Diversity of Bacterial Respiratory Systems

Volume II

Editor

Christopher J. Knowles, Ph.D.

Senior Lecturer in Biochemistry
Biological Laboratory
University of Kent
Canterbury, Kent
United Kingdom

CRC Press
Taylor & Francis Group
Boca Raton London New York

CRC Press is an imprint of the
Taylor & Francis Group, an **informa** business

T0174885

CRC Press
Taylor & Francis Group
6000 Broken Sound Parkway NW, Suite 300
Boca Raton, FL 33487-2742

First issued in paperback 2020

ISBN-13: 978-1-315-89240-5 (hbk)
ISBN-13: 978-0-367-65747-5 (pbk)

Library of Congress Cataloging in Publication Data

Main entry under title:

The Diversity of bacterial respiratory systems.
   Bibliography: p.
   Includes index.
   1. Microbial respiration. 2. Bacteria—
Physiology. I. Knowles, C. J. [DNLM:
1. Bacteria—Physiology. 2. Cell membrane—
Physiology. 3. Respiration. QW52.3 D618]
QR89.D58   589.9'01'2   79-17010
ISBN 0-8493-5399-8 (Volume I)
ISBN 0-8493-5400-5 (Volume II)

A Library of Congress record exists under LC control number: 79017010

Visit the Taylor & Francis Web site at
http://www.taylorandfrancis.com

and the CRC Press Web site at
http://www.crcpress.com

PREFACE

Although wide differences occur in the composition and function of mitochondrial respiratory systems, there is also a distinct and fundamental similarity between them, whether they originate from plants, animals, or microorganisms. However, in bacteria, respiratory systems vary enormously in both composition and function from the very simple to complex mitochondrial-like systems, depending on the degree of evolutionary sophistication of the organism and the type of habitat in which they exist. For example, there are bacteria that respire with oxygen, nitrate, fumarate, or sulfur compounds as electron acceptors, whereas mitochondrial systems respire only to oxygen. Some bacteria even require reversal of electron transfer against the normal electrochemical gradient in order to grow.

It is the aim of this book to present reviews on a wide range of aspects of bacterial respiratory systems. Because of the on-going publication elsewhere of reviews on bacterial respiration, a "blanket" coverage of the field has not been attempted. Rather, a range of topics have been selected, either because they are of special current interest, they have not been reviewed recently, or they have never been reviewed.

C. J. Knowles

THE EDITOR

Christopher J. Knowles, Ph.D., is Senior Lecturer in Biochemistry in the Biological Laboratory of the University of Kent, Canterbury, England. Dr. Knowles received his B.Sc. in chemistry from the University of Leicester in 1964 and his Ph.D. in biochemistry in 1967. From September 1967 to September 1969 he was a Postdoctoral Fellow of the American Heart Foundation at Dartmouth Medical School, Hanover, New Hampshire, U.S.A. In 1969 he returned to Britain as a Science Research Council Postdoctoral Fellow for one year at the University of Warwick. In October 1970 he was appointed Lecturer in Biochemistry at the University of Kent and promoted to Senior Lecturer in October 1977.

CONTRIBUTORS

Assunta Baccarini-Melandri, Ph.D.
Assistant Professor of Plant Physiology
University of Bologna
Bologna, Italy

Werner Badziong, Dr. rer. nat.
Research Associate
Fachbereich Biologie
Philipps-Universität Marburg
Auf den Lahnbergen
Marburg/Lahn
Federal Republic of Germany

Philip D. Bragg, Ph.D.
Professor of Biochemistry
University of British Columbia
Vancouver, British Columbia
Canada

Jan William Drozd, Ph.D.
Fermentation and Microbiology
  Division
Shell Research Limited
Shell Biosciences Laboratory
Sittingbourne Research Center
Sittingbourne, Kent
United Kingdom

I. John Higgins, Ph.D.
Senior Lecturer in Biochemistry and
  Microbiology
Biological Laboratory
University of Kent
Canterbury, Kent
United Kingdom

Peter Jurtshuk, Jr., Ph.D.
Professor of Biology
University of Houston
Houston, Texas

Christopher J. Knowles, Ph.D.
Senior Lecturer in Biochemistry
Biological Laboratory
University of Kent
Canterbury, Kent
United Kingdom

Wil N. Konings, Ph.D.
Associate Professor of Microbiology
Department of Microbiology
Biological Center
University of Groningen
Groningen
The Netherlands

Achim Kröger, Dr. phil.
Akademischer Rat
Institut für Physiologische Chemie
Universität München
München
Federal Republic of Germany

Paul A. M. Michels, Ph.D.
Research Fellow
Department of Microbiology
Biological Center
University of Groningen
Groningen
The Netherlands

Oense M. Neijssel, Ph.D.
Lecturer
Laboratorium voor Microbiologie
Universiteit van Amsterdam
Amsterdam
The Netherlands

Jae Key Oh, Ph.D.
Research Associate
Department of Microbiology
University of Manitoba
Winnipeg, Manitoba
Canada

L. F. Oltmann, Ph.D.
Research Fellow
Biological Laboratory
Free University
Amsterdam
The Netherlands

Robert K. Poole, Ph.D.
Lecturer
Department of Microbiology
Queen Elizabeth College
University of London
Campden Hill, London
United Kingdom

Irmelin Probst, Ph.D.
Research Associate
Institut für Mikrobiologie
Universität Göttingen
Göttingen
Federal Republic of Germany

Belinda Seto, Ph.D.
Senior Staff Fellow
Laboratory of Biochemistry
National Heart, Lung, and Blood
 Institute
National Institutes of Health
Bethesda, Maryland

Adrian H. Stouthamer, Ph.D.
Professor of Microbiology
Biological Laboratory
Free University
Amsterdam
The Netherlands

Isamu Suzuki, Ph.D.
Professor and Head
Department of Microbiology
University of Manitoba
Winnipeg, Manitoba
Canada

David W. Tempest, D.Sc.
Professor
Laboratorium voor Microbiologie
Universiteit van Amsterdam
Amsterdam
The Netherlands

Rudolf K. Thauer, Dr. rer. nat.
Professor of Microbiology
Fachbereich Biologie
Philips-Universität Marburg
Auf den Lahnbergen
Marburg/Lahn
Federal Republic of Germany

Jan van't Riet, Ph.D.
Senior Lecturer in Biochemistry
Biochemical Laboratory
Free University
Amsterdam
The Netherlands

Ralph S. Wolfe, Ph.D.
Professor of Microbiology
Department of Microbiology
University of Illinois at Urbana-
 Champaign
Urbana, Illinois

Tsan-yen Yang, Ph.D.
Postdoctoral Fellow
Johnson Research Foundation
University of Pennsylvania
School of Medicine
Philadelphia, Pennsylvania

Davide Zannoni, Ph.D.
Assistant Professor of Plant
 Biochemistry
University of Bologna
Bologna, Italy

TABLE OF CONTENTS

Volume I

TABLE OF CONTENTS

Volume II

Chapter 1

BACTERIAL ELECTRON TRANSPORT TO FUMARATE

A. Kröger

TABLE OF CONTENTS

I. INTRODUCTION

Until about 1970, it was generally believed that anaerobic nonphotosynthetic bacteria could synthesize ATP only by substrate level phosphorylation reactions and that the occurrence of electron transport phosphorylation was restricted to aerobic bacteria.[1] The reduction of nitrate as a known exception from this rule was thought to be a property of aerobic organisms only. In the meantime, it has been discovered that strictly anaerobic bacteria can derive the ATP required for growth from the reduction of fumarate, with either succinate or propionate as the end product. This was surprising in view of the redox potential of the fumarate/succinate couple which is nearly 400 mV more negative than that of nitrate/nitrite.

This chapter will concentrate mainly on two aspects:

1. The role of fumarate reduction in anaerobic metabolism
2. Fumarate reduction as a system of electron transport phosphorylation

Other aspects have been extensively discussed in recent reviews.[2-5]

II. THE ROLE OF FUMARATE IN ANAEROBIC METABOLISM

Hydrogen transfer is a general feature of aerobic and anaerobic metabolism. Substrate level phosphorylation, for instance, is often coupled to the oxidation of a carbon compound by the pyridine nucleotides. Under aerobic conditions, the reduced pyridine nucleotides are reoxidized by oxygen as an abundantly available acceptor with a sufficiently high redox potential (Table 1). An acceptor with similar properties is not at the disposal of anaerobic bacteria under most conditions. Although abundantly available from water dissociation, protons are rarely used as acceptors of reducing equivalents because of the low redox potential of the couple H^+/H_2 under the usually pH-neutral growth conditions. The production of molecular hydrogen from NADH and protons is an endergonic reaction which can occur only at the expense of other exergonic processes. CO_2 is expected to be available under most of the growth conditions of bacteria. However, the redox potential of bicarbonate with formate as the product of two-electron reduction is about as negative as that of protons. The redox potentials of the couples HCO^-_3/methane and HCO^-_3/acetate are distinctly more positive than those of H^+/H_2. Nevertheless, only methanobacteria and some other specialized organisms seem to be able to use CO_2 as a hydrogen acceptor and to derive useful energy from its reduction to methane or acetate.[2] In the fixation of CO_2 which is observed with many enteric bacteria fermenting sugars with succinate as one of the end products (see Figure 4B and C), CO_2 is not used as a true hydrogen acceptor. The fixed CO_2 is conserved in the carboxyl group of succinate and not further reduced. Sulfate and elemental sulfur can be used as hydrogen acceptors by certain specialized bacteria.[2] Sulfide is the end product in both cases. The redox potentials of sulfate and sulfur are in the same range as that of CO_2. In contrast to the acceptors mentioned above, nitrate exhibits an exceedingly high redox potential. Correspondingly a great variety of different bacteria have developed the ability to reduce nitrate either to nitrite or some other compounds.[2,4]

Most of the known anaerobic bacteria synthesize their hydrogen acceptors from the growth substrates. Pyruvate is a well-known example of a hydrogen acceptor which is reduced to lactate during the fermentation of glucose by many bacteria (see Table 1). Acrylate and crotonate appear to be more suitable hydrogen acceptors than pyruvate because their redox potentials are positive enough to oxidize most of the carbon com-

Table 1
REDOX COUPLES INVOLVED
IN BACTERIAL ENERGY
METABOLISM

	$E'_o{}^2$ (mV)
H^+/H_2	−420
HCO^-_3/formate	−416
NAD^+/NADH	−320
HCO^-_3/acetate	−280
S/HS^-	−274
HCO^-_3/CH_4	−244
SO_4^{-}/HS^-	−220
Pyruvate/lactate	−197
Dihydroxyacetone-P/glyerol-1-P	−190
Oxaloacetate/malate	−172
Acrylate/propionate	−30
Crotonate/butyrate	−30
Fumarate/succinate	+30
Nitrate/nitrite	+420
O_2/H_2O	+815

pounds involved in metabolism. The CoA-esters of acrylate and crotonate are well-known hydrogen acceptors in anaerobic bacteria that form fatty acids as end products.[2] The enzymes catalyzing the reduction reactions are soluble in contrast to fumarate reductase that is bound to the membrane.[2,6] Although the redox potential span between NADH and the CoA-esters of acrylate and crotonate is sufficient to allow the synthesis of 1 ATP/2e⁻, electron transport phosphorylation has not been shown to be coupled to the reduction of these compounds.[2]

The redox potential of the couple fumarate/succinate is even more positive than that of acrylate/propionate (see Table 1). Therefore, fumarate is a suitable hydrogen acceptor in anaerobic metabolism. In the absence of acceptors with more positive redox potentials, succinate cannot further be metabolized and is therefore a true end product of fermentation. Fumarate reduction occurs more frequently in bacteria than fatty acid fermentation and is about as widespread as nitrate reduction.[4] The formation of propionate proceeds via fumarate and not with acrylate as the intermeditate in most of the bacteria. The reason for the widespread occurrence of fumarate reduction resides probably in its availability from a variety of organic compounds, like amino acids which are likely to be present in many biotopes, and in its ability of being coupled to electron transport phosphorylation. In the following, examples of metabolic pathways involving fumarate reduction are given.

While growth of *Escherichia coli*[7,8] and *Vibrio succinogenes*[9] with fumarate is sustained only in the presence of a hydrogen donor (see Table 2) *Proteus rettgeri*[10] (see Figure 1A) and *Clostridium formicoaceticum*[11] (see Figure 1B) can grow with fumarate as the only energy substrate. In both pathways, part of the fumarate is oxidized, and the hydrogen liberated in these branches of metabolism is used for the reduction of fumarate to succinate. With *C. formicoaceticum*, 1 mol acetate and 2 mol CO_2 are the end products of the oxidation of 1 mol fumarate, while 2 mol fumarate are oxidized to give 1 mol succinate and 4 mol CO_2 with *P. rettgeri*. In the latter case, the acetyl-CoA is used to form citrate which is then oxidized via part of the citric acid cycle. Malate is fermented in the same way as fumarate by *P. rettgeri*. In contrast, *C. formicoaceticum* produces mainly acetate and CO_2 from malate. In this pathway, part of the acetate is formed by CO_2 reduction.[11]

The pathway of fermentation of citrate by growing *P. rettgeri* (see Figure 2) is similar to that of fumarate (see Figure 1A) in that part of the citrate is again oxidized via part of the citric acid cycle.[10] Fumarate is produced from oxaloacetate which is formed from citrate by the citrate lyase reaction. The formation of fumarate from oxaloacetate is a general process in those pathways by which glucose or C_3-compounds are fermented, with either succinate or propionate as end products (see Figures 3 and 4). These pathways may differ in the reactions by which oxaloacetate is formed. The pathways of fermentation of glycerol, lactate, and glucose[12,13] by growing propionic acid bacteria are given in Figures 3 and 4A. Propionic acid bacteria synthesize oxaloacetate from pyruvate by means of transcarboxylation.[12] In this reaction, methylmalonyl-CoA serves as the carboxyl donor to give propionyl-CoA. ATP is not required in this reaction. The succinate formed by fumarate reduction is subsequently converted to propionate.[12] The intermediates of this sequence are not given in the figures. In the first step of the sequence, succinyl-CoA is formed from succinate by transacylation with propionyl-CoA. This step does not require ATP. With glycerol as the growth substrate, the hydrogen required for the reduction of oxaloacetate is generated in the Emden-Meyerhof pathway alone, while with lactate and glucose, additional hydrogen is provided in the branch leading to acetate.

In the fermentation of glucose by growing *Bacteroides fragilis* (see Figure 4B), only part of the succinate is converted to propionate (not shown).[14,15] This means that the fermentation is associated with the fixation of CO_2. The CO_2 is incorporated into oxaloacetate which is synthesized from phosphoenolpyruvate with the concomitant phosphorylation of ADP. In further contrast to propionibacteria, the degradation of pyruvate to acetyl-CoA is not associated with NAD reduction, but yields formate as an end product instead.

Anaerobically growing enterobacteria form variable amounts of succinate from glucose.[16,17] As an example, the fermentation pathway of glucose in the presence of bicarbonate by *P. rettgeri*[10] is in Figure 4C. Oxaloacetate is probably synthesized from phosphoenolpyruvate without recovery of the phosphate bond. This is assumed because other enterobacteria were found to synthesize C_4 from C_3-compounds only via the phosphoenolpyruvate carboxylase reaction.[18-20] Part of the acetyl-CoA which is formed via pyruvate-formatelyase, is reduced to ethanol in *P. rettgeri*. In *E. coli*, malate is apparently not an obligatory intermediate of fumarate production, since a mutant that is deficient of malate dehydrogenase was found to synthesize fumarate from oxaloacetate via aspartate.[19]

III. GROWTH YIELDS

E. coli can grow in defined media at the expense of the reduction of fumarate by molecular hydrogen with succinate as the only product[7,8] (see Table 2). The same is true for *V. succinogenes*[9] and a strain of *Desulfovibrio*.[21] As the ATP required for growth cannot be derived from substrate level phosphorylation under these conditions, it is clear that the ATP must be formed by the electron transport phosphorylation coupled to fumarate reduction. A similar situation is valid for *V. succinogenes*, when growing on fumarate and formate with succinate and bicarbonate as the products (see Table 2). These cases demonstrate unambiguously that fumarate reduction can yield ATP via electron transport phosphorylation.

In the other cases in Table 2, the situation is complicated by the occurrence of substrate level phosphorylation reactions in addition to possible electron transport phosphorylation with fumarate. In these cases, the cell yields per mole of substrate (Y_s) have to be measured. Division of Y_s by the ATP yield per mole of substrate (n) which is expected from the corresponding metabolic pathway gives the cell yield per mole of

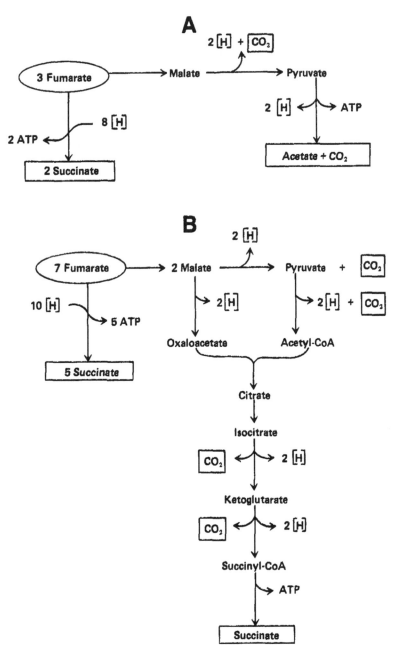

FIGURE 1. Pathway of fermentation of fumarate by (A) *C. formicoaceticum*[11] and (B) *P. rettgeri*.[10]

ATP (Y_s/n). As Y_s/n is believed to be a constant which is independent of the bacterial strain and the growth substrate under certain conditions[22-24] it is often possible to decide from growth yields whether fumarate reduction is associated with phosphorylation or not. The special implications of this method are discussed in detail by Neijssel and Tempest in Volume I, Chapter 1. The method is applied here to cases in which the growth reactions as well as the pathways of fermentation are known. The cell yields were obtained with batch cultures and the given numbers were not extrapolated for the "energy of maintenance".

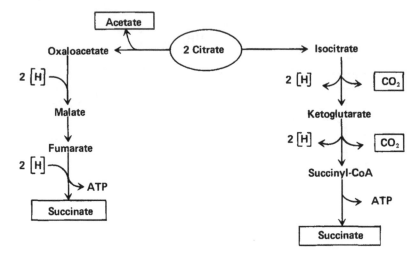

FIGURE 2.   Pathway of citrate fermentation by *P. rettgeri.*[10]

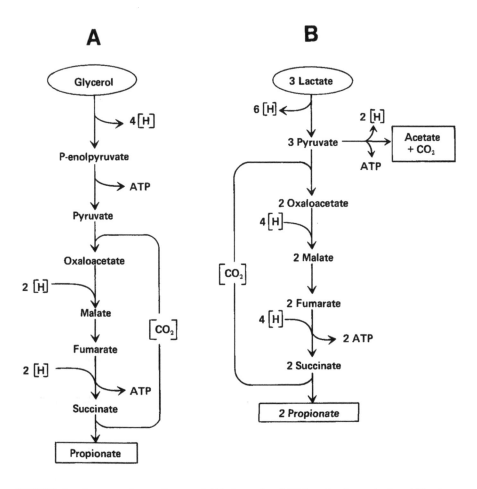

FIGURE 3.   Fermentation pathways of (A) glycerol and (B) lactate of propionic acid bacteria.[12,13]

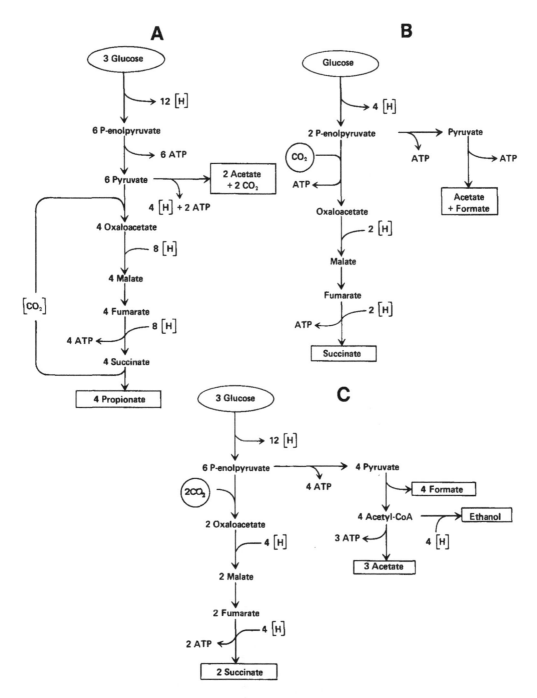

FIGURE 4.   Pathways of glucose fermentation by (A) propionic acid bacteria,[12,13] (B) *B. fragilis*,[14,15] and (C) *P. rettgeri*.[10]

According to the pathway of Figure 1B, the fermentation of fumarate by *P. rettgeri* yields 1 and 5 mol ATP per 7 mol fumarate by substrate level and electron transport phosphorylation respectively. The cell yield obtained with pyruvate (6.3 g cells per mole pyruvate) is used for judging the ATP yield with fumarate as growth substrate. Pyruvate is fermented in a simple pathway to give formate and acetate, and it is evident that 1 mol ATP per mole pyruvate is formed. The cell yield per mole ATP (Y_a/n) with

Table 2
GROWTH REACTIONS AND GROWTH YIELDS OF BACTERIA UNDER METABOLIC CONDITIONS INVOLVING FUMARATE REDUCTION

Bacterial strain	Growth reaction	Medium	Y_s (grams cells per mole substrate)	Ref.	Figure	n (moles ATP per mole substrate)		Y_s/n (grams cells per mole ATP)	$\Delta G'_s$ (kcal/mol ATP)
						SLP	ETP		
Escherichia coli	H_2 + fumarate→succinate	Complex	4.8	7,8	—	0	1	4.8	−20.5
Vibrio succinogenes	formate + fumarate + H_2O→HCO_3^- + succinate	Complex	4	9	—	0	1	4	−20.3
Proteus rettgeri	7 fumarate + 8 H_2O→6 succinate + $4HCO_3^-$ + $2H^+$	Mineral	4.7	10	1B	1/7	5/7	5.5	−17.6
P. rettgeri	2 citrate + 2 H_2O→2 succinate + acetate + $2HCO_3^-$ + H^+	Mineral	7.8	10	2	½	½	7.8	−18.5
P. rettgeri	Pyruvate + H_2O→acetate + formate + H^+	Mineral	6.3	10	—	1	0	6.3	−12
P. freudenreichii	3 lactate→2 propionate + acetate + HCO_3^- + H^+	Mineral	8.1	13	3B	⅓	⅔	8.1	−13.7
P. freudenreichii	Glycerol→propionate + H^+ + H_2O	Mineral	24	13	3A	1	1	12	−18
Clostridium formicoaceticum	3 fumarate + 4 H_2O→2 succinate + acetate + 2 HCO_3^- + H^+	Complex	6	11	1A	⅓	⅔	6	−16.4
P. freudenreichii	3 glucose→4 propionate + 2 acetate + $2HCO_3^-$ + $8H^+$	Mineral	65	13	4A	8/3	4/3	16	−18.8
Bacteroides fragilis	Glucose + HCO_3^-→succinate + acetate + formate + 3 H^+ + H_2O	Defined	53	14,15	4B	3	1	13	−16
P. rettgeri	3 glucose + $2HCO_3^-$→2 succinate + 3 acetate + ethanol + 4 formate + 9 H^+ + H_2O	Mineral	24	10	4C	7/3	⅔	8	−20

Note: The values of n (mol ATP/mol substrate) were taken from the corresponding metabolic pathways (Figures 1 to 4). SLP = substrate level phosphorylation, and ETP = electron transport phosphorylation with fumarate as acceptor.

fumarate is consistent with that of pyruvate fermentation only if phosphorylation coupled to fumarate reduction is taken into account (5.5 g cells per mole ATP) and would be five times greater on the basis that growth is sustained by substrate level phosphorylation alone. This was interpreted to indicate that ATP is gained from fumarate reduction in the fermentation of fumarate by *P. rettgeri*.[10] The same is likely to be true for the fermentation of citrate by *P. rettgeri*, where Y_s/n is 7.8 g cells per mole ATP, assuming that half of the ATP is formed by fumarate reduction (see Figure 2).

The pathway of fermentation of lactate by *Propionibacterium freudenreichii* (Figure 3B) indicates that per 3 mol lactate, 1 mol ATP is gained from substrate level phosphorylation and that an additional 2 mol ATP could be generated by fumarate reduction. Thus, n could maximally be 1, and the corresponding value of Y_s/n is 8.1. This number is greater than that measured for growth of *P. rettgeri* on pyruvate and therefore suggests that phosphorylation is coupled to fumarate reduction in *P. freudenreichii* growing on lactate. The relatively high cell yield of *P. freudenreichii* growing on glycerol (see Figure 3A) also suggests phosphorylative fumarate reduction. On the basis that the reduction of fumarate yields ATP in *C. formicoaceticum* growing on fumarate (see Figure 1), a value Y_s/n of 6.0 is obtained which is nearly equal to that measured with *P. rettgeri* growing on pyruvate. In view of the fact that *C. formicoaceticum* is grown on a complex medium in contrast to *P. rettgeri*, a greater value would be expected.[24] On the other hand, a value of Y_s/n of 18 g cells per mole ATP would be obtained on the basis of the assumption that growth of *C. formicoaceticum* on fumarate is sustained by substrate level phosphorylation alone. As similarly high values of Y_s/n have been found in some cases on one hand, and values lower than six were measured in spite of a complex medium on the other (see Table 2), it cannot be decided whether fumarate reduction is phosphorylative in *C. formicoaceticum* or not. A similar situation obtains from the growth yields measured with *Veillonella alcalescens* growing on lactate and malate.[25]

The fermentation of glucose by *P. freudenreichii* and *B. fragilis* can yield 4 mol ATP per mole glucose according to Figures 4A and B, while with *P. rettgeri* maximally 3 mol ATP per mole glucose can be obtained (see Figure 4C). The possible contribution of fumarate reduction to ATP formation is one third with *P. freudenreichii* and even lower with the other two strains. In view of the limited accuracy of the method of determining ATP yields from growth yields, it is not possible under these conditions to decide whether fumarate reduction contributes to ATP synthesis or not. As *P. freudenreichii* and *P. rettgeri* appear to synthesize ATP by fumarate reduction under other metabolic conditions, it is likely that they do so also with glucose as the growth substrate. *B. fragilis* can perform fumarate reduction only when grown with hemin.[14] From comparison of cell yields of *B. fragilis* grown on glucose in the presence and absence of hemin, it was concluded that fumarate reduction is phosphorylative in this bacterium.

The free energy that is required for the synthesis of 1 mol ATP ($\Delta G_o'$/mol ATP) was calculated as the ratio of the free energy change per mole of substrate of each reaction in Table 2 and the amount of ATP synthesized per mole of substrate (n). This number varies between 12 and 20.5 kcal/mol ATP and is close to the free energy of hydrolysis of ATP under cellular conditions (about −12 kcal/mol,[5]). This shows that the yields of ATP of the growth reactions given in Table 2 approach the energetic limit.

In summary, fumarate reduction appears to be coupled to phosphorylation in most of the bacteria investigated so far. However, it is possible that fumarate serves merely as hydrogen acceptor which drives substrate level phosphorylation reactions in some bacterial species.

IV. ENERGY TRANSFER

Phosphorylation coupled to fumarate reduction was also demonstrated directly. The endogenous ADP and AMP of spheroplasts of *V. succinogenes* which were grown on formate and fumarate were found to be phosphorylated as a function of the reduction of fumarate by formate.[5,26] The $P/2e^-$ ratio was 0.4—1. Phosphorylation, but not fumarate reduction, was fully inhibited by 10 μmol CCCP per gram protein.[5] This amount is five times greater than that required for full inhibiton of the mitochondrial oxidative phosphorylation.[27] Reddy and Peck[28] used a membrane preparation of *V. succinogenes* which were grown on formate and fumarate. This preparation catalyzed the ADP-dependent esterification of inorganic phosphate in the presence of glucose and hexokinase as a function of the reduction of fumarate by molecular hydrogen. $P/2e^-$ ratios of up to 0.54 were observed with some preparations. Pentachlorophenol caused nearly full inhibition of phosphorylation, while electron transport was stimulated by about 15%. The amount of pentachlorophenol used (220 μmol/g protein) was about 70 times greater than that required for uncoupling of the mitochondrial oxidative phosphorylation.[29] Phosphorylation was also inhibited by gramicidin which is known to collapse the electric potential across the membrane by facilitating the diffusion of Na^+ and K^+. In a similar experiment with a membrane preparation from *Desulfovibrio gigas*, $P/2e^-$ ratios of up to 0.9 were measured.[30] A membranous preparation of *E. coli* grown on glycerol and fumarate was reported to catalyze fumarate reduction by glycerol-1-phosphate. The phosphorylation coupled to this reaction ($P/2e^- = 0.1$) was decreased by 60% by the presence of 15 μmol CCCP per gram protein.[31]

The reduction of fumarate was shown to be associated with the electrogenic liberation of protons into the external medium with *V. succinogenes*,[4,5,26] *E. coli*, and *Klebsiella pneumoniae*.[32] The ratios H^+/e^- were 1.0—1.6. With *V. succinogenes*, the reduction of fumarate by formate was found to be associated with an uptake of protons from the cytoplasm which was dependent on the presence of K^+ in the external medium and was abolished by FCCP.[5]

These results indicate that fumarate reduction can drive energy-dependent reactions similar to those involved in oxidative phosphorylation of mitochondria and aerobic bacteria. It was further shown that fumarate reduction can also drive the active transport of amino acids, sugars, carboxylic acids, and several other metabolites.[3] It is likely that the phosphorylation that is coupled to fumarate reduction is catalyzed by the Mg^{2+}-dependent ATPase which is present in the membrane of all the bacteria examined so far.[2,28]

V. THE ELECTRON TRANSPORT CHAIN

Fumarate reduction is catalyzed by an electron transport chain that is localized in the cytoplasmic membrane of the bacteria.[4,5] Depending on the organism and on the growth conditions, molecular hydrogen, formate, NADH, lactate, glycerol-1-phosphate or malate (see Table 1) may serve as hydrogen donors for fumarate reduction.[4,5] The chain consists of at least one membrane-bound dehydrogenase that is specific for one of the donors, a naphthoquinone (MK or DMK), and fumarate reductase. Most of the bacteria catalyze fumarate reduction with more than one substrate and catalyze the reduction of other acceptors, like oxygen and nitrate in addition to fumarate. The quinones are not bound to protein, but can easily be extracted with hydrocarbons after lyophilization of the membrane preparation. The extraction causes inhibition of fumarate reduction by the donors, and the activity can be restored by reinsertion of the quinone into the membrane.[5] The naphthoquinones cannot be replaced by ubiquinone, the redox potential of which is too positive to allow oxidation by fumarate.[5] Inspection of the redox reactions of bacterial MK showed that the quinone is converted to the corresponding hydroquinone upon reduction by the donor substrates and that the hydroquinone is reoxidized to the quinone by fumarate.[33-36]

The function of the quinones appears to be similar to that of ubiquinone in the respiratory chain.[5,35,37] They collect the reducing equivalents liberated by the various dehydrogenases and transport them to fumarate reductase or to other acceptor-activating enzymes (reductases), like nitrate reductase and cytochrome oxidase. The transport is probably done by diffusion of the quinone in the lipid phase of the membrane. It is possible that the dehydrogenases as well as the reductases are randomly distributed on the surface of the membrane because the distances between the enzymes could be bridged by diffusion of the quinone. From this concept, it follows that the electron transport with fumarate as the acceptor may not be spatially separated from the respiratory chain or the pathway of nitrate reduction in the membrane of bacteria which exhibit all three activities. It is rather likely that fumarate reduction is only one process catalyzed by an electron transport system consisting of various dehydrogenases and reductases which are functionally linked by a diffusing quinone. The dehydrogenases and reductases probably form protein complexes with other electron transport components like cytochromes and iron-sulfur proteins, the function of which is to catalyze electron transport between the enzymes and the quinone.

The validity of this concept with respect to fumarate reduction is supported by the results which were obtained from studying the electron transport of *V. succinogenes* which catalyzes the reduction of fumarate by formate (see Figure 5). The system consists of the formate dehydrogenase complex, the fumarate reductase complex, and MK.[4,5,36,38,39] Electron transport from formate to fumarate is dependent on the presence of MK, whereas the oxidation of formate by artificial acceptors and the reduction of fumarate by artificial donors are not.[36] After solubilization of the membrane by detergents, the enzyme catalyzing the dehydrogenation of formate was separated from a fumarate reducing enzyme. After purification, formate dehydrogenase was found to consist of two identical subunits of 110,000 mol wt., each of which contained 1 atom of molybdenum and about 20 iron-sulfur groups.[39] Molybdenum is probably the primary acceptor of the reducing equivalents of formate, and the iron-sulfur groups may serve as mediators between molybdenum and the electron acceptor of formate dehydrogenase. Cytochrome b (-200 mV) appears to be the acceptor of formate dehydrogenase. This cytochrome is associated with the isolated formate dehydrogenase and is reduced by formate in the absence of other redox components. Fumarate reductase can be isolated in two different states.[46] In the smaller state, the enzyme (about 110,000 mol wt.) consists of two different subunits of 80,000 and 30,000 mol wt. and contains 8.4 μmol/g protein of covalently bound FAD and about 10 times as much iron-sulfur. The greater subunit contains the FAD which is linked at position 8α to the N-3 of a histidine residue of the apoprotein.[40] In the greater state, the enzyme is more stable; it consists of 3 subunits of 80,000, 30,000, and 25,000 mol wt. and contains cytochrome b in addition to FAD and iron-sulfur. Part of the cytochrome b can be reduced by succinate. This reaction as well as the activities of reduction of ferricyanide and methylene blue are inhibited by p-CMS which reacts with the iron-sulfur groups.[36,38] In contrast, the activities of fumarate reduction by artificial donors are not inhibited by p-CMS. This indicates that the iron-sulfur mediateq the electron transport between cytochrome b (-20 mV) and the FAD which is the primary donor of the reducing equivalents to fumarate. The iron-sulfur associated with formate dehydrogenase is different from that present in fumarate reductase.[39] Thus, the electron transport chain catalyzing reduction of fumarate by formate in *V. succinoenes* is probably made up of only five different polypeptides, each of which carries at least one redox-active

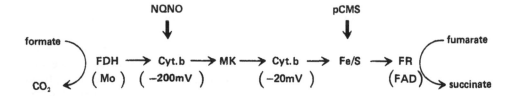

FIGURE 5.   Electron transport chain catalyzing fumarate reduction by formate of *V. succino-genes*.[36,38] FDH = formate dehydrogenase, Fe/S = iron-sulfur, FR = fumarate reductase. The *b*-cytochromes are characterized by their midpoint potentials (−200 and −20mV). NQNO = *n*-Nonyl-4-hydroxyquinoline-*N*-oxide, p-CMS = 4-chloromercuriphenylsulfonate.

prosthetic group. Based on the contents of molybdenum and covalently bound FAD of the membrane fraction, it was calculated that fumarate reductase is present in an about tenfold molar excess over formate dehydrogenase under the standard growth conditions of the bacteria. However, the contents vary as a function of the growth conditions.

It is likely that the chain catalyzing electron transport with fumarate in other bacteria is different from that in *V. succinogenes*. However, a detailed investigation on other systems of fumarate reduction is not available so far. Although *b* cytochromes seem to be involved in most cases, fumarate reduction can function in the absence of cytochromes in some bacteria.[4,5]

VI. SIDEDNESS OF FORMATE DEHYDROGENASE AND FUMARATE REDUCTASE IN *V. succinogenes*

The sidedness of formate dehydrogenase and fumarate reductase in the membrane was studied by measuring the enzymic activities with artificial redox dyes of cells before and after lysis (see Table 3). The activities of fumarate reductase were measured with anthrahydroquinone sulfonate as an impermeable donor or, in the reverse reaction, with succinate and ferricyanide as an impermeable acceptor. These activities were small with cells and were stimulated more than 20 times by lysis of the cells. This indicates that the fumarate reductase of cells is not accessible to impermeable dyes, unless the permeability barrier is removed. The activity of formate dehydrogenase as measured with the impermeable viologen dye diquat was high with cells and could not be further stimulated by lysis of the cells. This indicates that the formate dehydrogenase of cells is accessible to the impermeable dye.

In order to elucidate the sidedness of the sites of substrate reaction of the two enzymes, the activities were measured as functions of the concentrations of their substrates.[5] Fumarate reductase was assayed in the reverse reaction with succinate and methylene blue as permeable acceptor. The activities were related to the substrate concentrations according to the Michaelis equation with cells both before and after lysis. The K_M for formate was not altered by lysis of the cells, whereas the K_M for succinate was increased ten times. This result is consistent with the view that the sites of substrate interaction of the enzymes are oriented towards the same aspects of the membrane as the respective sites reacting with the impermeable dyes. The true K_M for succinate of fumarate reductase is measured only with lysed cells, whereas the apparent K_M obtained with intact cells implies the transport of succinate into the cytoplasm in addition.

The accessibility of the enzymes was also investigated with a particle preparation obtained with the French Press. These particles were reported to perform phosphoryl-

Table 3

ACTIVITIES OF FORMATE DEHYDROGENASE
AND FUMARATE REDUCTASE OF CELLS OF *V.
SUCCINOGENES* AND OF A MEMBRANE
PREPARATION THEREOF BEFORE AND   AFTER
LYSIS

| | Cells[a] | | Particles[a] | |
Activity	Intact	Lysed	Intact	Lysed
		μmol/min/g protein		
Anthrahydroquinone sulfonate Fumarate	0.07	2.8	4.5	4.4
Morfamquat Fumarate	—	—	11.2	12.6
Succinate Fe(CN)$_6^{3-}$	0.05	1.3	2.1	1.9
Formate Fumarate	—	1.5	0.47	1.9
Formate Methylene blue	7.1	6.7	2.8	9.6
Formate Diquat	0.86	0.82	0.77	2.3

Note: Cells were grown on formate and fumarate[36] and suspended in 0.5
M mannitol, 20 mM HEPES (N-2-hydroxyethylpiperazine-N-2-
ethansulfonate), and 50 mM KCl at pH 7.5 and 0°C. Lysis of the
cell was achieved by addition of 0.01% Triton® X-100. Particles
were prepared with the French press,[28] and lysis of the particles
was achieved either by freezing and thawing, by addition of 0.01%
Triton® X-100, or with phospholipase A$_2$. Nearly identical activities were measured after each of the treatments. The activities
were measured spectrophotometrically at 25°C.[36,38]

[a]   Expressed in μmol/min/g of protein of fumarate, succinate, or
formate.

ation of externally added ADP which is coupled to the reduction of fumarate by molecular hydrogen.[28] This indicates that the ATPase of a significant proportion of these
particles is exposed to the external medium. Experiments with ricoli showed that
treatment with the French Press causes an inversion of the cytoplasmic membrane.[41]
The activities of fumarate reductase with the impermeable dyes were not stimulated
by lysis of the particles (see Table 3). This indicates that the reactive sites of fumarate
reductase are exposed to the external medium (see Figure 6), in contrast to the situation
in cells. As, on the other hand, the reduction of fumarate by formate is increased
fourfold by lysis, this activity is obviously limited by the permeation of formate. It is
concluded that the active site of formate dehydrogenase faces the internal space of
part of the particles (see Figure 6). It is probable that the membrane of this part of
the particles is nearly impermeable to formate and that the activity observed with the

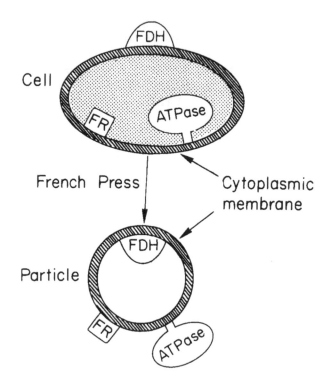

FIGURE 6.   Sidedness of formate dehydrogenase (FDH), fumarate reductase (FR), and ATPase in cells and French-press particles of *V. succinogenes*.

untreated preparation is mainly due to a second species which lacks the permeability barrier. This view is supported by the following observations:

1.   The degree of stimulation of formate oxidation by lysis of the particles is independent of the specific activity of the reaction after lysis and of the permeability of the acceptor (compare Lines 4, 5, and 6 of Table 3).
2.   Two types of particles could be separated by density gradient centrifugation, one showing a greater and the other a smaller degree of stimulation than the starting preparation. By this procedure particles were obtained, of which up to 80% appeared to be impermeable to formate as calculated from the degree of stimulation.
3.   The efficiency of the phosphorylation of external ADP which is coupled to the reduction of fumarate of French press particles is much lower with formate than with the permeable hydrogen as donor.[28]

   It is concluded that the sidedness of the active sites of formate dehydrogenase, fumarate reductase, and ATPase in the membrane of the particles is opposite to that of cells (see Figure 6) and that the permeation of formate through the membrane is slow, as compared with the activity of fumarate reduction by formate. In cells the active site of formate dehydrogenase faces the outside and that of fumarate reductase the inside of the membrane.
   These results suggest that the electrochemical proton potential across the membrane, generated as a function of electron transport from formate to fumarate, is built up according to the mechanism in Figure 7.[4,5] According to the chemiosmotic theory of

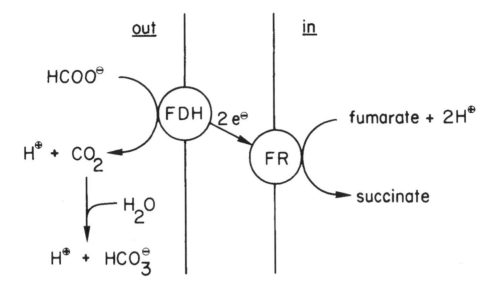

FIGURE 7.   Possible mechanism of generating an electrochemical proton potential across the bacterial membrane by the electron transport of *V. succinogenes*.[4,5] FDH = formate dehydrogenase, and FR = fumarate reductase.

Mitchell,[42,43] the proton potential represents the coupling device between electron transport and phosphorylation. The electron transport chain of *V. succinogenes* serves to generate an electrical field across the membrane by transporting only the electrons of formate from the outside to the inside of the membrane. The proton(s) abstracted from formate are liberated into the external medium, and the protons required for the reduction of fumarate are taken up from the cytoplasm. This view is consistent with the composition of the electron transport chain (see Figure 5) which prevents the transport of hydride ions from formate to MK and from MK to fumarate because formate dehydrogenase and fumarate reductase are linked to MK by electron carriers. The mechanism accounts for the observed H^+/e^- ratio, and therefore no other mechanism is required for explaining the generation of the proton potentia. However, it cannot be excluded that proton translocation by MK[4,5] may contribute to the generation of the proton potential.

Similar studies with *E. coli* showed that the fumarate reductase of cells is not accessible to impermeable dyes.[44] The same result was obtained for the formate dehydrogenase present in cells which were grown on glycerol plus nitrate. When *E. coli* were grown with glycerol, fumarate, and formate present, an enzyme catalyzing the oxidation of formate by viologen dyes ("formate hydrogen lyase") was induced which was accessible to impermeable dyes.[44] It may be speculated that this enzyme is responsible for growth of *E. coli* on formate and fumarate,[7,8] so that a similar mechanism of generation of the proton potential obtains in *E. coli* as in *V. succinogenes*. It is possible that a mechanism equivalent to that given in Figure 7 operates in fumarate reduction by molecular hydrogen in *V. succinogenes*[9] and *E. coli*,[7,8] if hydrogenase would face the external medium in intact cells.

In *C. formicoaceticum*, a soluble fumarate reductase is present in the periplasmic space of the bacterium.[45] It is possible that fumarate reduction is not coupled to phosphorylation (see above) and that fumarate serves merely as a hydrogen sink in this bacterium.

REFERENCES

1. Decker, K., Jungermann, K., and Thauer, R. K., Energy production in anaerobic organisms, *Angew. Chem. Int. Ed. Engl.*, 9, 138, 1970.
2. Thauer, R. K., Jungermann, K., and Decker, K., Energy conservation in chemotrophic anaerobic bacteria, *Bacteriol. Rev.*, 41, 100, 1977.
3. Konings, W. N. and Boonstra, J., Anaerobic electron transfer and active transport in bacteria, *Curr. Top. Membr. Transp.*, 9, 177, 1977.
4. Kröger, A., Phosphorylative electron transport with fumarate and nitrate as terminal hydrogen acceptors, in *Microbial Energetics* 27th Symp. Soc. General Microbiology, Haddock, B. A. and Hamilton, W. A., Eds., 1977, 61.
5. Kröger, A., Fumarate as terminal acceptor of phosphorylative electron transport, *Biochim. Biophys. Acta*, 505, 129, 1978.
6. Brockman, H. L. and Wood, W. A., Electron transferring flavoprotein of *Peptostreptococcus elsdenii* that functions in the reduction of acrylyl-CoA, *J. Bacteriol.*, 124, 1447, 1975.
7. Macy, J., Kulla, H., and Gottschalk, G., H_2-dependent anaerobic growth of *E. coli* on L-malate: succinate formation, *J. Bacteriol.*, 125, 423, 1976.
8. Bernhard, Th. and Gottschalk, G., Cell yields of *E. coli* during anaerobic growth on fumarate and molecular hydrogen, *Arch. Microbiol.*, 116, 235, 1978.
9. Wolin, M. J., Wolin, E. A., and Jacobs, N. J., Cytochrome-producing anaerobic vibrio, *Vibrio succinogenes* sp. n, *J. Bacteriol.*, 81, 911, 1961.
10. Kröger, A., Schmikat, M., and Niedermaier, S., Electron transport phosphorylation coupled to fumarate reduction in anaerobically grown *Proteus rettgeri*, *Biochim. Biophys. Acta*, 347, 273, 1974.
11. Dorn, M., Andreesen, J. R., and Gottschalk, G., Fermentation of fumarate and L-malate by *Clostridium formicoaceticum J. Bacteriol.*, 133, 26, 1978.
12. Wood, H. G., Transcarboxylase, *Enzymes*, 6, 83, 1972.
13. De Vries, W., van Wyk-Kapteyn, M. C., and Stouthamer, A. H., Generation of ATP during cytochrome-linked anaerobic electron transport in propionic acid bacteria, *J. Gen. Microbiol.*, 76, 31, 1973.
14. Macy, J., Probst, I., and Gottschalk, G., Evidence for cytochrome involvement in fumarate reduction and ATP synthesis by *Bacteroides fragilis* grown in the presence of hemin, *J. Bacteriol.*, 123, 436, 1975.
15. Macy, J., Ljungdahl, L. G., and Gottschalk, G., Pathway of succinate and propionate formation in *Bacteroides fragilis*, *J. Bacteriol.*, 134, 84, 1978.
16. Wood, W. A., Fermentation of carbohydrates and related compounds, *Bacteria*, 2, 59, 1961.
17. Hernandez, E. and Johnson, M. J., Anaerobic growth yields of *Aerobacter cloacae* and *E. coli, J. Bacteriol.*, 94, 991, 1967.
18. Kornberg, H. L., Anaplerotic sequences and their role in metabolism, *Essays Biochem.*, 2, 1, 1966.
19. Courtright, J. B. and Henning, U., Malate dehydrogenase mutants in *E. coli* K-12, *J. Bacteriol.*, 102, 722, 1970.
20. Theodore, T. S. and Englesberg, E., Mutant of *Salmonella typhimurium* deficient in the CO_2-fixing enzyme phosphoenolpyruvic carboxylase, *J. Bacteriol.*, 88, 946, 1964.
21. Badziong, W., Thauer, R. K., and Zeikus, J. G., Isolation and characterization of *Desulfovibrio* growing on hydrogen plus sulfate as sole energy source, *Arch. Microbiol.*, 116, 41, 1978.
22. Bauchop, T. and Elsden, S. R., The growth of microorganisms in relation to their energy supply, *J. Gen. Microbiol.*, 23, 457, 1960.
23. Stouthamer, A. H. and Bettenhausen, C., Utilization of energy for growth and maintenance in continuous and batch cultures of microorganisms, *Biochim. Biophys. Acta*, 301, 53, 1973.
24. Stouthamer, A. H., A theoretical study on the amount of ATP required for synthesis of microbial cell material, *Antonie van Leeuwenhoek J. Microbiol. Serol.*, 39, 545, 1973.
25. De Vries, W., Rietveld-Struijk, T. R., and Stouthamer, A. H., ATP formation associated with fumarate and nitrate reduction in growing cultures of *Veillonella alcalescens*, *Antonie van Leeuwenhoek J. Microb. Serol.*, 43, 153, 1977.
26. Kröger, A., The electron transport coupled phosphorylation of the anaerobic bacterium *Vibrio succinogenes*, in *Electron Transfer Chains and Oxidative Phosphorylation*, Quagliariello, E., et al. Eds., North-Holland, Amsterdam, 1975, 165.
27. Heytler, P. G., Uncoupling of oxidative phosphorylation by carbonyl cyanide phenylhydrazones, *Biochemistry*, 2, 357, 1963.
28. Reddy, C. A. and Peck, H. D., Electron transport phosphorylation coupled to fumarate reduction by H_2, and Mg^{2+}-dependent adenosine triphosphatase activity in extracts of the rumen anaerobe *Vibrio succinogenes*, *J. Bacteriol.*, 134, 982, 1978.

29. **Weinbach, E. C.**, The effect of pentachlorophenol on oxidative phosphorylation, *J. Biol. Chem.*, 210, 545, 1954.
30. **Barton, L. L., Le Gall, J., and Peck, H. D.**, Phosphorylation coupled to oxidation of hydrogen with fumarate in extracts of the sulfate reducing bacterium *Desulfovibrio gigas*, *Biochem. Biophys. Res. Commun.*, 41, 1036, 1970.
31. **Miki, K. and Lin, E. C. C.**, Anaerobic energy-yielding reaction associated with transhydrogenation from glycerol-3-phosphate to fumarate by an *Escherichia coli* system, *J. Bacteriol.*, 124, 1282, 1975.
32. **Brice, J. M., Law, J. F., Meyer, D. J., and Jones, C. W.**, Energy conservation in *Escherichia coli* and *Klebsiella pneumoniae*, *Biochem. Soc. Trans.*, 2, 523, 1974.
33. **Kröger, A. and Dadak, V.**, On the role of quinones in bacterial electron transport. The respiratory system of *B. megaterium*, *Eur. J. Biochem.*, 11, 328, 1969.
34. **Kröger, A., Dadak, V., Klingenberg, M., and Diemer, F.**, On the role of quinones in bacterial electron transport. Differential roles of ubiquinone and menaquinone in *Proteus rettgeri*, *Eur. J. Biochem.*, 21, 322, 1971.
35. **Kröger, A. and Klingenberg, M.**, Quinones and nicotinamide nucleotides associated with electron transfer, *Vitam. Horm. (Leipzig)*, 28, 533, 1970.
36. **Kröger, A. and Innerhofer, A.**, The function of menaquinone, covalently bound FAD and iron-sulfur protein in the electron transport of formate to fumarate of *Vibrio succinogenes*, *Eur. J. Biochem.*, 69, 487, 1976.
37. **Kröger, A. and Klingenberg, M.**, Further evidence for the pool function of ubiquinone as derived from the inhibition of electron transport by antimycin, *Eur. J. Biochem.*, 39, 313, 1973.
38. **Kröger, A. and Innerhofer, A.**, The function of the *b*-cytochromes in the electron transport from formate to fumarate in *Vibrio succinogenes*, *Eur. J. Biochem.*, 69, 497, 1976.
39. **Kröger, A., Winkler, E., Innerhofer, A., and Schagger, H.**, The formate dehydrogenase involved in electron transport from formate to fumarate in *Vibrio succinogenes*, *Eur. J. Biochem.*, 94, 465, 1979.
40. **Kenney, W. C. and Kroger, A.**, The covalently bound flavin of *Vibrio succinogenes* succinate dehydrogenase, *FEBS Lett.*, 73, 239, 1977.
41. **Futai, M.**, Orientation of membrane vesicles from *Escherichia coli* prepared by different procedures, *J. Membr. Biol.*, 15, 15, 1974.
42. **Mitchell, P.**, Coupling of phosphorylation to electron and hydrogen transfer by a chemi-osmotic type of mechanism, *Nature (London)* 191, 144, 1961.
43. **Mitchell, P.**, Chemiosmotic coupling in oxidative and photosynthetic phosphorylation, *Biol. Rev. Biol. Proc. Cambridge Philos. Soc.*, 41, 445, 1966.
44. **Jones, R. W. and Garland, P. B.**, Sites and specificity of the reaction of bipyridylium compounds with anaerobic respiratory enzymes of *Escherichia coli*, *Biochem. J.*, 164, 199, 1977.
45. **Dorn, M., Andreesen, J. R., and Gottschalk, G.**, Fumarate reductase of *Clostridium formicoaceticum*, a peripheral protein, *Arch. Microbiol.*, 119, 7, 1978.
46. **Unden, G., Hackenberg, H., and Kröger, A.**, *Biochim. Biophys. Acta*, in press, 1980.

Chapter 2

RESPIRATION WITH NITRATE AS ACCEPTOR

A. H. Stouthamer, J. van't Riet, and L. F. Oltmann

TABLE OF CONTENTS

I. INTRODUCTION

Many bacteria can utilize nitrate as terminal hydrogen acceptor instead of oxygen. This process, called nitrate respiration or dissimilatory nitrate reduction, occurs only under anaerobic conditions. The first step of nitrate reduction is a reduction of nitrate to nitrite. In a number of organisms, nitrite may also be used as terminal hydrogen acceptor under anaerobic conditions. In this latter case, nitrate is converted to gaseous products, such as nitrogen or nitrous oxide, a process called denitrification. The distribution of the capacity of dissimilatory nitrate reduction and of denitrification among various species of bacteria has been recently reviewed.[1-3] The properties of nitrate reductase, the composition of the electron transport chain towards nitrate, and energy production during nitrate respiration have also been treated in a number of recent reviews.[4-11] Denitrification was reviewed by Payne and Balderston.[12] This review will be devoted to some new developments.

II. PURIFICATION AND PROPERTIES OF NITRATE REDUCTASE

A. Solubilization of Nitrate Reductase from the Cytoplasmic Membrane

The different methods that have been used for the solubilization of nitrate reductase from the cytoplasmic membrane during enzyme isolation procedures have included:

Solubilization by detergents—Ionic[13-21] as well as nonionic[15,17,22-25] detergents have been employed for this purpose. The detergents substitute for membrane lipids whereby proteins are solubilized as protein-detergents micelles. A primary requirement for a suitable solubilizing agent, in addition to a good recovery, is a good selective solubilization of a specific enzyme under investigation. For example, after treatment of membrane vesicles of *Klebsiella aerogenes* with 1.5% deoxycholate,[15] 85% of nitrate reductase activity was solubilized, whereas only 30% of total membrane protein became soluble. The use of Triton® X-100, however, resulted in the solubilization of 90% of nitrate reductase activity and over 70% of membrane proteins. Addition of 0.3 M KNO_3, a chaotropic agent, made the solubilization with deoxycholate more reliable.[21] Other detergents that have also been used inlude sodium dodecylsulphate (SDS), employed for the solubilization of nitrate reductase from *Bacillus stearothermophilus* membranes,[17] as well as nonionic detergents like Triton® X-100[15,17,23-25] and Brij®-36T.[22]

Solubilization by organic solvent—This method results in the extraction of lipids from membranes into the organic phase, thus rendering a portion of the membrane proteins soluble. Using acetone Forget and co-workers[26-28] solubilized nitrate reductase from *Paracoccus denitrificans*, *Micrococcus halodenitrificans*, and *Escherichia coli* membranes, whereas Iida and Taniguchi[13] used isobutanol. Oltmann et al.[29] have used a repeated extraction of *Proteus mirabilis* membranes with tertiary amyl alcohol which resulted in the solubilization of other membrane-bound enzymes like tetrathionate reductase, chlorate reductase C, hydrogenase, and formate dehydrogenase as well.

Solubilization by heat treatment—Taniguchi and Itagaki[30] were the first to solubilize nitrate reductase from *E. coli* membranes with this method which was later repeated by MacGregor et al.[31] and Lund and DeMoss.[32] Since the presence of the protease inhibitor *p*-aminobenzamidine during heat treatment inhibited release of the enzyme from the membranes, MacGregor[33] concluded that heat treatment activates a protease which splits the catalytic part of nitrate reductase from the membranes. DeMoss[20] believes that proteolysis is not necessary, but may enhance the solubilization of nitrate reductase during heat treatment.

The different procedures used for the solubilization of nitrate reductase have led to the isolation of different nitrate reductase complexes. It has been shown, for instance, that nitrate reductase from *E. coli* can be solubilized while still complexed with formate dehydrogenase (formate being an electron donor for nitrate reductase in *E. coli*, see the section entitled *Escherichia coli*) and cytochrome *b*.[13,34] On the other hand, removal of the solubilizing conditions (e.g., detergents) prior to further purification may lead to hydrophobic aggregation with other solubilized membrane components,[15] such as cytochrome *b*, glycoproteins, etc.

B. Molecular Properties of Respiratory Nitrate Reductase

1. Purification and Subunit Structure

Several groups have purified nitrate reductase from *E. coli* to homogeneity.[18,25,28,30-32] Nitrate reductase from the related *K. aerogenes* was purified by van 't Riet and Planta[15,19] and from *P. mirabilis* by Oltmann et al.[35] Nitrate reductase from *Micrococci* was purified by Forget et al.[26,27] and by Lam and Nicholas.[14] Recently, a homogeneous enzyme has also been isolated from *Bacillus licheniformis*.[21]

Though differences in subunit structure have been published, t emerges from the literature that the minimal composition of respiratory nitrate reductase, which still reduces NO_3^- to NO_2^- with the use of artificial electron donors like reduced viologens, consists of only two polypeptides. Thus nitrate reductase from *E. coli*,[18,25,31,32] *K. aerogenes*,[19] *P. mirabilis*,[35] and *B. licheniformis*[21] consists of two polypeptide chains having molecular weights of about 150,000 (α-subunit) and 60,000 (β-subunit) which are generally present in an equimolar ratio. Of note is the observation that the α-subunits as well as the β-subunits of nitrate reductase from *E. coli*, *K. aerogenes*, and *B. licheniformis* comigrate upon sodium dodecylsulfate polyacrylamide gel electrophoresis.[162] If *E. coli* nitrate reductase, prior to further purification, is solubilized from the membranes with Triton® X-100, the purified enzyme usually also contains cytochrome *b*.[18,23,25] Since this association was not found for nitrate reductase from other organisms which had been solubilized with detergents,[19,21] it may be indicative of a closer connection of cytochrome *b* and nitrate reductase in *E. coli*. Alternatively if *E. coli* nitrate reductase is solubilized by heat treatment, cytochrome *b* is apparently not present in the purified enzyme,[30-32] and the β-subunit appears to have been split into several smaller polypeptides by heat-activated proteases.[31,32] DeMoss[20] has shown that treatment with trypsin removes a 15,000-dalton fragment from the β-subunit, leaving a 43,000-dalton portion of the β-subunit intact. van't Riet et al.,[21] have shown that proteolytic breakdown of *B. licheniformis* nitrate reductase also results in several distinct polypeptides, which are thought to have arisen from the β-subunit. *E. coli* K_{12} nitrate reductase[25] contains two forms of the β-subunit, differing in molecular weight by about 2,000 daltons and present in an equimolar ratio as compared with the α-subunit. In view of the presence of carbohydrate in *E. coli* K_{12} nitrate reductase,[36] the possibility that a portion of the β-subunits in Clegg's preparation[25] contains glyco-moieties should be considered. van't Riet and Planta[19] and Oltmann et al.[35] have shown the presence of another polypeptide in the 60,000-dalton region of nitrate reductase from *K. aerogenes* and *P. mirabilis*. The origin of these polypeptides is not known. The authors[19] conclude that this polypeptide (M_r 52,000) does not have a catalytic function in the reduction of nitrate by reduced benzylviologen. Its presence, however, makes the enzyme more stable upon storage. Since this polypeptide, which contains iron not liganded by acid-labile sulfide, has not been found in immunoprecipitates from Triton® X-100 extracted membranes, it may therefore be a breakdown product of one of the subunits of the native enzyme. The presence of this polypeptide in nitrate reductase, like the presence of cytochrome *b* in the *E. coli* enzyme gives rise to enzyme preparations which differ in molecular weight compared to the native enzyme containing only intact α- and β-subunits. Occasionally both entities have been found in isolated enzyme preparations.[18-20,35]

2. Molecular Weight, Association, and Dissociation

The monomeric form of isolated nitrate reductase, consists of one α- and one β-subunit and has a molecular weight of about 200,000 daltons.[15,19,21,25,27,32] The molecular weight is obviously increased with extra polypeptides, i.e., cytochrome b, being present.[19,25,35] In the absence of detergent, nitrate reductase from *E. coli* and *K. aerogenes* after isolation appear to have a much higher molecular weight (about 800,000 daltons).[15,19,21,25,30-32] Values of 320,000 and 400,000 daltons have also been found for *E. coli*[28] and for *K. aerogenes* nitrate reductase,[15] respectively. As has been shown,[15,19,25,32] these are tetrameric and dimeric forms of nitrate reductase. *P. mirabilis* nitrate reductase can also exist in these different aggregation forms.[35] van 't Riet and Planta[19] and Clegg[25] have shown that this aggregation phenomenon is governed by hydrophobic forces, since detergents bring about a reversible dissociation to the monomeric form even at relative high protein concentrations of 2 to 4 mg/mℓ. Lund and DeMoss[32] have shown that such a dissociation is also obtainable by dilution which does not exclude the possibility of hydrophobic forces governing the association-dissociation behavior. Interesting is the aforementioned observation of DeMoss[20] that trypsin splits a 15,000 dalton piece from the β-subunit, which concurrently results in the loss of the aggregative properties of the remaining enzyme. As DeMoss[20] suggests, this tryptic polypeptide of the β-subunit may at least be partially responsible for the binding of nitrate reductase to the cytoplasmic membrane. Nitrate reductase from *B. licheniformis*[21] and two types of *Micrococcus*[26,27] after isolation appear to be present in a monomeric form. Since van 't Riet et al.[21] applied the same procedure for the isolation of nitrate reductase from *B. licheniformis* and *K. aerogenes*, this property cannot be related to the solubilization method used. These authors suggest that nitrate reductase from *B. licheniformis* and the *Micrococci* only have a small hydrophobic surface, as compared with nitrate reductase from *E. coli* and *K. aerogenes*. This implies a more intense hydrophobic interaction of the enzyme from the latter organisms with the bacterial cytoplasmic membrane.[21]

C. The Interaction of Nitrate Reductase with the Cytoplasmic Membrane

1. The Localization of Nitrate Reductase in the Membrane

Two different approaches have been used to study the localization of nitrate reductase in cytoplasmic membranes. Nitrate reductase protein is labeled *in situ* or enzymic methods are employed for the localization of the active center of the enzyme (see next paragraph). Clegg and Boxer[37] have shown that tyr residues from the α-subunit of *E. coli* nitrate reductase can only be labeled by the[125]I/lactoperoxidase method from the cytoplasmic aspect of the membrane, whereas the β-subunit was not labeled at all from either side of the membrane. A similar result has been obtained by MacGregor and Christopher,[38] who used the transglutaminase system which labels glutamine residues. Presumably due to a lack of glutamine, the β-subunit was not labeled. Other possible explanations of the β-subunit not being accessible to th labeling systems used[37,38] may be that it is buried within the membrane, has a very hydrophobic surface, or is masked by the α-subunit. Recently Graham and Boxer,[39] by using the indirect immunofluorescence technique in combination with antibodies specific for the α-subunit, have shown that this subunit in *E. coli* is located solely on the cytoplasmic face of the membrane. In contrast to the results of MacGregor and Christopher,[38] Boxer and Clegg[37] found that cytochrome b can be labeled from the periplasmic aspect of the membrane. The latter authors therefore suggest that the cytochrome b-nitrate reductase complex has a transmembrane orientation. No conclusion, however, could be drawn regarding the transmembrane location of nitrate reductase itself. Wientjes et al.[40,41] have studied the localization of nitrate reductase in *K. aerogenes* and *B. licheniformis* membranes with the use of the ^{125}I/lactoperoxidase and immunofluorescence techniques. By the former method, they obtained similar results for *K. aerogenes* as for *E. coli*[37,38] nitrate reductase; however, in *B. licheniformis* both subunits were labeled from the cytoplasmic aspect of the membrane. By using antibodies against the distinct subunits in the indirect immunofluorescence technique, Wientjes et al.[40,41] showed that whereas no antigenic groups of nitrate reductase were exposed on the periplasmic side of *B. licheniformis* membranes, this was actually the case with *K. aerogenes*. Since antisera against *K. aerogenes* nitrate reductase subunits unfortunately cross-reacted in the micro-immunodiffusion assay, no conclusion could be drawn as to whether one or both of the subunits were located on the periplasmic side of the membrane. This result demonstrates that the catalytic unit of nitrate reductase (α- plus β-subunit) in *K. aerogenes* is a transmembrane protein. Since both the α- and the β-subunit are labeled by lactoperoxidase using isolated nitrate reductase, Wientjes et al.[40,41] assumed that no tyr residues are exposed on the periplasmic side of the membrane. Since *E. coli* nitrate reductase after solubilization from the membranes by heat treatmnt is present in a tetrameric form,[23,31,42] it has been suggested that this is the form of the enzyme *in situ*. Evidence for this suggestion, however, does not seem to be strong at the moment. If membrane lipids or other hydrophobic proteins like cytochrome *b* occupy the hydrophobic surfaces of the enzyme, nitrate reductase need not be self-associated within the membrane. In fact Enoch and Lester[43] have shown that isolated *E. coli* nitrate reductase containing cytochrome *b* is in a dimeric form, consisting of two α- and two β-subunits and four cytochrome *b* molecules. The 15,000-dalton product of tryptic digestion of the β-subunit[20] (see before) may therefore be responsible for hydrophobic interaction with the membrane lipids or cytochrome *b*. MacGregor[42] has shown that the presence of cytochrome *b* in *E. coli* membranes is necessary for the incorporation of nitrate reductase into the membranes. The high concentration of nitrate reductase found in the cytoplasmic membranes[31,43,44] (about 25% of the total membrane protein), however, may argue in favor of an associated form of nitrate reductase *in situ*.

2. The Localization of the Active Center of Nitrate Reductase in the Membranes

As will be discussed further on, nitrate reductase *in situ*, accepts electrons from a cytochrome *b*. The activity of the solubilized enzyme not containing cytochrome *b* can only be measured by using artificial electron donors like reduced viologens.[19] Kinetic evidence obtained with *K. aerogenes*[9,45] and *B. licheniformis*[21] nitrate reductase, based on the different inhibition character of azide with respect to reduced benzylviologen and nitrate, respectively, illustrates that the electron-accepting site and the electron-donating site (nitrate-reducing site) of the catalytic center are different. Jones and Garland[46] have shown that the electron-accepting site (from reduced viologens) of *E. coli* nitrate reductase is located on the cytoplasmic aspect of the membrane, whereas the nitrate-reducing site is on the periplasmic aspect of the membrane.[47,48] More recently, however, this location of the nitrate-reducing site of nitrate reductase in *E. coli* membranes has been doubted.[49] Differences in azide inhibition of nitrate reductase activity in cells and inverted membrane vesicles by varying pH have been interpreted in such a way that the nitrate reducing site is on the cytoplasmic aspect of the membrane.[49] Hence the location of the active center of nitrate reductase in *E. coli* appears to be the same as in *P. denitrificans*[50] and *B. licheniformis*.[41]

3. Summary

Summarizing the experimental evidence on the localization of respiratory nitrate reductase in the bacterial cytoplasmic membrane, (Figure 1) the following emerges:

1.   The active center of nitrate reductase is located on the cytoplasmic side of the membrane of the bacteria studied thus far.
2.   In *B. licheniformis*, both subunits are located on the cytoplasmic side, whereas in *E. coli* only the α-subunit can be detected on that side, the β-subunit not being accessible. In the latter organism cytochrome *b* is, however, on the periplasmic side, indicating a transmembrane location for the cytochrome *b* nitrate reductase complex. In *K. aerogenes*, nitrate reductase itself is a transmembrane protein, and the possibility cannot be excluded that both subunits are transmembrane.[41] The presence of carbohydrate in both subunits of this enzyme[51] is also considered in Figure 1.
3.   The different association-dissociation behavior of isolated nitrate reductase from *E. coli* and *K. aerogenes* on the one hand and of *B. licheniformis* on the other (see the section entitled Molecular Weight, Association, and Dissociation) coincides with the different locations of nitrate reductase in the membranes, the latter having a less intense interaction with the membrane.

The functional meaning of the differences in membrane localization of nitrate reductase may be related to the various ways in which nitrate reduction in these organisms is connected with oxidative phosphorylation (see the section entitled Energy Conservation during Nitrate Reduction).[5,7,9,41]

D. The Participation of Metals in Nitrate Reductase Activity
1. The Presence of Metals and Acid-Labile Sulfide in Nitrate Reductase
Per 200,000 daltons, *E. coli* nitrate reductase contains about 12 Fe-S groups[28,32] and one atom of molybdenum.[28,31,32] *B. licheniformis*[21] and *P. denitrificans*[26] nitrate reductase contain about eight Fe-S groups and one atom of molybdenum per enzyme molecule. *K. aerogenes* nitrate reductase contains the same amount of Fe-S groups[45] and molybdenum[44] as the latter enzymes. If the 52,000-dalton polypeptide is present, however, the latter enzyme contains also 12 iron atoms, 4 of which are not liganded by acid-labile sulfide. Sometimes lower metal contents have been found,[27,28,45] presumably due to losses caused by the purification method[27,45] or the treatment of the enzyme prior to metal determination.[44] These results and electron paramagnetic resonance (EPR) work to be discussed indicate the presence of two or three 4Fe-4S iron-sulfur clusters in nitrate reductase.

2. The Role of Iron-Sulfur Clusters and Molybdenum
Evidence for the participation of molybdenum in nitrate reductase activity has been obtained by growing organisms in the presence of WO_4^{--}, instead of MoO_4^{--}. Since W is incorporated into nitrate reductase which makes the enzyme inactive, these organisms do not exhibit nitrate respiration.[4] The role of molybdenum is confirmed by the inhibitory effect of molybdenum chelating agents like thiocyanate and dithiol on the enzyme activity.[14,16,45] Participation of Fe-S groups is indicated by the inhibitory effect of agents like bathophenathroline.[16,45] Unequivocal evidence for the participation of these metals in the intramolecular electron transfer was obtained by EPR studies on nitrate reductase from *M. dentrificans*,[52] *K. aerogenes*,[44,45] and *E. coli*.[53-55] Molybdenum resonances were studied at 80 to 120°K and those of iron-sulfur groups at 12 to 18°K. Table 1 summarizes the signals obtained thus far. At low temperature the oxidized enzyme in general shows a resonance at about g = 2.015, due to nonheme ferric iron.[44,75,52] This signal disappears upon reduction by dithionite[45,52,53,55] or NADH[44] (when membrane vesicles are used) with the concommitant appearance of rhombic signals of the Fe-S clusters. Two of the latter signals are found in *E. coli* and *M. denitrificans* nitrate reductase,[52,53,55] and only one is found in the *K. aerogenes* enzyme.[45]

Table 1

ELECTRON PARAMAGNETIC RESONANCE SIGNALS OF NITRATE REDUCTASE

Nitrate reductase from	g values				Reductor	Ref.
	12 to 18°K		77 to 120°K			
	Oxidized[a]	Reduced	Oxidized[a]	Reduced		
Micrococcus denitrificans	2.016	2.057,1.947,1.881 2.031,1.926 2.057,1.947,1.881	2.045[b],1.985	2.023,1.999 None	$Na_2S_2O_4$ $Na_2S_2O_4$ Hydrogenase + H_2	52
Escherichia coli	2.074,2.005	2.047,1.889,1.861 2.030,1.948	1.988	2.032,2.008	$Na_2S_2O_4$	53
Escherichia coli	2.015	2.041,1.945,1.921 2.033,1.888,1.870	2.006[b] 1.987,1.980,1.961 1.999,1.985,1.964 1.996,1.984,1.956	None		54,55
Klebsiella aerogenes	2.015 2.10,2.03	2.10,2.03 2.05,1.95,1.88	2.02[b] 1.98	2.02[b] 1.98	$Na_2S_2O_4$ $Na_2S_2O_2$	45
Klebsiella aerogenes (membranes)	2.04,2.00 2.02	2.01,2.00 2.05,1.95,1.88	2.04[b],2.02[b],2.00[b] 1.989,1.968	2.04[b],2.02[b],2.00[b] None	NADH NADH	44

[a] Enzyme isolated as such or reoxidized by NO_3^-.

[b] These signals are not due to molybdenum.[a]

Reoxidation by nitrate gives rise to the original signal at about g = 2.015. These results show clearly the participation of Fe-S centers in the nitrate-reducing activity of the enzyme. For *K. aerogenes*, nitrate reductase also an axial-symmetrical signal was found with lines at g = 2.10 and g = 2.03, due to the iron not liganded by acid-labile sulfur.[45] This signal was not influenced by reduction or reoxidation.

At higher temperatures (80 to 120°K), the Fe-S centers do not give resonances, and at least those signals with g_{av} at about 1.98[44,54,55] can be attributed to molybdenum.

The oxidation state of molybdenum may vary from (I) to (VI). The uneven states are paramagnetic and detectable by EPR, whereas the even oxidation states are diamagnetic and EPR-undetectable. Quantitation of the Mo(V) signal in the oxidized enzyme shows that only part of molybdenum, even in the presence of NO^-_3, is present as Mo(V).[52,54,55] In oxidized membrane vesicles of *K. aerogenes*, even only 2% of nitrate reductase molybdenum is detectable as Mo(V).[44] Vincent and Bray[55] described several EPR detectable forms of Mo(V) in *E. coli* nitrate reductase: a high- and a low-pH species, of which the low pH one is functional. Interaction with NO^-_3 and NO^-_2 give modified Mo(V) signals. Moreover these authors describe a nonfunctional species of Mo.

The events occurring upon reduction of the enzyme seem to be dependent on the composition and the way the enzyme is isolated. Mild reduction of nitrate reductase from *M. denitrificans*[52] by hydrogenase plus H_2 results in the disappearance of the Mo(V) signal, molybdenum being reduced to diamagnetic Mo(I). Reduction of this enzyme and also that of *E. coli*[53] by a slight excess of dithionite results also in the disappearance of the original Mo(V) signal, but gives rise to another signal (g values at 2.03 and 2.00) which Forget and DerVartanian[52,53] suggest to represent Mo(III). Both-mentioned enzyme preparations do not contain cytochrome *b*. Vincent and Bray[55] performed similar experiments with *E. coli* nitrate reductase, containing cytochrome *b*. In this case, the Mo(V) signal disappeared. Bosma et al.,[44] working under near physiological conditions with *K. aerogenes* membranes, also showed the disappearance of the Mo(V) signal upon NADH reduction. In the mentioned studies, the Mo(V) signals reappear upon reoxidation of the samples by NO^-_3.

3. On the Mechanism of Molybdenum Participation

The aforementioned experiments do not elucidate in detail the mechanism of enzyme action. Though the iron-sulfur clusters have also to be considered, almost nothing is known about the interaction of these groups with molybdenum. On the other hand, some remarks can be made on the action of molybdenum itself. For the reduction of NO^-_3 to NO^-_2, two electrons and two protons are needed. The involvement of protons is also reflected in the EPR signals of Mo(V) which show splittings due to interaction with protons.[44,54,55] Since all experimental evidence points to the presence of one atom of molybdenum per enzyme molecule, the metal should transfer two electrons to NO^-_3. Molybdenum therefore could function as a Mo(III)/Mo(V) or a Mo(IV)/Mo(VI) redox couple. Most experimental evidence[44] now points to a Mo (IV)/Mo(VI) redox couple (see Stiefel).[56] Some remarks have to be made on the occurrence of the Mo(V) signal in isolated oxidized enzyme which maximally counts for 25% of the molybdenum present. Upon reduction of the enzyme, Mo(IV) is present,[44,54,55] and the results might indicate a Mo(IV)/Mo(V) redox couple functioning in the reduction of NO^-_3. Perhaps two one-electron transfers could occur sequentially or a two-electron transfer followed by a rapid intermolecular equilibration of molybdenum and iron-slfur (see Vincent and Bray.[55]) However, some evidence is presented indicating that NO^-_3 has accepted only one electron. This evidence follows from the presence of some unknown signals in *E. coli* nitrate reductase after a dithionite reduction/nitrate oxidation cycle.[53,55]

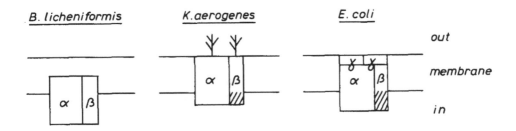

FIGURE 1. A schematic representation of our present (still fragmentary) knowledge on the localization of respiratory nitrate reductase in the cytoplasmic membrane in some bacteria (see text). α and β denote the subunits of the enzyme, while γ represents cytochrome b. The hatched area in the β-subunit represents that part which contains no tyr residues, but reacts with β-subunit specific antiserum.[164] Only in *Klebsiella aerogenes* is the presence of carbohydrate in the subunits accounted for. For *Escherichia coli*, the presence of carbohydrate in nitrate reductase[36,165] is more difficult to reconcile with the depicted structure.

These signals are suggested to be due to interaction of NO with Fe-S centers[53] or heme-iron.[55] In this connection, some earlier and more recent studies on reduction of NO_3^- with inorganic molybdenum complexes have to be mentioned. Reduction of NO_3^- with Mo(V)-complexes results in the formation of NO_2 (one electron transfer).[57,58] NO_2 can be further reduced and/or reacts with H_2O, giving rise to NO and also NO_2^- in substoichiometric quantities.[57] On the other hand, Mo(III)-complexes reduce NO_3^- stoichiometrically to NO_2^-, Mo(III) being oxidized to Mo(V).[59,60] Though these studies have not been possible with Mo(IV)-complexes which are very labile, they suggest that molybdenum can transfer either one or two electrons to NO_3, giving rise to different products. By analogy the occurrence of signals of NO metal cmplexes in nitrate reductase suggest that during the reduction-oxidation cycle partly one-electron transfers to NO_3^- have taken place.[53,55] Two reasons could be mentioned for that. Firstly, during the isolation procedure, part of the enzymes molybdenum may, dueto conformational stress, have lost the possibility to be oxidized to the hexavalent state. Nonfunctional molybdenum is also present in other molybdenum containing enzymes (see Bray).[61] The very low content (2%) of Mo(V) in oxidized membrane vesicles,[44] as compared with oxidized isolated nitrate reductases[52,53,55] (up to 25%), is in favor of this proposition. The second reason may reside in the use of dithionite as a reductor. van't Riet et al.[19,45] have shown that isolated *K. aerogenes* nitrate reductase hardly produces any NO_2^- from NO_3^- with dithionite, though dithionite per se is able to reduce the metal groups in the enzyme,[45] which on turn can be reoxidized by NO_3^-. Only the addition of viologens results in the production of NO_2^-. Thus, by using dithionite as the sole reductor, it seems possible that NO_3^- is reducd to other products than NO_2^-. Further kinetic EPR studies will be needed to elucidate the mechanisms by which the Fe-S centers and molybdenum function in nitrate reductase and whether "transient" Mo(V) signals between Mo(IV) and Mo(VI) can be found.

4. The Molybdenum Cofactor

Since the classic work of Nason's group,[62-64] it is well established that molybdenum is present as a cofactor in several molybdo-enzymes and is presumably a peptide having a molecular weight of about 1000 daltons. Possibly the molybdenum cofactor is identical with a small molybdenum-binding molecule which was observed in the supernatant fraction of a cell-free extract of *E. coli* grown in the presence of ^{99}Mo.[65] This cofactor can be exchanged between several enzymes and thus appears not to be very specific. Recent results obtained by Brill and co-workers[66,67] indicate the presence of an Fe-Mo cofactor in nitrogenase. Nitrogenase without this cofactor cannot be activated with the molybdenum cofactor isolated from xanthine oxidase and assimilatory nitrate reductase from *Neurospora crassa.*[67] The molybdenum cofactor from xanthine oxidase, however, is able to activate the assimilatory nitrate reductase of a mutant of *N. crassa*, indicating that nitrate reductase and xanthine oxidase share a common molybdenum cofactor which is different from the molybdenum cofactor in nitrogenase.[67] These results are confirmed by studies with chlorate resistant mutants. Chlorate-resistant mutants are deficient in nitrate reductase, but generally are affected in the formation of a number of other enzymes[47,68] as well. A number of chlorate resistant mutants are affected in the formation of the molybdenum cofactor for nitrate reductase.[47,68] It has been demonstrated that nitrogenase activity is not affected in chlorate resistant mutants of *E. coli* [69,70] (containing the *nif* genes of *K. pneumoniae*), *Rhizobium* sp.,[71] and *Azospirillum* sp.[72] Mutants of *E. coli* carrying the *chlD* gene form nitrate reductase, nitrogenase, and formate dehydrogenase only after growth in media with very high concentrations of molybdate.[69,73] Since *chlD* mutants are not affected in the uptake of molybdate,[73] it is evident that nitrate reductase and nitrogenase share at least one molybdenum-processing function.[69]

The presence or absence of iron in the Mo cofactor of respiratory nitrate reductase has not been determined as yet. The Mo cofactor can be extracted from several enzymes by acid treatment.[62-64] van 't Riet et al.[21] have shown that the Mo cofactor of respiratory nitrate reductase both from *B. licheniformis* and *K. aerogenes* is also removed from the enzyme by sodium dodecylsulfate treatment, indicating that the Mo cofactor is not covalently bound to one of the subunits. Also interesting is the observation that molybdenum is dissociated from the cofactor upon sodium dodecylsulfate electrophoresis in the absence, but not in the presence, of β-mercaptoethanol, suggesting that thiol groups or reducing conditions per se bind Mo to the cofactor. Wientjes and van 't Riet[163] have shown that antiserum against reductase from *K. aerogenes* like that from *B. licheniformis* contains antibodies against the Mo cofactor which precipitates antigenic material from bacteria grown in such a way that no active enzyme or subunits were detectable. Presumably, because of the antigeneity of the Mo cofactor, other molybdo-enzymes, e.g. formate dehydrogenase, are also precipitated by nitrate reductase antiserum. This may indicate that the Mo cofactors associated with various enzymes of an organism are similar if not identical entities. Considerable genetic evidence is available in support of this proposition.[4,68]

III. RESPIRATORY CHAIN TO NITRATE

The organization of the respiratory chain to nitrate has been reviewed several times recently.[4-6] The discussion in this chapter will therefore be restricted to only a few organisms.

A. *Escherichia coli*

Based on a comparison of the rates of nitrate reduction with membrane vesicles with formate or NADH as hydrogen donors, respectively, it is generally agreed that formate is the preferential electron donor for nitrate reduction in *E. coli.*[4] This conclusion, however, cannot be maintained at the moment. For the related *K. aerogenes*, it has been shown that NADH is the most important hydrogen donor for nitrate reduction.[4] This conclusion followed from the fermentation balance for anaerobic growth of *K. aerogenes* with glucose and nitrate.[4] Similar studies have now been performed with *E. coli.*[74] The results showed that during anaerobic growth of *E. coli* with glucose and nitrate, no ethanol is formed as a fermentation product, indicating that all the NADH produced in the glycolytic system is oxidized by nitrate. At the end of the growth period, considerable amounts of formate were still present. Subsequently the formate was slowly oxidized with nitrate as hydrogen acceptor.[74] As a consequence, in vivo NADH is the most important hydrogen donor for nitrate reduction in *E. coli*.

The formate dehydrogenase of *E. coli* has been purified and was found to contain a cytochrome *b* subunit plus molybdenum and selenium.[43] Nitrate reductase also contains a cytochrome *b* subunit (the γ subunit) and molybdenum.[18] The cytochrome *b* involved in nitrate reduction is different from the cytochrome *b* components of the aerobic respiratory chain. On base of these data, the aerobic respiratory chain and the respiratory chain to nitrate are shown in Figure 2, which is a modification of earlier versions in recent reviews.[5,6] As mentioned before, the most recent experiments indicate that nitrate is reduced at the cytoplasmic side of the membrane.[49] These results indicate furthermore that in contrast with earlier statements,[46-48] nitrate reductase is not proton translocating and also that nitrate should cross the cytoplasmic membrane before reduction.

B. *Klebsiella aerogenes*

The respiratory chain to oxygen and nitrate for *K. aerogenes* is shown in Figure 2.[4] It has not been studied in this organism at which side of the cytoplasmic membrane nitrate is reduced.

C. *Paracoccus denitrificans*

The electron transport chain to oxygen, nitrate and nitrite is shown in Figure 2.[9,76,77] Evidence has been obtained that nitrate is reduced at the cytoplasmic side of the membrane.[36] Nitrite reductase, which is identical with cytochrome *cd*,[14,78] is an inducible enzyme.[14,79] The enzyme is localized at the periplasmic side of the cytoplasmic membrane.[80] For *Pseudomonas aeruginosa* it had been shown earlier that the terminal reductase in nitrite respiration is located in the periplasmic space.[81] This indicates that in both organisms nitrite is reduced at the outer side of the cytoplasmic membrane. With *Paracoccus denitrificans* reduction of nitrite can be obtained with ascorbate plus N,N,N′,N′-tetramethyl-p-phenylenediamine (TMPD), as hydrogen donor.[80] Ascorbate-TMPD directly donates electrons to cytochrome *c*. In this reduction, four protons are used from the medium per mol nitrite reduced, indicating that nitrite is reduced to nitrogen[80] ($NO_2^- + 4 H^+ + 3e^- \rightarrow \frac{1}{2} N_2 + 2 H_2O$). Nitrous oxide, which is generally considered to be an intermediate in the reduction of nitrite to itrogen,[1,2,12] is not reduced with ascorbate + TPMD. On the other hand nitrous oxide gives oxidation of cytochrome *c* and, furthermore, the reduction of nitrous oxide is strongly inhibited by antimycin A.[82] These observations indicate that reduced cytochrome *c* is the direct electron donor for the reduction of nitrous oxide. It is generally accepted that nitrous oxide is an obligatory free intermediate in the reduction of nitrite.[1,2,12] In our opinion however the evidence can also be explained in a different way. It might be that enzyme bound intermediates of nitrite reduction may be released as nitrous oxide under various circumstances, e.g., in the presence of acetylene,[83,84] and that different pathways exist for the reduction of this enzyme bound intermediate and that of exogenous nitrous oxide. We can speculatively represent the pathway of denitrification as in Figure 3; in which nitrous oxide is an intermediate in denitrification but not an obligatory free intermediate.

The reduction of nitrous oxide is strongly inhibited by azide, whereas the reduction of nitrite is not or is scarcely influenced.[85-88] Recently, it has been shown for *P. aeruginosa* that nitrite is reduced to nitrogen in the presence of azide.[89] These observations were explained by the assumption that nitrite released the inhibitory effect of azide on the reduction of nitrous oxide. However the explanation given in Figure 3 seems more likely. In *Paracoccus denitrificans* nitrite was also reduced to nitrogen in the presence of azide with ascorbate + TPMD as hydrogen donor with the consumption of four

Escherichia coli

NADH $\xrightarrow{\ 1\ }$ fp   cyt b_{556} $\longrightarrow$ cyt o $\longrightarrow$ O_2

CoQ (2)

succinate $\longrightarrow$ fp   cyt b_{558} $\longrightarrow$ cyt d $\longrightarrow$ O_2

NADH $\longrightarrow$ fp

CoQ $\longrightarrow$ cyt b_{NR} $\longrightarrow$ nitrate

formate $\longrightarrow$ cyt $b_{\overline{FDH}}$

Klebsiella aerogenes

formate

NADH $\xrightarrow{\ 1\ }$ fp $\longrightarrow$ CoQ$_8$ $\longrightarrow$ cyt b_{559} $\xrightarrow{2}$ cyt o $\longrightarrow$ O_2

cyt b_{559} $\longrightarrow$ cyt b_{563} $\longrightarrow$ cyt a_1 $\longrightarrow$ cyt d $\longrightarrow$ O_2

succinate $\longrightarrow$ fp

nitrate reductase (2)

nitrate

Paracoccus denitrificans

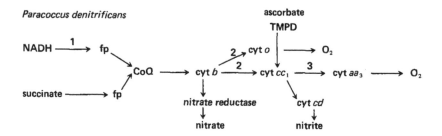

Thiobacillus denitrificans

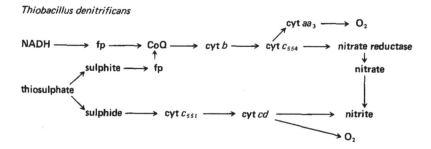

FIGURE 2.   The respiratory chain of a number of selected bacteria to oxygen, nitrate and nitrite. fp, flavoprotein: CoQ, coenzyme Q (ubiquinone). Phosphorylation sites in the respiratory chains of *Escherichia coli*, *Klebsiella aerogenes*, and *Paracoccus denitrificans* are indicated by 1, 2, and 3, respectively.

protons from the outer medium.[80] High concentrations of uncouplers inhibit the reduction of nitrous oxide, but not the formation of nitrogen from nitrite in *Pseudomonas denitrificans*,[88] which can also be explained by the scheme given in Figure 3. However in a strain of the *Alcaligenes/Achromobacter* group uncouplers completely

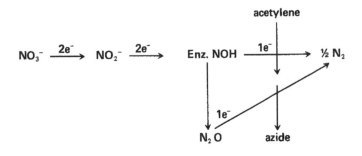

FIGURE 3.   Hypothetical scheme for the pathway of denitrification.
Enz. NOH denotes an enzyme-bound intermediate. The sites of inhibition by acetylene and azide are given by arrows.

inhibited the reduction of nitrous oxide, but in this case nitrite was only reduced to
nitrous oxide in the presence of the inhibitor.[90] More recent work with high concentrations of uncoupler show that both nitrite and nitrous oxide reduction are inhibited.[91]
It is thus evident that although some data can be explained well by the pathway of
denitrification shown in Figure 3, more work is necessary to see whether this scheme
is valid under all circumstances and in all species.

D. *Thiobacillus denitrificans*

The respiratory chain of this organism to oxygen, nitrate, and nitrite has been studied intensively by Nicholas and associates.[92-95] *T. denitrificans* is an obligate chemolith-
otroph which can derive the energy needed for growth by performing the reaction

$$5S_2O_3^{2-} + 8NO_3^- + H_2O \longrightarrow 10SO_4^{2-} + 4N_2 + 2H^+$$

The respiratory chain in this organism is in Figure 2. An unusual point in the respiratory chain of this organism is that cytochrome *c* is involved in nitrate reduction. Normally nitrate reductase accepts electrons from cytochrome *b* instead of cytochrome *c*.[4]
In *T. denitrificans*, however, nitrate reduction is strongly inhibited by antimycin A
which inhibits electron transfer between cytochrome *b* and cytochrome *c*.[94] It is evident
from Figure 2 that separate chains for electron transport exist for sulfide and sulfite
which both arise from thiosulfate. Electrons from sulfite are transferred to nitrate,
whereas electrons from sulfide are transferred to nitrite.[95] Different *c*-type cytochromes are involved in these two branches of the respiratory chain.

IV. ENERGY CONSERVATION DURING NITRATE REDUCTION

The $\Delta G_o'$ of the reaction

$$NADH + NO_3^- + H^+ \longrightarrow NAD^+ + NO_2^- + H_2O \text{ is } -34 \cdot 1 \text{ kcal/mol}$$

Consequently this reaction may be associated with the formtion of ATP. That nitrate
reduction is indeed associated with energy production is indicatd from four types of
experiments.

1.   The demonstration of ATP formation with cell-free extracts during electron
     transfer from various substrates to nitrate
2.   The extrusion of protons during the reduction of nitrate by whole cells
3.   The influence of nitrate on molar growth yields

4.   The energization by nitrate reduction of the active uptake of solutes in bacterial cells and membrane vesicles

A. Oxidative Phosphorylation Coupled to Reduction of Nitrate by Cell-Free Extracts and Whole Cells

In cell-free preparations, generally low P/O ratios are obtained.[96] It has been concluded that these values are not representative of those in growing cells.[8,96-98] The same applies to the P/O ratios determined in intact resting cells.[99] The P/2e$^-$ ratios determined for oxidative phosphorylation coupled to nitrate respiration are given in Table 2. Also in the case of nitrate respiration low P/2e$^-$ values are obtained, and furthermore in all cases, these values are lower than those for electron transfer to oxygen. In *P. denitrificans*, the P/2e$^-$ ratios for electron transfer to nitrate are 0.9 and 0.06, with NADH and succinate respectively as electron donors.[79] This was one of the arguments used for the location of the phosphorylation sites in the respiratory chain (Figure 2).

B. Extrusion of Protons During Nitrate Respiration

According to the chemiosmotic hypothesis of the mechanism of oxidative phosphorylation, oxidation of substrates must be accompanied by the extrusion of protons. It is implicit in this hypothesis that H$^+$/2e$^-$ quotients reflect the efficiency of oxidative phosphorylation. H$^+$/nitrate ratios have been measured for *E. coli* and *Klebsiella pneumoniae* for the oxidation of endogenous substrate; values of 4.20 and 3.85 were found, respectively.[109] These values are about the same as the H$^+$/O ratios for these organisms. This indicates that the efficiency of oxidative phosphorylation with oxygen and nitrate is about the same, which had been concluded earlier for, the related *K. aerogenes* from the measurement of molar growth yields.[110,111]

For *E. coli* nitrate-dependent proton translocation associated with the oxidation of various substrates added to starved cells yielded H$^+$/2e$^-$ ratios of four for L-malate and two for the oxidation of glycerol, succinate and D-lactate.[47] It was concluded that the site of formate oxidation is on the inner aspect of the cytoplasmic membrane and that the H$^+$/nitrate ratio is greater than two.[47] On basis of these results and the data on the localization of nitrate reductase in the membrane (see Section I.C.) the scheme shown in Figure 4 has been proposed for the functional organization of the redox carriers involved in the nitrate reductase system of *E. coli*.[5,48]

Studies on proton translocation coupled to nitrate reduction in *P. denitrificans* yield a very complicated picture. Orginally, no proton extrusion could be demonstrated with nitrate and nitrous oxide.[80] In recent experiments however it could be shown that when thiocyanate, which is normally present in the incubation medium for experiments on proton extrusion, is omitted proton extrusion during the reduction of nitrate and nitrous oxide occurs.[82] The latter observations are more in accordance with our knowledge of the electron transfer pathways to nitrate and nitrous oxide. During nitrite reduction proton extrusion occurs with a H$^+$/nitrite ratio of 6.4.[80] Recently it has been found that in *P. denitrificans* under anaerobic conditions in the presence of nitrate no detectable proton motive force was generated.[112] The occurrence of oxidative phosphorylation coupled to nitrate respiration[79] (Table 2) on the one hand and the absence of the generation of a proton motive force[112] are seemingly contradictory results. A possible explanation might be that nitrite is a potent uncoupling agent.[113] Nitrite has a lowering effect on H$^+$/O ratios and increases the rate of decay of the proton pulse curve. High concentrations can even completely circumvent the extrusion of protons. Therefore the occurrence of oxidative phosphorylation during reduction of nitrate or nitrite will depend on the extent of the uncoupling effect of nitrite under the prevailing conditions. Since nitrite has a potent uncoupling effect, the accumulation of nitrite

33

Table 2
OXIDATIVE PHOSPHORYLATION COUPLED TO REDUCTION OF NITRATE IN CELL-FREE EXTRACTS AND IN WHOLE CELLS OF VARIOUS ORGANISMS

Organism	Hydrogen donor cell-free extracts	P/2e⁻	Ref.
Pseudomonas	Succinate	0.25	100
denitrificans	Lactate	0.3	101
Pseudomonas	Lactate	0.3	102
aeruginosa			
Pseudomonas	NADH	0.52	103, 104
saccharophila	succinate	0.09	103, 104
Escherichia coli	NADH	0.55	105
	Glycerol-3-phosphate	0.20—0.30	106
Paracoccus deni-	NADH	0.9	79
trificans	Succinate	0.06	79
Nitrobacter agilis	NADH	0.7	107
	Whole cells		
Proteus mirabilis	Endogenous	0.37	99
Pseudomonas	Endogenous	0.50	108
aeruginosa			

during growth will have a strong influence on molar growth yields (see the section entitled Influence of Nitrate Respiration on Molar Growth Yields).

The data discussed above and those discussed in the section entitled *Paracoccus denitrificans* lead to the picture of nitrate and nitrite respiration in Figure 5. In this figure, the extrusion of three to four protons per pair of electrons and per phosphorylation site is assumed which is in agreement with the experimental evidence for this organism.[77,114,115] An important point is that the four protons used in the reduction of nitrite do not give a contribution to the formation of the proton motive force. The two protons used in the reduction of nitrate to nitrite, however, do give a contribution to the formation of the protonmotive force. This difference is due to the difference in localization of nitrate and nitrite reductase. During reduction of nitrate to nitrogen, the nitrite produced has to move to the other side of the membrane for further reduction. It is uncertain whether the nitrite is transferred as undissociated acid or not; in the first case, a proton is cotransported. Another uncertain factor is the transport of nitrate across the membrane. In bacteria evidence for active transport systems for nitrate uptake is completely absent. Since the influence of these processes on the proton balance is unknown, they will be neglected for the moment. We expect a H⁺/nitrite ratio between 5 and 8 (Figure 5) which is in good agreement with the experimental value of 6.4.[80] From the data in Figures 2 and 5, we may conclude that one ATP is formed in the electron transport from NADH to nitrate and two in the electron transport from NADH to nitrite. From a comparison of the respiratory chains of *E. coli* (Figures 2 and 4) and of *P. denitrificans* (Figures 2 and 5), it is evident that in *E. coli* two phosphorylation sites are present between NADH and nitrate against only one in *P. denitrificans*. The explanation for this difference is still lacking. One might speculate that the difference is due to a difference in the transmembrane character of the nitrate reductases in these organisms (see the section entitled The Interaction of Nitrate Reductase with the Cytoplasmic Membrane). In Figure 5, the branch of the respiratory chain to cytochrome aa_3 has been omitted because it is known that cells grown under anaerobic conditions in the presence of nitrate only contain very small amounts of this cytochrome.[14,116]

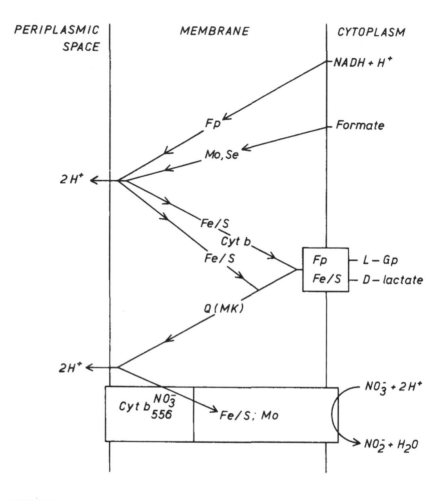

FIGURE 4.   Proposed functional organization of the redox carriers for electron transport with nitrate as terminal hydrogen acceptor in *Escherichia coli*. The scheme is a revision of the version of Haddock and Jones[5] to account for the new findings of Jones et al.[49] Mo, molybdenum cofactor: Se, selenium-containing polypeptide: α, β, subunits of nitrate reductase: Fp, flavoprotein, Fe/S, iron-sulfur protein: Q, coenzyme Q (ubiquinone): and MK, menaquinone.

C. Influence of Nitrate Respiration of Molar Growth Yields

The amount of bacterial dry weight formed during growth is directly proportional to the amount of ATP which can be obtained from the catabolism of the energy-yielding substrate.[117] The relation between growth and energy production has been discussed in a number of recent reviews.[8,9] The influence of nitrate on molar growth yields has also been reviewed recently.[4,8,9] Some results are in Table 3. For comparison aerobic and anaerobic growth yields are included. In all cases, the molar growth yield for a certain substrate and organism is higher when nitrate is present than when no hydrogen acceptor is available. On the other hand, the molar growth yields with oxygen as hydrogen acceptor are always higher than those with nitrate. With the enteric bacteria, the increase in the molar growth yields in the presence of nitrate is due to two effects: (1) occurrence of oxidative phosphorylation and (2) formation of more acetate. The molar growth yield for aerobic growth is higher than for anaerobic growth in the presence of nitrate, since in the latter case the citric acid cycle does not function.[129,130] On the basis of these molar growth yields, it has been concluded that the efficiency of

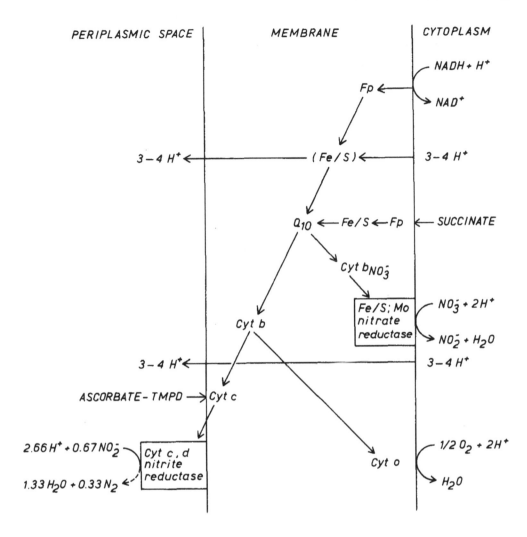

FIGURE 5. Proposed functional organization of the redox carriers for electron transport with nitrate, nitrite, and oxygen in cells of *Paracoccus denitrificans* grown anaerobically with nitrate. Abbreviations as in Figure 4. The location of nitrate reductase and nitrite reductase shown is only of importance to demonstrate the site where the respective substrates are reduced, since the exact location of these enzymes in the membrane is not known. In the scheme, the transfer of two electrons to nitrite reductase allows the reduction of 0.67 mol nitrite to nitrogen. The formation of nitrogen is shown with a broken line to denote that it is likely that in addition to nitrite reductase, other electron carriers are involved in the reduction of nitrite to nitrogen.

oxidative phosphorylation of electron transport to oxygen and nitrate is about the same.[110,111]

It is very remarkable that in a number of anaerobic organisms, nitrate respiration occurs.[4] In this paper, some data will be reported on *Propionibacterium pentosaceum*. *Propionibacterium acidi-propionici*, and *Veillonella alcalescens*. With *P. pentosaceum*, several values for the molar growth yield are given. The growth curve obtained with anaerobic cultures growing in the presence of nitrate could be differentiated into various phases.[122] In the first period, a large part of the lactate was converted into pyruvate which was subsequently used in the later periods. It could be shown that the citric acid cycle functioned under anaerobic conditions in the presence of nitrate. The molar growth yields indicated a P/2e⁻ ratio of 2 for electron transfer from NADH to nitrate and of 1 for electron transfer from lactate to nitrate. Similar studies with a

Table 3
INFLUENCE OF NITRATE ON MOLAR GROWTH YIELDS IN BATCH CULTURES OF VARIOUS ORGANISMS

Organism	Energy source	Hydrogen acceptor	Molar growth yield (g/mol⁻)	Ref.
Klebsiella aerogenes [a]	Glucose	None	26.1	118
		Nitrate	45.5	110
		Oxygen	72.7	118
	Mannitol	None	21.8	110
		Nitrate	50.6	110
		Oxygen	95.5	118
Proteus mirabilis [a]	Glucose	None	14.0,[a] 49.0[d]	75,[a] 119[d]
		Nitrate	30.1,[a] 93.0[d]	75, 119
		Oxygen	58.1	75
Citrobacter sp. [b]	Glucose	None	45	120
		Nitrate	65.8	120
		Oxygen	96.1	120
Escherichia coli [b]	Formate	Nitrate	7.8	121
	Glucose	None	24	74
		Nitrate	46	74
		Oxygen	160	74
Propionibacterium pentosaceum [b]	Lactate	None	12.9	122
		Nitrate	22—52[c]	122
	Glycerol	None	26.3	122
		Nitrate	62—81[c]	122
Propionibacterium acidi-propionici [b]	Lactate	None	5.3	123
		Nitrate	10.8	123
	Glucose	None	54.2	123
		Nitrate	53.9	123
Pseudomonas aeruginosa [a]	Succinate	Nitrate	22.5	108,124
		Oxygen	34	108,124
Bacillus licheniformis [d]	Glucose	None	16.1	119
		Nitrate	53.4	119
Pseudomonas denitrificans [a]	Glutamate	Nitrate	33.3	125
		Oxygen	64	125
	Aspartate	Nitrate	19.3	125
		Oxygen	36.3	125
	Alanine	Nitrate	12.3	125
		Oxygen	31.4	125
Clostridium perfringens [b]	Glucose	None	38.3	126
		Nitrate	45	126
Clostridium tertium [b]	Glucose	None	34.1	127
		Nitrate	46	127
Veillonella alcalescens [b]	Lactate	None	8.9	128
		Nitrate	19.8—22.9	128
	Pyruvate	None	Poor growth	128
		Nitrate	25.5	128
	Citrate	None	19.3	128
		Nitrate	25.3	128

[a]   Minimal salts medium
[b]   Complex medium
[c]   For explanation, see text.
[d]   pH auxostat cultures.

strain of *P. acidi-propionici* yielded much lower growth yields for anaerobic growth in the presence of nitrate with lactate.[123] Probably, in this organism, the citric acid cycle does not function under anaerobic conditions with nitrate. The molar growth yields with *V. alcalescens* indicate that 1 mole of ATP is generated in the electron transport from pyruvate, NADH, and NADPH to nitrate, whereas ATP is not produced in the electron transport from lactate to nitrate.[128] In Clostridia oxidative phosphorylation during nitrate respiration is absent.[126,127] In these organisms, reduced ferredoxin is the hydrogen donor for nitrate reduction.[131] The molar growth yields are increased by the presence of nitrate which is due to an increase in the proportion of fermentation products which can participate in substrate-level phosphorylation. During growth in the presence of nitrate, more acetate and less ethanol and butyrate are formed from glucose than during growth in its absence.[126,127,132]

From these data, it is clear that the efficiency of oxidative phosphorylation coupled with nitrate respiration can have very different values. Most probably these variations are due to differences in the complexity of the respiratory chain in these organisms.

The data reported in Table 3 were all performed with batch cultures. Relatively few studies have been performed with continuous cultures. Some results are reported in Table 4. The Y^{MAX}_{el} values for growth with nitrate or nitrite as hydrogen acceptor are lower than those for growth with oxygen. In *P. denitrificans*, the Y^{MAX}_{el} value for growth with gluconate as carbon and energy source and nitrate as hydrogen acceptor was dependent on the nature of the growth-limiting factor.[133] With nitrate as growth-limiting factor, the Y^{MAX}_{el} value was higher than with gluconate as growth-limiting factor. This is due to the fact that in the latter cultures more nitrite accumulates which acts as an uncoupling agent.[113] Consequently anaerobic growth with nitrite as hydrogen acceptor is only possible when nitrite is the growth-limiting factor. Originally it was assumed that the toxic effect of nitrite was exerted on phosphorylation at Site I.[133] However, this explanation was found to be incorrect.[113] The great difference in Y^{MAX}_{el} with oxygen and nitrate as hydrogen acceptor is largely due to uncoupling by the nitrite that accumulated during nitrate respiration. The toxic effect of nitrite has also been found in yield studies with batch cultures. Y_{ATP} in anaerobic cultures of *K. aerogenes*[8,110] and in *P. pentosaceum*[122] was found to be dependent on the concentration of nitrite in the medium. However, species differences exist in the sensitivity towards nitrite, since concentrations which have a drastic effect on *K. aerogenes*, *P. pentosaceum*, and *P. denitrificans* have a very limited influence on *V. alcalescens*.[128] Similarly no difference is found between thiosulfate- and nitrate-limited chemostat cultures of *T. denitrificans*.[134] From the chemostat studies with *T. denitrificans* it has been concluded that 4 to 5 mol ATP and 6 to 7 ATP are produced per mol thiosulfate with nitrate and oxygen as hydrogen acceptor, respectively.[134] This corresponds to a Y^{MAX}_{ATP} of about two, which compares favorably with an earlier estimate of 1.75 for Y_{ATP} at $\mu = 0.03^{h-1}$.[135]

Molar growth yields of *Pseudomonas denitrificans* for chemostat cultures with glutamate as the carbon and energy source in the presence of nitrate, nitrite or nitrous oxide were found to be 28.6 g/mol nitrate, 16.9 g/mol nitrite and 8.8 g/mol nitrous oxide after correction for the requirement of maintenance energy.[136] The energy yield was proportional to the oxidation number of the nitrogen in the hydrogen acceptor. It was therefore concluded that oxidative phosphorylation coupled to reduction of nitrate, nitrite, and nitrous oxide have similar efficiencies that are all lower than the efficiency of oxidative phosphorylation with oxygen. In *Paracoccus denitrificans* Y^{MAX}_{el} for growth with nitrite is higher than that for growth with nitrate (Table 4), which is as expected from our knowledge of the respiratory chain to nitrate and nitrite (Figures 2 and 5). Good growth in chemostat cultures with nitrous oxide as hydrogen acceptor could not be obtained for *P. denitrificans*.[82] In this respect there seems to be a difference between *Pseudomonas denitrificans* and *Paracoccus denitrificans*. However, absence of growth with nitrous oxide was also observed for a strain of *P. aeruginosa*.[137]

Table 4

INFLUENCE OF NITRATE ON MOLAR GROWTH YIELDS IN CHEMOSTAT
CULTURES OF VARIOUS ORGANISMS

Organism	Hydrogen donor	Limiting factor	Hydrogen acceptor	Y^{MAX}_{sub} (g/mol)	Y^{MAX}_{el}	Ref.
Pseudomonas denitrificans	Glutamate	Glutamate	Nitrate	n.r.[a]	4.5	125
	Glutamate	Glutamate	Oxygen	n.r.	7.7	125
Paracoccus denitrificans	Succinate	Succinate	Nitrate	35.2	3.7	133
	Succinate	Nitrite	Nitrite	38.4	4.50	133
	Succinate	Succinate	Oxygen	40.2	8.55	114
	Gluconate	Gluconate	Nitrate	79.7	5	133
	Gluconate	Nitrate	Nitrate	71.4	7	133
	Gluconate	Gluconate	Oxygen	77.8	10.85	114
Thiobacillus denitrificans	Thiosulphate	Thiosulphate	Oxygen	14.7	n.r.	134
	Thiosulphate	Nitrate	Nitrate	11.4	n.r.	134
	Tetrathionate	Nitrate	Nitrate	21.5	n.r.	134

Note: Y^{MAX}_{sub} is the molar growth yield for the indicated substrate corrected for energy of maintenance. Y^{MAX}_{el} is growth yield per grams equivalent of electrons transferred to the indicated hydrogen acceptor corrected for energy of maintenance.

[a] n.r. = not reported.

D. Active Transport Associated with Nitrate Respiration

The influence of nitrate respiration on the active accumulation of solutes has been treated in some recent reviews.[10,11] Anaerobic active transport of the lactose in whole cells of *E. coli* was found to be strongly increased by the presence of formate and nitrate.[138] This increase was only obtained with cells which had been grown anaerobically in the presence of nitrate. Active transport of amino acids can also be obtained under these circumstances.[139] Similar results have been obtained with membrane vesicles.[140] Formate is a good electron donor for the accumulation, but NADH is not, since NADH cannot enter the membrane vesicles. When the cells are grown anaerobically with glycerol and nitrate, active transport in membrane vesicles can also be stimulated by L-α-glycerol phosphate plus nitrate.[140] In membrane vesicles of *V. alcalescens*, the active accumulation of L-glutamate can occur in the presence of L-lactate as electron donor and nitrate as hydrogen acceptor.[141]

V. REGULATION OF THE FORMATION OF RESPIRATORY NITRATE REDUCTASE

In general, microorganisms which have the ability for nitrate respiration form respiratory nitrate reductase in the absence of oxygen, when nitrate is present. However, this observation does not give direct information about the mechanism of the regulation process. It might suggest an involvement of nitrate and oxygen in the induction and repression of the biosynthesis of nitrate reductase, but there are many observations which prove that the mechanism is not that simple.

A mutant of *E. coli* was isolated which has an unidentified block in the respiratory chain.[142] This mutant becomes sensitive to chlorate under aerobic conditions, proving the formation of nitrate reductase in the presence of oxygen. Similar mutants have been isolated from *E. coli* K12.[143] The latter mutants, which also form nitrate reductase during aerobic growth, exhibit a defect in the biosynthesis of ubiquinone.[144] It can be concluded from these facts that oxygen itself cannot repress the synthesis of nitrate reductase directly.

A number of special conditions for the biosynthesis of nitrate reductase in the absence of nitrate have been reported. They give evidence that it is not exclusively nitrate that induces the synthesis of the reductase. Under anaerobic conditions, nitrite or azide can induce nitrate reductase as well.[145-147] Furthermore it was found in *E. coli*, as well as in *P. mirabilis*, that the formation of nitrate reductase is temporary derepressed just after the shift of an aerobic culture to anaerobic growth conditions, irrespective of the presence of nitrate.[146,148] Analysis of cytoplasmic membrane proteins of a chlorate-resistant mutant of *P. mirabilis* which is supposed to be defective in the acquisition of the molybdenum cofactor suggested the presence of inactive nitrate reductase after anaerobic growth in the absence of nitrate.[147] A similar effect is observed in the wild type strain of *P. mirabilis* during anaerobic growth in the presence of tungstate. In spite of the absence of nitrate, the bacterial cells grown under this condition contain a relatively high quantity of inactive nitrate reductase which can be reactivated within a few minutes upon addition of molybdate.[150]

So far, we have described a number of unusual inducing conditions for nitrate reductase. Notwithstanding their peculiar character, they should fit into a current regulation model. Evidently, they are incompatible with the hypothesis which introduces the redox potential of the medium as regulating factor for the formation of reductases[130,151] and also with the hypothesis which introduces a nitrate-sensitive and a redox-sensitive repressor.[148] These hypotheses are at variance with, for instance, the formation of nitrate reductase under aerobic conditions in respiratory mutants and cannot be brought into line with the derepression of the enzyme by azide[145] or tungstate[150] under anaerobic conditions.

More in accordance with the experimental data appears to be the regulation model introduced by De Groot and Stouthamer.[146,152] They introduced the electron flow through the respiratory chain, as a regulating factor in the formation of nitrate reductase and other reductases. Under conditions which make the electron flow to a reductase impossible, repression of its formation occurs, for instance, by a regulation mechanism known as "autogenous regulation of gene expression".[153] The introduction of the respiratory chain in the regulation model for nitrate reductase appears to be a reasonable supposition for other reasons too.

As demonstrated recently in *P. mirabilis*, alterations in the growth conditions which induce different anaerobic respiration routes, e.g., to nitrate, tetrathionate, or fumarate, are accompanied with rather drastic alterations in the composition of the cytochrome *b* pool.[154] Furthermore Wallace and Young[155] reported an alteration in the ubiquinone/menaquinone ratio in *E. coli* during induction of nitrate reductase. So the induction of nitrate reductase should not be seen as a single or unique event in the bacterial cell. The formation of an operational anaerobic respiratory system requires, beyond the formation of an active reductase, the synthesis of appropriate electron carriers, and the coregulation of the formation of all composing parts. In Enterobacteriaceae a special cytochrome *b* functions in the electron transport to nitrate reductase (see the sections entitled The Interaction of Nitrate Reductase with the Cytoplasmic Membrane and *Escherichia coli*). In several cases, this cytochrome has been copurified with nitrate reductase (see section entitled The Interaction of Nitrate Reductase with the Cytoplasmic Membrane) and some authors consider it to be an integral part of the reductase. For this reason, MacGregor[23] suggested in 1976 that this cytochrome *b* is involved in the regulation mechanism for nitrate reductase. In *P. mirabilis* this cytochrome has been designated as cytochrome b_{559},[146,158] and its synthesis always parallels the formation of nitrate reductase, even in the case of unbalanced synthesis of high quantities of the enzyme which occurs after induction by azide or tungstate.[146,150] Based on the coregulation of this cytochrome and nitrate reductase, a model for the regulation of the formation of nitrate reductase in *P. mirabilis* was introduced.[149] This model, which might be considered as an elaboration of the regulation model of De Groot and Stouthamer,[146,152] is (slightly modified) illustrated in Figure 6. It introduces two repressors, as in an earlier model,[148] for the biosynthesis of nitrate reductase, namely cytochrome b_{559} and the nitrate reductase molecules themselves or their precursors. The argument for two repressors came from the observation of a temporary derepression of nitrate reductase after a shift from aerobic to anaerobic growth and in addition from the observation that the formation of nitrate reductase under the peculiar inducing conditions, namely addition of azide or tungstate, is still subject to repression by oxygen.[146,150] The supposition that nitrate reductase or a soluble precursor of the enzyme which is found in *E. coli* [42] functions as an autogenous repressor for its own biosynthesis is based on the fact that it appears to be a plausible common target for the known inducers, namely nitrate (substrate), nitrite (product), azide (inhibitor), and finally tungstate or mutation in molybdenum acquisition (interference with Mo cofactor of the enzyme). Interaction of nitrate, nitrite, or azide with the reductase or the absence of the Mo cofactor may introduce conformational shifts of the reductase molecule, rendering this molecule unfit for its repressor function. Under anaerobic conditions when no biosynthesis of nitrate reductase occurs, the biosynthesis of cytochrome b_{559} is supposed to be repressed also by nitrate reductase molecules. During aerobic respiration neither nitrate, nor azide or tungstate bring about formation of nitrate reductase and cytochrome b_{559}. Since a direct interaction of oxygen with the repressor function of the nitrate reductase molecule should be excluded on the basis of the formation of nitrate reductase in the presence of oxygen in respiratory mutants, the formation of a nitrate respiratory system may be repressed under these conditions by the same factors which regulate the formation of the components of the respiratory chain. It might be an autogenous mechanism too, as supposed in the given regulation model. Cytochrome b_{559} or its apoprotein may repress its own biosynthesis and the biosynthesis of nitrate reductase as well, when its assemblage into the respiratory chain is impossible because of electron transport to oxygen.

The regulation model just described also included a feed-back control of the amount of nitrate reductase and cytochrome b_{559} which will be synthesized during derepression by nitrate. When the intracellular nitrate concentration is lowered sufficiently by the operative nitrate reductase system, the repressor function of the freshly synthesized nitrate reductase, being not eliminated by interaction with nitrate off, will switch of further synthesis. On the other hand, the model predicts an unbalanced continuation of the formation of reductase and cytochrome after induction by azide and tungstate. Indeed an unbalanced formation of up to fourfold greater quantiies of nitrate reductase and cytochrome b_{559} have been reported to occur under these circumstances.[146,150] A similar, also predictable, unbalanced formation of nitrate reductase has been reported for heme-deficient mutants of *Staphylococcus aureus*[157] and *E. coli*.[42,158] On the other hand, other *hem* mutants of *P. mirabilis*[146] and *E. coli*[57] stop or slow down the synthesis of nitrate reductase upon dilution of functional cytochromes after omission of the heme-precursor δ-aminolevulinic acid from the growth medium. This different behavior of various *hem* mutants is still an unexplaned discrepancy. There is one conclusion from experimental results with *hem* mutants which seems to be at variance with the regulation model as described here. Recently MacGregor and Bishop[158] conclude from the repression of the formation of nitrate reductase under aerobic conditions in the heme-exhausted cells that cytochromes are not responsible for nitrate reductase repression by air. They conclude that it should be the oxygen level itself which is responsible for the repression. However, such a direct interaction is incompatible with the observed aerobic synthesis of nitrate reductase in respiratory mutants of *E. coli* and *E. coli* K12.[142-144] When a repressor function of nonfunctional, heme-deficient apoprotein of cytochrome *b* is assumed, these data are still in accordance with the regulation model in Figure 6.

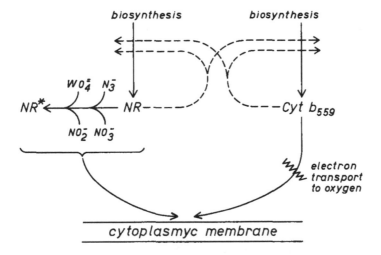

FIGURE 6.   Model for the repression /derepression mechanism for nitrate
reductase in *Proteus mirabilis*. The biosynthesis of both nitrate reductase and
cytochrome b_{559} are repressed by either nitrate reductase molecules (NR) or
cytochrome b_{559} molecules. NR* represents nitrate reductase molecules which
are conformationally modified by interaction with nitrate, nitrite, or azide,
or on account of the absence of a suitable molybdenum cofactor. These modified reductase molecules do not function as repressor.

In conclusion, we can state that the regulation hypothesis of Oltmann et al. is in
accordance with experimental data known at the moment. As a second advantage, this
model can be adjusted to the regulation of other anaerobic respiration routes without
major alterations, as shown by the authors in the original paper.[149] On the other hand,
the model is only a working hypothesis. It shifts part of the regulation problems to
the also unknown regulation mechanism for the formation of cytochromes, and it
ignores the problem of the physical separation on the DNA of a number of genes
which are involved in the formation of a nitrate respiration system. Furthermore the
presented regulation scheme is hardly supported by experimental data from bacteria
which do not belong to the family of the Enterobacteriaceae simply because compara-
ble data from those bacteria are not available. Meanwhile it is clear that there are
differences between the nitrate respiration systems in the various types of microorganisms which may also give rise to variations in the regulation mechanism. For instance,
oxygen sensitivity of nitrate respiration appeared to be less in *B. licheniformis*.[160] Denitrifying bacteria isolated from activated sludge were found to synthesize respiratory
nitrate reductase in air-saturated culture.[161] At the moment, experimental data for a
more universal regulation model for the formation of respiratory nitrate reductase in
bacteria are lacking.

VI. SUMMARY AND FINAL CONCLUSIONS

During the last years, great progress has been made in the characterization of nitrate
reductase and in our knowledge on its localization in the cytoplasmic membrane. A
molybdenum cofactor of low molecular weight has been shown to be involved in enzymatic activity. This molybdenum cofactor has not yet been isolated in pure state,
and our knowledge of the molybdenum cofactor of nitrate reductase lags far behind
that of the molybdenum cofactor of nitrogenase. Our knowledge of energy formation
associated with nitrate reduction has also increased strongly during the last years. This
knowledge has been obtained largely from a combination of yield studies in chemostat

cultures, studies on proton extrusion during nitrate reduction, and studies on the composition of the respiratory chains to nitrate, nitrite, and other nitrogenous oxides. In all cases, the efficiency of energy generation associated with nitrate respiration was found to be less than that for respiration with oxygen. Our present knowledge indicates that there are large species specific differences in the efficiency of phosphorylation associated with nitrate reduction which may be due to diversity in bacterial respiratory chains and in the localization of the nitrate reductase in the cytoplasmic membrane. The observation that nitrite — the reduction product of nitrate — is a very potent uncoupling agent has important implications. The influence of nitrate on molar growth yields is very complicated, since it is dependent on the efficiency of oxidative phosphorylation associated with nitrate respiration and on the sensitivity towards nitrite. Species-specific differences also exist for the property.

In a number of organisms, nitrate has been shown to be reduced at the inner aspect of the cytoplasmic membrane. Virtually nothing is known about the uptake process for nitrate by bacteria. In denitrifying organisms, nitrite is reduced at the outer aspect of the cytoplasmic membrane. Again nothing is known about the transport of nitrite from the cytoplasm to the periplasmic space to achieve further reduction of nitrite or release of toxic nitrite to the medium.

Due to these factors, many uncertainties are present in our knowledge of energy generation associated with nitrate respiration. There is great uncertainty about the pathway of denitrification. In this area, many contradictory results have been published. In this paper, the proposal has been made that only enzyme-bound intermediates occur during the reduction of nitrite to nitrogen. Nitrous oxide, generally considered to be an obligate intermediate in denitrification, is proposed to be reduced by a pathway other than via nitrite. Great species differences are present in the pathway of denitrification and in the enzymes involved. Part of the discrepancies in the literature may be due to this factor.

Our knowledge of the regulation of the formation of nitrate reductase still shows many gaps. More knowledge of the bacterial respiratory chain and the mechanism of its adaptability seems indispensible for the further elucidation of this regulation.

Much further work is necessary to clarify the many points on which uncertainties still exist.

REFERENCES

1. **Pichinoty, F.,** La reduction bacterienne des composes oxygenes mineraux de l'azote, *Bull. Inst. Pasteur Paris,* 71, 317-395, 1973.
2. **Payne, W. J.,** Reduction of nitrogenous oxides by microorganisms, *Bacteriol. Rev.,* 37, 409, 1973.
3. **Hall, J. B.,** Nitrate-reducing bacteria, in *Microbiology—1978,* Schlessinger, D., Ed., American Society for Microbiology, Washington, D.C., 1978, 296.
4. **Stouthamer, A. H.,** Biochemistry and genetics of nitrate reductase in bacteria, *Adv. Microb. Physiol.,* 14, 315, 1976.
5. **Haddock, B. A. and Jones, C. W.,** Bacterial respiration, *Bacteriol. Rev.,* 41, 47, 1977.
6. **Thauer, R. K., Jungermann, K., and Decker, K.,** Energy conservation in chemotrophic anaerobic bacteria, *Bacteriol. Rev.,* 41, 100, 1977.
7. **Kröger, A.,** Phosphorylative electron transport with fumarate and nitrate as terminal hydrogen acceptors, *Symp. Soc. Gen. Microbiol.,* 27, 61, 1977.
8. **Stouthamer, A. H.,** *Yield-Studies in Micro-Organisms,* Meadowfield Press, Durham, England, 1976.
9. **Stouthamer, A. H.,** Energetic aspects of the growth of micro-organisms, *Symp. Soc. Gen. Microbiol.,* 27, 285, 1977.

10. **Konings, W. N.,** Active transport of solutes in bacterial membrane vesicles, *Adv. Microb. Physiol.,* 15, 175, 1977.

11. **Konings, W. N. and Boonstra, J.,** Anaerobic electron transfer and active transport in bacteria, *Curr. Top. Membr. Transp.,* 177, 1977.

12. **Payne, W. J. and Balderston, W. L.,** Denitrification, in *Microbiology—1978,* Schlessinger, D., Ed., American Society of Microbiology, Washington, D.C., 1978, 339.

13. **Iida, K. and Taniguchi, S.,** Studies on nitrate reductase system of *Escherichia coli.* 1. Particulate electron transport system to nitrate and its solubilization, *J. Biochem. (Tokyo),* 46, 1041, 1959.

14. **Lam, Y. and Nicholas, D. J. D.** A nitrate reductase from *Micrococcus denitrificans, Biochim. Biophys. Acta,* 178, 225, 1969.

15. **van't Riet, J. and Planta, R. J.,** Purification and some properties of the membrane-bound respiratory nitrate reductase of *Aerobacter aerogenens, FEBS Lett.,* 5, 249, 1969.

16. **Radcliffe, B. C. and Nicholas, D. J. D.,** Some properties of a nitrate reductase from *Pseudomonas denitrificans, Biochim. Biophys. Acta,* 205, 273, 1970.

17. **Kiszkiss, D. F. and Downey, R. J.,** Localization and solubilization of the respiratory nitrate reductase of *Bacillus stearothermophilus, J. Bacteriol.,* 109, 803, 1972.

18. **Enoch, H. G. and Lester, R. L.,** The role of a novel cytochrome *b*-containing nitrate reductase and quinone in the in vitro reconstruction of formate-nitrate reductase activity of *E. coli, Biochem. Biophys. Res. Commun.,* 61, 1234, 1974.

19. **van't Riet, J. and Planta, R. J.,** Purification, structure and properties of the respiratory nitrate reductase of *Klebsiella aerogenes, Biochim. Biophys. Acta,* 379, 81, 1975.

20. **DeMoss, J. A.,** Limited proteolysis of nitrate reductase purified from membranes of *Escherichia coli, J. Biol. Chem.,* 252, 1696, 1977.

21. **van't Riet, J., Wientjes, F. B., Van Doorn, J., and Planta, R. J.,** Purification and characterization of the respiratory nitrate reductase of *Bacillus licheniformis, Biochim. Biophys. Acta,* 576, 347, 1979.

22. **Villarreal-Moguel, E. I., Ibarra, V., Ruiz-Herrera, J., and Gitler, C.,** Resolution of the nitrate reductase complex from the membrane *Escherichia coli, J. Bacteriol.,* 113, 1264, 1973.

23. **MacGregor, C. H.,** Anaerobic cytochrome b_i in *Escherichia coli:* association with and regulation of nitrate reductase, *J. Bacteriol.,* 121, 1111, 1975.

24. **MacGregor, C.H.,** Synthesis of nitrate reductase components in chlorate resistant mutants of *Escherichia coli, J. Bacteriol.,* 121, 1117, 1975.

25. **Clegg, R. A.,** Purification and some properties of nitrate reductase (EC 1.7.99.4) from *Escherichia coli* K12, *Biochem. J.,* 153, 533, 1976.

26. **Forget, P.,** Les nitrate-reductase bacteriennes. Solubilisation, purification et proprietes de l'enzyme A de *Micrococcus denitrificans, Eur. J. Biochem.,* 18, 442, 1971.

27. **Rosso, J. P., Forget P., and Pichinoty, F.,** Les nitrate-reductases bacteriennes. Solubilization, purification et proprietes de l'enzyme A de *Micrococcus halodenitrificans, Biochim. Biophys. Acta,* 321, 443, 1973.

28. **Forget, P.,** The bacterial nitrate reductases. Solubilization, purification and properties of the enzyme A of *Escherichia coli* K. 12, *Eur. J. Biochem.,* 42, 325, 1974.

29. **Oltmann, L. F., Schoenmaker, G. S., and Stouthamer, A. H.,** Solubilization and purification of a cytoplasmic membrane bound enzyme catalyzing tetrathioriate and thiosulphate reduction in *Proteus mirabilis, Arch. Microbiol.,* 98, 19, 1974.

30. **Taniguchi, S. and Itagaki, E.,** Nitrate reductase of nitrate respiration type fron *E. coli.* I. Solubilization and purification from the particulate system with molecular characterization as a metalloprotein, *Biochim. Biophys. Acta,* 44, 263, 1960.

31. **MacGregor, C. H., Schaitman, A., Normansell, D. E., and Hodgins, M. G.,** Purification and properties of nitrate reductase from *Escherichia coli* K12, *J. Biol. Chem.,* 249, 5321, 1974.

32. **Lund, K. and DeMoss, J. A.,** Association-dissociation behavior and sub-unit structure of heat-released nitrate reductase from *E. coli, J. Biol. Chem.,* 251, 2207, 1976.

33. **MacGregor, C. H.,** Solubilization of *Escherichia coli* nitrate reductase by a membrane-bound protease, *J. Bacteriol.,* 121, 1102, 1975.

34. **Itagaki, E., Fujita, T., and Sato, R.,** Solubilization and properties of formate dehydrogenase and cytochrome b_i from *Escherichia coli, J. Biochem. (Tokyo),* 52, 131, 1962.

35. **Oltmann, L. F., Reijnders, W. N. M., and Stouthamer, A. H.,** Characterization of purified nitrate reductase A and chlorate reductase C from *Proteus mirabilis, Arch. Microbiol.,* 111, 25, 1976.

36. **Forget, P. and Rimassa, R.,** Evidence for the presence of carbohydrate units in the nitrate reductase A of *Escherichia coli* K12, *FEBS Lett.,* 77, 182, 1977.

37. **Boxer, D. H. and Clegg, R. A.,** A transmembrane-location for the proton-translocating reduced ubiquinone-nitrate reductase segment of the respiratory chain of *Escherichia coli, FEBS Lett.,* 60, 54, 1975.

38. **MacGregor, C. H. and Christopher, A. R.**, Assymmetric distribution of nitrate reductase subunits in the cytoplasmic membrane of *Escherichia coli*. Evidence derived from surface labelling studies with transglutaminase, *Arch. Biochem. Biophys.* 185, 204, 1978.

39. **Graham, A. and Boxer, D. H.**, Immunochemical localisation of nitrate reductase in *Escherichia coli*, *Biochem. Soc. Trans.*, 6, 1210, 1978.

40. **Wientjes, F. B., Kolk, A. H. J., and van't Riet, J.**, *Abstr. Commun. Meet. Fed. Eur. Biochem. Soc.*, 11, 662, 1977.

41. **Wientjes, F. B., Kolk, A. H. J., and van't Riet, J.**, Respiratory nitrate reductase: its localisation in the cytoplasmic membrane of *Klebsiella aerogenes* and *Bacillus licheniformis*, *Eur. J. Biochem.*, 95, 61, 1979.

42. **MacGregor, C. H.**, Biosynthesis of membrane-bound nitrate reductase in *Escherichia coli* : Evidence for a soluble precursor, *J. Bacteriol.*, 126, 122, 1976.

43. **Enoch, H. G. and Lester, R. L.**, The purification and properties of formate dehydrogenase and nitrate reductase from *Escherichia coli*, *J. Biol. Chem.*, 250, 6653, 1975.

44. **Bosma, J. H., Wever, R., and van't Riet, J.**, Electron paramagnetic resonance studies on membrane-bound respiratory nitrate reductase of *Klebsiella aerogenes*, *FEBS Lett.*, 90, 107, 1978.

45. **van't Riet, J., van Ee, J. H., Wever, R., van Gelder, B. F., and Planta, R. J.**, Characterization of the respiratory nitrate reductase of *Klebsiella aerogenes* as a molybdenum-containing iron-sulfur enzyme, *Biochim. Biophys. Acta*, 405, 306, 1975.

46. **Jones, R. W. and Garland, P. B.**, Sites and specificity of the reaction of bipyridylium compounds with anaerobic respiratory enzymes of *Escherichia coli*. Effects of permeability barriers imposed by the cytoplasmic membrane, *Biochem. J.*, 164, 199, 1977.

47. **Garland, P. B., Downie, J. A., and Haddock, B. A.**, Proton translocation and the respiratory nitrate reductase of *Escherichia coli*, *Biochem. J.*, 152, 547, 1975.

48. **Garland, P. B., Clegg, R. A., Boxer, D. H., Downie, J. A., and Haddock, B. A.**, Protontranslocating nitrate reductase of *Escherichia coli*, in *Electron Transport Chains and Oxidative Phosphorylation*, Quagliariello, E., Papa, S., Palmieri, F., Slater, E. C., and Siliprandi, N., Eds., North-Holland, Amsterdam, 1975, 351.

49. **Jones, R. W., Ingledew, W. J., Graham, A., and Garland, P. B.**, Topography of nitrate reductase of the cytoplasmic membrane of *Escherichia coli*. The nitrate reducing site, *Biochem. Soc. Trans.*, 6, 1287, 1978.

50. **John, P.**, Aerobic and anaerobic bacterial respiration monitored by electrodes, *J. Gen. Microbiol.*, 98, 231, 1977.

51. **Wientjes, F. B., Abraham, P., and van't Riet, J.**, Localization of nitrate reductase subunits in bacterial membranes, *Abstr. 11th Congr. Biochem.*, p. 340, 1979.

52. **Forget, P., and DerVartanian, D. V.**, The bacterial nitrate reductases: EPR studies on nitrate reductase A from *Micrococcus denitrificans*, *Biochim. Biophys. Acta*, 255, 600, 1972.

53. **DerVartanian, D. V. and Forget, P.**, The bacterial nitrate reductases, EPR studies on the enzyme A of *Escherichia coli* K12, *Biochim. Biophys. Acta*, 379, 74, 1975.

54. **Bray, R. C., Vincent, S. P., Lowe, D. J., Clegg, R. A., and Garland, P. B.**, Electron paragmagnetic resonance studies on the molybdenum of nitrate reductase from *E. coli* K12, *Biochem. J.*, 135, 201, 1976.

55. **Vincent, S. P. and Bray, R. C.**, Electron-paramagnetic-resonance studies on nitrate reductase from *Escherichia coli* K12, *Biochem. J.*, 171, 639, 1978.

56. **Stiefel, E. I.**, Proposed mechanism for the action of Molybdenum in enzymes: coupled proton and electron transfer, *Proc. Natl. Acad. Sci. U.S.A.*, 70, 988, 1973.

57. **Guymon, E. P. and Spence, J. T.**, The reduction of nitrate by Molybdenum (V.), *J. Phys. Chem.*, 70, 1964, 1966.

58. **Garner, C. D., Hyde, M. R., Mabbs, F. E., and Routledge, V. I.**, Possible model reactions for the nitrate reductases, *Nature (London)*, 252, 579, 1974.

59. **Ketchum, P. A., Taylor, R. C., and Young, D. C.**, Model reaction for biological reduction of nitrate involving Mo(III) Mo(V), *Nature (London)*, 259, 202, 1976.

60. **Ketchum, P. A.**, Reactions of Molybdenum with nitrate and naturally produced Phenolates, *Adv. Chem. Ser.*, 162, 408, 1977.

61. **Bray, R. C.**, Molybdenum iron-sulfur flavin hydroxylases and related enzymes, in *The Enzymes* Vol. 12, Boyer, P. D., Ed., Academic Press, New York, 1976, 229.

62. **Ketchum, P. A., Cambier, H. Y., Frazier, W., Madansky, C., and Nason, A.**, In vitro assembly of *Neurospora* assimilatory nitrate reductase from protein subunits of a *Neurospora* mutant and the xanthine oxidizing or aldehyde oxidase systems of higher animals, *Proc. Natl. Acad. Sci. U.S.A.*, 66, 1016, 1970.

63. Nason, A., Lee, K. Y., Pan, S. S., Ketchum, P. A., Lamberti, A., and de Vries, J., In vitro formation of assimilatory reduced nicotinamide dinucleotide phosphate nitrate reductase from a *Neurospora* mutant and component molybdenum enzymes, *Proc. Natl. Acad. Sci. U.S.A.*, 68, 3242, 1971.

64. Lee, K. Y., Pan, S. S., Erickson, R., and Nason, A., Involvement of molybdenum and iron in the *in vitro* assembly of assimilatory nitrate reductase utilizing *Neurospora* mutant nit-1, *J. Biol. Chem.*, 249, 3941, 1974.

65. Dubordieu, M., Andrade, E., and Puig, J., Molybdenum and chlorate resistant mutants in *Escherichia coli* K12, *Biochem. Biophys. Res. Commun.*, 70, 766, 1976.

66. Shah, V. U. and Brill, W. J., Isolation of an iron-molybdenum cofactor from nitrogenase, *Proc. Natl. Acad. Sci. U.S.A.*, 74, 3249, 1977.

67. Pienkos, P. T., Shah, V. U., and Brill, W. J., Molybdenum cofactors from molybdoenzymes and in vitro reconstitution of nitrogenase and nitrate reductase, *Proc. Natl. Acad. Sci. U.S.A.*, 74, 5468, 1977.

68. Haddock, B. A., The isolation of phenotypic and genotypic variants for the functional characterization of bacterial oxidative phosphorylation, *Symp. Soc. Gen. Microbiol.*, 27, 95, 1977.

69. Kennedy, C. and Postgate, J. R., Expression of *Klebsiella pneumoniae* nitrogen fixation genes in nitrate reductase mutants of *Escherichia coli*, *J. Gen. Microbiol.*, 98, 551, 1977.

70. Skotnicki, M. L. and Rolfe, B. G., Interaction between the nitrate respiratory system of *Escherichia coli* K12 and the nitrogen fixation genes of *Klebsiella pneumoniae*, *Biochem. Biophys. Res. Commun.*, 78, 726, 1977.

71. Pagan, J. D., Snowcroft, W. R., Dudman, W. F., and Gibson, A. H., Nitrogen fixation in nitrate reductase-deficient mutants of cultured Rhizobia, *J. Bacteriol.*, 129, 718, 1977.

72. Magelhães, L. M. S., Neyra, C. A., and Döbereiner, J., Nitrate and nitrite reductase negative mutants of N₂-fixing *Azospirillum* spp., *Arch. Microbiol.*, 117, 247, 1978.

73. Glaser, J. H. and DeMoss, J. A., Phenotypic restoration by molybdate of nitrate reductase activity in *chlD* mutants of *Escherichia coli*, *J. Bacteriol.*, 108, 854, 1971.

74. Ishimoto, M. and Yamamoto, J., Cell growth and metabolic products of *Escherichia coli* in nitrate respiration, *Z. Allg. Mikrobiol.*, 17, 309, 1977.

75. Stouthamer, A. H. and Bettenhaussen, C. W., Influence of hydrogen acceptors on growth and energy production of *Proteus mirabilis*, *Antonie van Leeuwenhoek J. Microbiol. Serol.*, 38, 81, 1972.

76. John, P. and Whatley, F. R., *Paracoccus denitrificans* and the evolutionary origin of the mitochondrion, *Nature (London)*, 254, 495, 1975.

77. van Verseveld, H. W. and Stouthamer, A. H., Electron transport chain and coupled oxidative phosphorylation in methanol-grown *Paracoccus denitrificans*, *Arch. Microbiol.*, 118, 13, 1978.

78. Newton, N., The two haem nitrite reductase of *Micrococcus denitrificans*, *Biochim. Biophys. Acta*, 185, 316, 1969.

79. John, P. and Whatley, F. R., Oxidative phosphorylation coupled to oxygen uptake and nitrate reduction in *Micrococcus denitrificans*. *Biochim. Biophys. Acta*, 216, 342, 1970.

80. Meyer, E. M., van der Zwaan, J. W., and Stouthamer, A. H., Location of the proton consuming site in nitrite reduction and stoichiometries for proton pumping in *Paracoccus denitrificans*, *FEMS Microbiol. Lett.*, 5, 369, 1979.

81. Wood, P. M., Periplasmic location of the terminal reductase in nitrite respiration, *FEBS Lett.*, 92, 214, 1978.

82. Boogerd, F. C., van Verseveld, H. W., and Stouthamer, A. H., unpublished results.

83. Balderston, W. L., Sherr, B., and Payne, W. J., Blockage by acetylene of nitrous oxide reduction in *Pseudomonas perfectomarinus*, *Appl. Env. Microbiol.*, 31, 504, 1976.

84. Yoshinari, T. and Knowles, R., Acetylene inhibition of nitrous oxide reduction by denitrifying bacteria, *Biochem. Biophys. Res. Commun.*, 69, 705, 1976.

85. Allen, M. B. and van Niel, C. B., Experiments on bacterial denitrification, *J. Bacteriol.*, 64, 397, 1952.

86. Kluyver, A. J. and Verhoeven, W., Studies on true dissimilatory nitrate reduction. II. Mechanism of denitrification, *Antonie van Leeuwenhoek J. Microbiol. Serol.*, 20, 241, 1954.

87. Pichinoty, F. and d'Ornano, L., Recherches sur la reduction du protoxyde l'azote par *Micrococcus denitrificans*, *Ann. Inst. Pasteur Paris*, 101, 418, 1961.

88. Sacks, L. E. and Barker, H. A., Substrate oxidation and nitrous oxide utilization in denitrification, *J. Bacteriol.*, 64, 247, 1952.

89. Sidransky, E., Walter, B. and Hollocher, T. C., Studies on the differential inhibition by azide on the nitrite/nitrous oxide level of denitrification, *Appl. Env. Microbiol.*, 35, 247, 1978.

90. Matsubara, T. and Mori, T., Studies on denitrification. IX. Nitrous oxide, its production and reduction to nitrogen, *J. Biochem. (Tokyo)*, 64, 863, 1968.

91. Walter, B., Sidransky, E., Kristjansson, J. K., and Hollocher, T. C., Inhibition of denitrification by uncouplers of oxidative phosphorylation, *Biochemistry*, 17, 3039, 1978.

92. **Aminuddin, M. and Nicholas, D. J. D.**, Sulphide oxidation linked to the reduction of nitrate and nitrite in *Thiobacillus denitrificans*, *Biochim. Biophys. Acta*, 325, 81, 1973.
93. **Aminuddin, M. and Nicholas, D. J. D.**, Electron transfer during sulphide and sulphite oxidation in *Thiobacillus denitrificans*, *J. Gen. Microbiol.*, 82, 115, 1974.
94. **Sawhney, V. and Nicholas, D. J. D.**, Sulphite- and NADH-dependent nitrate reductase from *Thiobacillus denitrificans*, *J. Gen. Microbiol.*, 100, 49, 1977.
95. **Sawhney, V. and Nicholas, D. J. D.**, Sulphide-linked nitrite reductase from *Thiobacillus denitrificans* with cytochrome oxidase activity: purification and properties, *J. Gen. Microbiol.*, 106, 119, 1978.
96. **Harold, F. M.**, Conservation and transformation of energy by bacterial membranes, *Bacteriol. Rev.*, 36, 172, 1972.
97. **Stouthamer, A. H.**, Determination and significance of molar growth yields, *Meth. Microbiol.*, 1, 629, 1969.
98. **van Verseveld, H. W. and Stouthamer, A. H.**, Oxidative phosphorylation in *Micrococcus denitrificans*. Calculation of the P/O ratio in growing cells, *Arch. Microbiol.*, 107, 241, 1976.
99. **van der Beek, E. G. and Stouthamer, A. H.**, Oxidative phosphorylation in intact bacteria, *Arch. Microbiol.*, 89, 327, 1973.
100. **Ohnishi, T.**, Oxidative phosphorylation coupled with nitrate respiration with cell free extracts of *Pseudomonas denitrificans*, *J. Biochem. (Tokyo)*, 53, 71, 1963.
101. **Ohnishi, T. and Mori, T.**, Oxidative phosphorylation coupled with denitrification in intact cell systems, *J. Biochem. (Tokyo)*, 48, 406, 1960.
102. **Yamanaka, T., Ota, T., and Okunuki, K.**, Oxidative phosphorylation. I. Evidence for phosphorylation coupled with nitrate reduction in a cell-free extract of *Pseudomonas aeruginosa*, *J. Biochem. (Tokyo)*, 51, 253, 1962.
103. **Ishaque, M., Donawa, A., and Aleem, M. I. H.**, Oxidative phosphorylation in *Pseudomonas saccharophila* under autotrophic and heterotrophic growth conditions, *Biochem. Biophys. Res. Commun.*, 44, 244, 1971.
104. **Ishaque, M., Donawa, A., and Aleem, M. I. H.**, Energy coupling mechanisms under aerobic and anaerobic conditions in autotrophically grown *Pseudomonas saccharophila*, *Arch. Biochem. Biophys.*, 159, 570, 1973.
105. **Ota, A., Yamanaka, T., and Okunuki, K.**, Oxidative phosphorylation coupled with nitrate respiration. II. Phosphorylation coupled with anaerobic nitrate reduction in a cell-free extract of *Escherichia coli*, *J. Biochem. (Tokyo)*, 55, 131, 1964.
106. **Miki, K. and Lin, E. C. C.**, Electron transport chain from glycerol-3-phosphate to nitrate in *Escherichia coli*, *J. Bacteriol.*, 124, 1288, 1975.
107. **Sewell, D. L. and Aleem, M. I. H.**, NADH-linked oxidative phosphorylation in *Nitrobacter agilis*. *Bacteriol. Proc.*, p. 169, 1974.
108. **van Hartingsveldt, J. and Stouthamer, A. H.**, Properties of a mutant of *Pseudomonas aeruginosa* affected in aerobic growth, *J. Gen. Microbiol.*, 83, 303, 1974.
109. **Brice, J. M., Law, J. F., Meyer, D. J., and Jones, C. W.**, Energy conservation in *Escherichia coli* and *Klebsiella pneumoniae*, *Biochem. Soc. Trans.*, 2, 523, 1974.
110. **Hadjipetrou, L. P. and Stouthamer, A. H.**, Energy production during nitrate respiration by *Aerobacter aerogenes*, *J. Gen. Microbiol.*, 38, 29, 1965.
111. **Stouthamer, A. H.**, Mutant strains of *Aerobacter aerogenes* which require both methionine and lysine for aerobic growth, *J. Gen. Microbiol.*, 46, 389, 1967.
112. **Kell, D. B., John, P., and Ferguson, S. J.**, The protonmotive force in phosphorylating membrane vesicles from *Paracoccus denitrificans*. Magnitude, sites of generation and comparison with the phosphorylation potential, *Biochem. J.*, 174, 257, 1978.
113. **Meijer, E. M., van der Zwaan, J. W., Wever, R., and Stouthamer, A. H.**, Anaerobic respiration and energy conservation in *Paracoccus denitrificans*. Functioning of iron-sulfur centers and the uncoupling effect of nitrite, *Eur. J. Biochem.*, 96, 69, 1979.
114. **Meijer, E. M., van Verseveld, H. W., van der Beek, E. G., and Stouthamer, A. H.**, Energy conservation during aerobic growth in *Paracoccus denitrificans*, *Arch. Microbiol.*, 112, 25, 1977.
115. **Lawford, H. G.**, Energy-transduction in the mitochondrion-like bacterium *Paracoccus denitrificans* during carbon- or sulphate-limited aerobic growth in continuous culture, *Can. J. Biochem.*, 56, 13, 1978.
116. **Scholes, P. B. and Smith, L.**, Composition and properties of the membrane-bound respiratory chain system of *Micrococcus denitrificans*, *Biochim. Biophys. Acta*, 153, 363, 1968.
117. **Bauchop, I. and Elsden, S. R.**, The growth of microorganisms in relation to their energy supply, *J. Gen. Microbiol.*, 23, 457, 1960.
118. **Hadjipetrou, L. P., Gerrits, J. P., Teulings, F. A. G., and Stouthamer, A. H.**, Relation between energy production and growth of *Aerobacter aerogenes*, *J. Gen. Microbiol.*, 36, 139, 1964.

119. Oltmann, L. F., Schoenmaker, G. S., Reijnders, W. N. M., and Stouthamer, A. H., Modification of the pH-auxostat culture method for the mass cultivation of bacteria, *Biotechnol. Bioeng.*, 20, 921, 1978.

120. Kapralek, F., The physiological role of tetrathionate respiration in growing *Citrobacter, J. Gen. Microbiol.*, 71, 133, 1972.

121. Yamamoto, I. and Ishimoto, M., Anaerobic growth of *Escherichia coli* on formate by reduction of nitrate, fumarate, and trimethylamine N-oxide, *Z. Allg. Mikrobiol.*, 17, 235, 1977.

122. van Gent-Ruijters, M. L. W., de Vries, W., and Stouthamer, A. H., Influence of nitrate on fermentation pattern, molar growth yields and synthesis of cytochrome *b* in *Propionibacterium pentosaceum, J. Gen. Microbiol.*, 88, 36, 1975.

123. Kaneko, M. and Ishimoto, M., Effect of nitrate reduction on metabolic products and growth of *Propionibacterium acidi-propionici, A. Allg. Mikrobiol.*, 17, 211, 1977.

124. van Hartingsveldt, J. and Stouthamer, A. H., unpublished results.

125. Koike, J. and Hattori, A., Growth yield of a denitrifying bacterium, *Pseudomonas denitrificans*, under anaerobic and denitryfying conditions, *J. Gen. Microbiol.*, 88, 1, 1975.

126. Hasan, S. M. and Hall, J. B., The physiological function of nitrate reduction in *Clostridium perfringens, J. Gen. Microbiol.*, 87, 120, 1975.

127. Hasan, S. M. and Hall, J. B., Dissimilatory nitrate reduction in *Clostridium tertium, Z. Allg. Mikrobiol.*, 17, 501, 1977.

128. de Vries, W., Rietveld-Struijk, T. R. M., and Stouthamer, A. H., ATP formation associated with fumarate and nitrate reduction in growing cultures of *Veillonella alkalescens, Antonic van Leeuwenhoek J. Microbiol. Serol.*, 43, 153-167, 1977.

129. Forget, P. and Pichinoty, F., Le cycle tricarboxylique chez *Aerobacter aerogenes, Ann. Inst. Pasteur Paris*, 112, 261, 1967.

130. Wimpenney, J. W. T., and Cole, J. A., The regulation of metabolism in facultative bacteria. III. The effect of nitrate, *Biochim. Biophys. Acta*, 148, 233, 1967.

131. Chiba, S. and Ishimoto, M., Ferredoxin-linked nitrate reductase from *Clostridium perfringens, J. Biochem. (Tokyo)*, 73, 1315, 1973.

132. Ishimoto, M., Umeyama, M., and Chiba, S., Alteration of fermentation products from butyrate to acetate by nitrate reduction in *Clostridium perfringens, Z. Allg. Mikrobiol.*, 14, 115, 1974.

133. van Verseveld, H. W., Meijer, E. M., and Stouthamer, A. H., Energy conservation during nitrate respiration in *Paracoccus denitrificans, Arch. Microbiol.*, 112, 17, 1977.

134. Justin, P. and Kelly, D. P., Growth kinetics of *Thiobacillus denitrificans* in anaerobic and aerobic chemostat culture, *J. Gen. Microbiol.*, 107, 123, 1978.

135. Timmer-ten Hoor, A., Energetic aspects of the metabolism of reduced sulphur compounds in *Thiobacillus denitrificans, Antonie v. Leeuwenhoek J. Microbiol. Serol.*, 42, 483, 1976.

136. Koike, I. and Hattori, A., Energy yield of denitrification: an estimate from growth yield in continuous cultures of *Pseudomonas denitrificans* under nitrate-, nitrite-, and nitrous oxide-limited conditions, *J. Gen. Microbiol.*, 88, 11, 1975.

137. St. John, R. T. and Hollocher, T. C., Nitrogen 15 tracer studies on the pathway of denitrification in *Pseudomonas aeruginosa, J. Biol. Chem.*, 252, 212, 1977.

138. Konings, W. N. and Kaback, H. R., Anaerobic transport in *Escherichia coli* membrane vesicles, *Proc. Natl. Acad. Sci. U.S.A.*, 70, 3376, 1973.

139. Boonstra, J., Sips, H. J., and Konings, W. N., Active transport by membrane vesicles from anaerobically grown *Escherichia coli* energized by electron transfer to ferricyanide and chlorate, *Eur. J. Biochem.*, 69, 35, 1976.

140. Boonstra, J., Huttunen, M. T., Konings, W. N., and Kaback, H. R., Anaerobic electron transport in *Escherichia coli* membrane vesicles, *J. Biol. Chem.*, 250, 6792, 1975.

141. Konings, W. N., Boonstra, J., and de Vries, W., Amino acid transport in membrane vesicles of the obligately anaerobic *Veillonella alkalescens, J. Bacteriol.*, 122, 245, 1975.

142. Simoni, R. D. and Schallenberger, M. K., Coupling of energy to active transport of amino acids in *Escherichia coli, Proc. Natl. Acad. Sci. U.S.A.*, 69, 2663, 1972.

143. Giordano, G., Rosset, R., and Azoulay, E., Isolation and study of mutants of *Escherichia coli* K12 that are sensitive to chlorate and derepressed for nitrate reductase, *FEMS Microbiol. Lett.*, 2, 21, 1977.

144. Azoulay, E., Giordano, G., Guillet, L., Rosset, R., and Haddock, B. A., Properties of *Escherichia coli* K12-mutants that are sensitive to chlorate when grown aerobically, *FEMS Microbiol Lett.*, 4, 235, 1978.

145. Hackenthal, E. and Hackenthal, R., Die Spezifitat der Nitratreduktase-Induktion bei *Bacillus cereus, Arch. Pharm. Exp. Path.*, 254, 56, 1966.

146. De Groot, G. N. and Stouthamer, A. H., Regulation of reductase formation in *Proteus mirabilis*. II. Influence of growth with azide and of haem deficiency on nitrate reductase formation, *Biochim. Biophys. Acta*, 208, 414, 1970.

147. **Chippaux, M. and Pichinoty, F.**, Les nitrate-reductases bacteriennes. V. Induction de la biosynthese de l'enzyme A par l'azoture, *Arch. Mikrobiol.*, 71, 361, 1970.

148. **Showe, M. K. and DeMoss, J. A.**, Localization and regulation of synthesis of nitrate reductase in *Escherichia coli*, *J. Bacteriol.*, 95, 1305, 1968.

149. **Oltmann, L. F., Reijnders, W. N. M., and Stouthamer, A. H.**, The correlation between the protein composition of cytoplasmic membranes and the formation of nitrate reductase A, chlorate reductase C and tetrathionate reductase in *Proteus mirabilis* wild type and some chlorate resistant mutants, *Arch. Microbiol.*, 111, 37, 1976.

150. **Oltmann, L. F., Claassen, V. P., Kastelein, P., Reijnders, W. N. M., and Stouthamer, A. H.** Influence of tungstate on the formation and activities of four reductases in *Proteus mirabilis*, Identification of two molybdoenzymes: chlorate reductase and tetrathionate reductase, *FEBS Lett.*, 106, 43, 1979.

151. **Wimpenny, J. W. T.**, Oxygen and carbon dioxide as regulators of microbial growth and metabolism, *Symp. Soc. Gen. Microbiol.*, 19, 161, 1969.

152. **De Groot, G. N. and Stouthamer, A. H.**, Regulation of reductase formation in *Proteus mirabilis*. III. Influence of oxygen, nitrate and azide on thiosulfate reductase and tetrathionate reductase formation, *Arch. Microbiol.*, 74, 326, 1970.

153. **Goldberger, R. F.**, Autogenous regulation of gene expression, *Science*, 183, 810, 1974.

154. **van Wielink, J. E., Leeuwerik, F. J., Oltmann, L. F., and Stouthamer, A. H.**, The composition of the cytochrome *b*-pool in *Proteus mirabilis* in connection with different growth conditions. Characterization by means of potentio., meteric titrations, submitted.

155. **Wallace, B. J. and Young, J. G.**, Role of quinones in electron transport to oxygen and nitrate in *Escherichia coli*. Studies with *ubi A⁻ men A⁻* double quinone mutant, *Biochim. Biophys. Acta*, 461, 84, 1977.

156. **van der Beek, E. G.**, Oxidative Phosphorylation and Electron Transport in *Proteus mirabilis*, Ph.D. thesis, Free University, Amsterdam, 1976.

157. **Burke, K. A. and Lascelles, J.**, Nitrate reductase activity in heme-deficient mutants of *Staphoylococcus aureus*, *J. Bacteriol.*, 126, 225, 1976.

158. **MacGregor, C. H. and Bishop, C. W.**, Do cytochromes function as oxygen sensors in the regulation of nitrate reductase biosynthesis?, *J. Bacteriol.*, 131, 372, 1977.

159. **Kemp, M. B., Haddock B. A., and Garland, P. B.**, Synthesis and sidedness of membrane-bound respiratory nitrate reductase (EC 1.7.99.4) in *Escherichia coli* lacking cytochromes, Biochem. J., 148, 329, 1975.

160. **Schulp, J. A. and Stouthamer, A H.**, The influence of oxygen, glucose and nitrate upon the formation of nitrate reduction and the respiratory system in *Bacillus licheniformis*, *J. Gen. Microbiol.*, 64, 195, 1970.

161. **Krul, J. M. and Veeningen, R.**, The synthesis of the dissimilatory nitrate reductase under aerobic conditions in a number of denitrifying bacteria, isolated from activated sludge and drinking water, *Water Res.*, 11, 39, 1977.

162. **Boxer, D. H. and van't Riet, J.**, unpublished observations.

163. **Wientjes, F. B. and van't Riet, J.**, unpublished results.

164. **Graham, A., Wientjes, F. B., and van't Riet, J.**, unpublished results.

165. **Graham, A. and Boxer, D. H.**, personal communication.

Chapter 3

THE STICKLAND REACTION

B. Seto

TABLE OF CONTENTS

I. INTRODUCTION

In 1925, Quastel et al.[1] demonstrated that *Bacterium coli (Escherichia coli)* could be cultivated under anaerobic conditions, provided that pairs of hydrogen donors and acceptors, such as lactate and fumarate or glycerol and aspartate, were supplied. Undoubtedly, catalysis of these coupled oxidation-reduction reactions yielded the energy needed to support growth of the organism. An extension of these studies on anaerobic energy-yielding coupled reactions was undertaken by Stickland in 1934[2] and later by Woods in 1936.[3] They showed that the obligate anaerobe, *Clostridium sporogenes*, required a nutrient medium containing mixtures of amino acids for growth. Catabolism of these mixtures involved the oxidation of a particular amino acid and the reduction of another as shown by the following general equation:

$$R^1CHNH_2COOH + R^2CHNH_2COOH + H_2O \longrightarrow R^1COCOOH$$

$$+ R^2CH_2COOH + 2NH_3 \qquad (1)$$

This type of reaction became known as a Stickland reaction.

By coupling each half reaction to the oxidation or reduction of dyes, such as benzyl viologen, methyl viologen, cresyl blue, and others as redox indicators, Stickland[2] concluded that alanine, valine, and leucine served as hydrogen donors for *C. sporogenes*, whereas glycine, proline, and hydroxyproline were hydrogen acceptors in the coupled oxidation-reduction of amino acid pairs. Kocholaty and Hoogerheide[4] and Woods[3] reported that ornithine and tryptophan could be utilized either as a hydrogen donor or as acceptor.

Early investigations of the Stickland reactions were performed with whole-cell suspensions of *C. sporogenes*. The characterization of these systems has been extensively reviewed by Nisman.[5] In the present review, emphasis will be on the enzymatic systems that catalyze the Stickland reactions, the physical and chemical properties of these enzymes, and their roles in energy metabolism.

II. OCCURRENCE OF THE STICKLAND REACTION

In addition to *C. sporogenes* which was studied by Stickland, a number of other anaerobic bacteria, most of which belong to the family Clostridiae,[6-11] have been found to catalyze Stickland reactions (Table 1).

III. THE STICKLAND REACTION

The efforts of the initial investigators of the Stickland reaction were directed primarily toward the chemical identification of reaction products. For example, Stickland[12] showed that alanine and glycine interacted to form acetic acid, ammonia, and carbon dioxide (Reaction 2) in the presence of *C. sporogenes* as follows:

$$CH_3CHNH_2COOH + 2CH_2NH_2COOH + 2H_2O \longrightarrow$$

$$3CH_3COOH + 3NH_3 + CO_2 \qquad (2)$$

Alternatively, the oxidative deamination of alanine could be coupled to the reduction of proline to form δ-aminovalerate.[13] With suspensions of *Clostridium caproicum* or *Clostridium valerianicum*, Cohen-Bazire et al.[14] showed that the reduction of proline can also be coupled to the oxidation of valine, leucine, or isoleucine, forming the respective volatile fatty acids, isobutyric acid, isovaleric acid, and valeric acid.

Table 1
DISTRIBUTION OF THE STICKLAND
REACTION AMONG CLOSTRIDIAE

Clostridia capable of carrying out Stickland reactions:

C. acetobutylicum	C. ghonii
C. aerofoetidum	C. lentoputresin
C. bifermentans	C. mitelmanii
C. botulinum	C. saprotoxicum
C. butyricum	C. sordelii
C. caproicum	C. sporogenes
C. carnofoetidum	C. sticklandii
C. histolyticum	C. valerianicum
C. indolicus	

Clostridia incapable of carrying out Stickland reactions:

C. iodophilum	C. tetani
C. propionicum	C. tetanomorphum
C. saccharobutyricum	C. welchii
C. teras	

Earlier attempts to study the Stickland-type reactions in cell extracts failed for the most part, probably because of the oxygen sensitivity of the enzymes.[15] In fact, the addition of thiol compounds was found to reactivate *C. sprogenes* cells exposed to air. Recognizing the probable involvements of SH groups in the enzyme systems, Stadtman[16] discovered that dimercaptans such as 1,3-dimercaptopropanol would not only protect the enzymes from oxidation, but could also function as artificial electron donors. This facilitated study of the reductive half-reactions. In the following sections, discussions will be confined to two of the enzyme systems that transfer reducing equivalents to amino acid acceptors in the Stickland reactions: glycine reductase and proline reductase.

A. Reduction of Glycine
1. Chemical Reaction

Enzyme preparations, subsequently referred to as the glycine reductase complex from *Clostridium sticklandii*, have been shown to catalyze the reductive deamination of glycine in the presence of 1,3-dimercaptopropanol with concomitant esterification of ADP to ATP as illustrated in Equation 3:[17]

$$\text{Glycine} + R(SH)_2 + Pi + ADP \longrightarrow \text{acetate} + NH_3 + R\begin{matrix} S \\ | \\ S \end{matrix}$$

$$+ \text{ATP} \qquad (3)$$

This reaction represents a type of anaerobic respiration which generates a high-energy compound (ATP) at the expense of the substrate (glycine) and intracellular reducing equivalents. The stoichiometry of one ATP per mole of acetate formed was determined by chemical balance studies as in Table 2. The esterification of adenine nucleotide is a specific phenomenon and other base nucleotides are ineffective. The enzymatic mechanism of deamination has not been established, but studies with [2-C¹⁴]glycine showed that the carbon skeleton of glycine is preserved in the product, acetate. Based on radioisotope dilution studies[18] in which unlabeled glycollic acid did not dilute the radioactive acetate formed from glycine, it was concluded that glycollic acid is not a free intermediate. Barnard and Akhtar[19] studied the stereochemistry of the reaction and concluded that the reaction proceeds with an inversion of configuration at the meth-

Table 2
ATP SYNTHESIS DURING REDUCTION OF GLYCINE TO
ACETATE AND AMMONIA

Experiment number	Glycine decomposed	(SH) oxidized	PO_4 uptake	P_{10} min formed[a]	Acetate formed	NH_3 formed
	(μmol)	(μequiv)	(μmol)	(μmol)	(μmol)	(μmol)
1	1.01	—	1.07	0.98	1.01	1.23[b]
2	—	2.52[c]	1.25	0.89	1.21	—

[a]   Amount of phosphate released when heated for 10 min at 100°C in 1 N Acid.
[b]   The enzyme preparation used to measure NH_3 formation had been precipitated with
       saturated Na_2SO_4 and redissolved in buffer to lower its $(NH_4)_2SO_4$ concentration.
[c]   The amount of dimercaptopropanol oxidized was measured in incubation mixtures
       reduced to one half the usual volume; all components were added in proportionally
       smaller amounts except for enzyme and glycine.

Reproduced from Stadtman, T. C., Elliot, P., and Tiemann, L., *J. Biol. Chem.*, 231, 961, 1958. With permission.

ylene carbon of glycine as in Figure 1. These investigators prepared enantiomeric stereospecifically tritiated and deuteriated glycines via the serine transhydroxymethylase reaction which catalyzes the exchange of the hydrogen on the α carbon of glycine with the medium, either ^3H_2O or D_2O. [^{14}C]glycine was added to each sample of stereoisomer. These stereospecifically labeled glycines were then used as substrates for glycine reductase reaction. The product of the reaction, acetate, was converted to L-malate for configurational analyses. As in Table 3, the product derived from the 2R-glycine enantiomer retained the majority of its tritium.

Glycine appears to be the only substrate deaminated by glycine reductase.[18] No utilization of a number of substrate analogues and derivatives, such as glycine anhydride, glycine methyl ester, glycylglycine, glycollic acid, glyoxylic acid, etc. was detected. The nature of the electron donor is not as restrictive, and a number of dimercaptans are satisfactory donors. These are, however, artificial in vitro electron donors and presumably do not have any physiological significance except perhaps as chemical models. Using French press extracts of *C. sticklandii* and *Clostridium lentoputrescens*, Stadtman demonstrated that NADH can function as the electron donor for glycine reduction.[20] This NADH-linked reaction is completely inhibited by 1×10^{-4} M arsenite, suggesting that vicinal -SH groups are involved. An additional requirement for FMN was exhibited by partially purified protein fractions, suggesting the mediation of a flavoprotein as electron carrier between NADH and glycine reductase.

2. Metal Requirements

In the course of the purification of components of the glycine reductase complex, Turner and Stadtman[21] observed that glycine reductase activity of cells of *C. sticklandii* harvested at different stages of growth varied considerably. There was a marked decrease with the age of the culture, suggesting that an essential substance is depleted from the medium. Prompted by the findings of Andreesen and Ljungdahl[22] that the synthesis of formate dehydrogenase in *Clostridium thermoaceticum* can be sustained throughout the log phase of growth by the addition of selenite and molybdate, Turner and Stadtman undertook similar nutritional studies with *C. sticklandii*. Their results showed that the addition of selenite resulted in increased glycine reductase activity, whereas added molybdate showed no effect. Subsequently, these investigators demonstrated that one of the components of the glycine reductase complex is a selenoprotein. This protein will be described in the section entitled Selenoprotein.

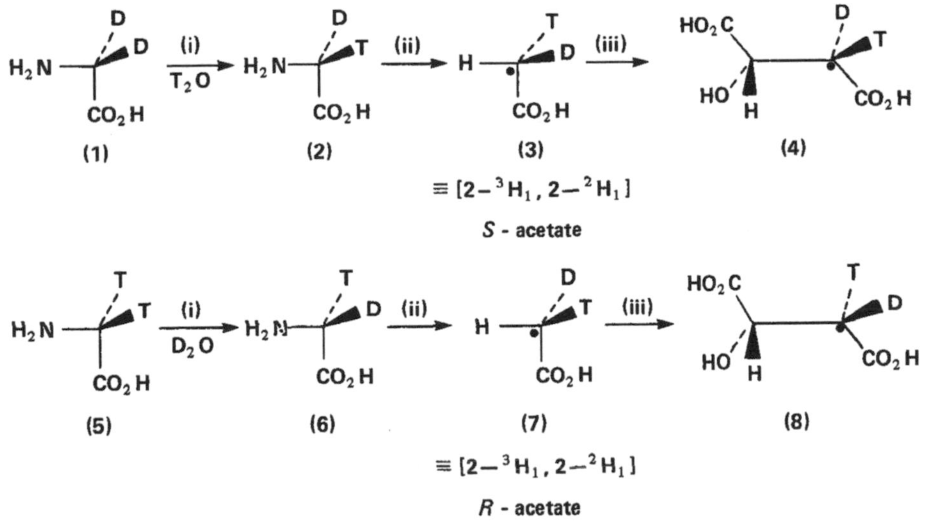

FIGURE 1. Stereochemistry of the glycine reductase reaction. (i) Serine transhydroxymethylase, (ii) glycine reductase, and (iii) malate synthetase. (Reproduced from Barnard, G. and Akhtar, M., *J. Chem. Soc. Chem. Commun.*, p. 980, 1975. With permission.)

Table 3
STEREOCHEMISTRY OF THE GLYCINE REDUCTASE REACTION

Source of acetate		$^3H/^{14}C^a$	Retention of ^2H (%)
$[2-^3H_1,2-^2H_1]$-2S-glycine	Starting malate	1.835	
	Equilibrated malate	0.546	29.7
$[^3H_2]$-glycine	Starting malate	3.781	
	Equilibrated malate	1.893	50.1
$[2-^3H_1,2-^2H_1]$-2R-glycine	Starting malate	1.848	
	Equilibrated malate	1.438	77.8

* Various samples of acetate, as indicated, were converted into malate. Aliquots of the malate samples were then treated with fumarase to equilibrate the C-3 pro-R hydrogen with protons of the medium. $^3H/^{14}C$ ratios refer to the recrystallized malate.

Reproduced from Barnard, g. and Akhtar, M., *J. Chem. Soc. Chem. Commun.*, p. 980, 1975. With permission.

A strict growth requirement for a selenium supplement in the culture medium was reported for *C. sporogenes*[23] when glycine, but not proline, was used as the obligatory electron acceptor in the Stickland reaction. Although the glycine reductase enzymes have not been investigated in detail from this organism, both the nutritional requirement and preliminary ^{75}Se-labeling experiments[24] suggest that a selenoprotein component is involved.

3. Assays of Glycine Reductase
a. Oxidation of Redox Indicator Dyes
In the early experiments, the rate of glycine-dependent oxidation of reduced dyes in the presence of cell suspensions was followed colorimetrically. Glycine oxidizes a variety of dyes, including leucomethylviologen and leucobenzylviologen.[2,12,13,25] Since this assay was developed prior to the isolation of the glycine reductase complex, it is difficult to assess whether these dyes were primary electron donors.

b. Production of [¹⁴C]Acetate from [¹⁴C]Glycine²⁶

This assay is most commonly performed with dithiothreitol as the electron donor in the presence of ADP, Pi, Mg²⁺, and [1-¹⁴C]glycine. With intact cells, ADP is omitted. The product, [1-¹⁴C]acetate can be separated from the substrate [1-¹⁴C]glycine either by Dowex® 50 H⁺ chromatography or by extraction with ethyl acetate.²⁶ This assay requires the presence of all the protein components of the glycine reductase complex. The activity of each individual protein component can be determined in the presence of excess amounts of the complementary protein.

c. Tritium Exchange Assay²⁷

Protein B (a component of the glycine reductase complex that contains an essential carbonyl group) catalyzes tritium-hydrogen exchange between [2-³H]glycine and H₂O or ³H₂O and glycine, presumably via a Schiff base intermediate. This exchange serves as an independent assay for protein B. When [2-³H]glycine is used, radioactivity of ³H₂O is determined following removal of residual labeled substrate by Dowex® chromatography. Alternatively, following incubation of unlabeled glycine in ³H₂O, radioactivity of [2-³H]glycine can be determined after repeated drying of the sample to remove ³H₂O.

4. Purification and Properties of the Glycine Reductase Complex
a. Selenoprotein

Stadtman and Elliot²⁸ first obtained glycine reductase activity in cell extracts of *Clostridium* strain HF (later named *C. sticklandii*) prepared by sonication or alumina grinding. Subsequently, other enzyme systems involved in this type of reaction were purified and characterized.

The glycine reductase system was separated into two distinct protein fractions by treatments with ammonium sulfate, calcium phosphate gel, and acidic pH.¹⁸,²⁹ One of these fractions contained a very acidic protein, referred to as Protein A or selenoprotein. Stadtman and co-workers²¹,²⁶ took advantage of this property and further purified this protein by DEAE cellulose chromatography. The protein can be eluted from DEAE cellulose with 1 *M* sodium chloride or 1 *M* potassium phosphate. An additional property which is useful for purification purposes is the presence of a selenohydryl group in the fully reduced protein. Selenohydryl groups have been shown to have a higher affinity for mercury than sulfhydryl groups.³⁰ Thus, affinity chromatography on mercuriagarose was exploited and the highly purified selenoprotein eluted with 100 m*M* mercaptoethanol was suitable for chemical analyses.

The glycine reductase selenoprotein is a small, heat stable, acidic protein. The molecular weight, as determined by three different methods, ranged from 12,000 to 16,000 (Table 4). The apparent discrepancy in these results can be explained by the significant carbohydrate content of this protein. Cone et al.²⁶ showed that as much as 16% of the dry weight of the selenoprotein may be carbohydrate. A high molecular weight estimate for glycoproteins is common based on the observation of Braun and Schroeder.³¹ They demonstrated that carbohydrates of glycoproteins bind less sodium dodecyl sulfate than the polypeptide, resulting in retarded electrophoretic mobilities.

One of the striking features of this small acidic protein is the presence of a catalytically essential selenium moiety. This was suggested by the results of the nutritional studies with trace elements and further substantiated by incorporation experiments with ⁷⁵Se. Chromatography profiles of the protein isolated from ⁷⁵Se-labeled bacteria showed coincidence of enzymatic activity with radioactivity. Moreover, these two parameters remained at a constant ratio through several different chromatographic procedures.²¹ However, the definitive identification of the protein-bound selenium moiety

Table 4
SUMMARY OF PHYSICAL AND CHEMICAL PARAMETERS
OF SELENOPROTEIN A

Parameter	Value
Molecular weight	
Gel filtration (native protein)	16,200
Sodium dodecyl sulfate gel electrophoresis	14,500
Amino acid composition	12,000
Carbohydrate	16% (w/w)
E_{277} (1% protein, 1-cm light path)	
Based on Lowry assay (average value, five preparations)	2.22
Based on dry weight (average value, three preparations)	2.16
Elemental selenium[a]	
(g·atom/mol protein, average value of three preparations)	0.66
Selenocysteine (residues/mol protein)	1
Half cystine (residues/mol protein)	2
Tyrosine (residues/mol protein	1
Phenylalanine (residues/mol protein)	6
NH₂-terminal	Blocked
COOH-terminal	Aspartic acid

[a] Determined by atomic absorption spectrophotometry and fluorescence assay by the diaminonaphthalene procedure.

Reproduced from Cone, J. E., Martin del Rio, R., and Stadtman, T. C., *J. Biol. Chem.*, 252, 5337, 1977. With permission.

was hampered by its extreme lability to air oxidation. It was only recently that Cone et al.[32] indirectly identified the organoselenium moiety as selenocysteine. They overcame the instability problem by preparing alkyl derivatives, e.g., carboxymethyl-, carboxyethyl-, and aminoethyl-, of the selenoprotein. The acid hydrolysis products of these derivitized proteins were then compared with various alkylated derivatives of selenocysteine. All the chemical parameters studied support the contention that the ⁷⁵Se-labeled moiety obtained from a reduced and alkylated protein sample is indistinguishable from the corresponding alkyl derivative of authentic selenocysteine. Though the identification is an indirect one, their work represents a breakthrough in the field of selenium biochemistry. Subsequently, the selenium moieties of other selenoproteins, such as mammalian glutathione peroxidase[33,34] and bacterial formate dehydrogenase,[35,36] have been shown to be selenocysteine residues.

Some spectral properties of the glycine reductase selenoprotein are shown in Figure 2A. The selenide anion of the selenocysteine residue can be generated at pH 7 to 8 by KBH_4 reduction, and it exhibits a characteristic electronic absorption spectrum with a maximum at 238 nm. The chromophore is extremely oxygen sensitive and disappears immediately upon exposure to air. After alkylation the selenoprotein no longer exhibits this chromophore upon KBH_4 reduction. Based on the absorbance at 240 nm and the molar extinction coefficient of 5000 $M^{-1}cm^{-1}$ reported for the selenide anion of selenocysteine,[37,38] Cone et al.[26] calculated 0.99 mol selenocysteine per 12,000 daltons of selenoprotein.

Another notable feature of the absorption spectrum of selenoprotein A is the number of distinct maxima between 250 to 270 nm due to the presence of phenylalanine. The absorption due to tyrosine is also discernible in the spectrum. Consistent with these spectral properties are the results of the amino acid analyses which showed a

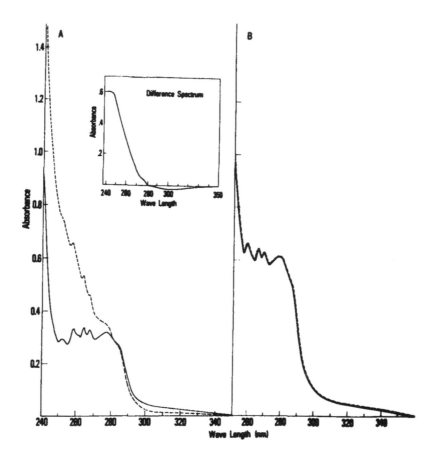

FIGURE 2.   Absorption spectra of the reduced and the alkylated selenoprotein. (A)
The selenoprotein absorption spectrum was recorded before (——) and after (- - - -)
the addition of KBH$_4$. The inset shows the difference spectrum of reduced versus oxidized protein. (B) The absorption spectra of the alkylated selenoprotein before (——)
and after (- - - -) the addition of KBH$_4$. (Reproduced from Cone, J. E., Martin del
Rio, R., and Stadtman, T. C., *J. Biol. Chem.*, 252, 5337, 1977.

ratio of five to six phenylalanine residues to one tyrosine residue and the absence of
tryptophan.

In addition to a single residue of selenocysteine, the reduced selenoprotein contains
two residues of sulfur cysteine. The structural or spatial relationship of these groups
has not been elucidated. However, this problem must be studied with consideration of
the biological role of the selenoprotein. Though it has been known that the reductive
deamination of glycine to acetate requires the participation of the selenoprotein, its
precise role during catalysis is not clear. In view of the nature of the reaction catalyzed
by the glycine reductase complex and the molecular properties of the selenoprotein,
an electron transport carrier function for the selenoprotein appears to be an attractive
possibility. In this case, an oxidation-reduction reaction involving disulfide (S—S) or
mixed selenosulfide (Se—S) at the active center is entirely possible. An alternative role
would involve phosphate transfer associated with the concomitant formation of ATP
during glycine reduction. No phosphoprotein intermediate in the glycine reductase system has been detected, despite the unequivocal evidence for phosphate esterification.
The recent discovery of Pigiet and Conley[39] of a cysteine S-thiol phosphate in an electron carrier protein, thioredoxin, suggests a new possible function of this protein,
namely, in a phosphate transfer reaction. It is tempting to postulate that an analogous

function be ascribed for the glycine reductase selenoprotein. The conclusion must await the demonstration of either cysteinyl-S-phosphate or selenocysteinyl-Se-phosphate.

b. Glycine Reductase Protein B

The reductive deamination of glycine to acetate and the simultaneous formation of ATP requires at least three distinct protein components, one of which is the selenoprotein described in the previous section. The remaining two proteins, protein B and fraction C, exist in a tightly associated membrane-bound complex. They are collectively termed glycine reductase fraction because they cochromatograph on DEAE-cellulose prior to detergent solubilization. In contrast to the acidic selenoprotein which requires 1 M phosphate buffer for elution from DEAE-cellulose, glycine reductase can be eluted with 0.25 M phosphate buffer.[21] Recently, Tanaka and Stadtman were able to solubilize the membrane bound complex by detergent treatment with a mixture of 1.5% Lubrol® WX and 0.5% sodium deoxycholate.[27] Subsequently, protein B and fraction C can be separated and purified individually by repeated chromatography on DEAE-cellulose® and Sephacryl® S-200. Protein B purified by these procedures is approximately 90% homogeneous based on polyacrylamide gel electrophoresis.

The molecular weight of protein B is estimated by gel filtration on Sephacryl® S-200 to be 200,000 to 250,000. It contains a catalytically essential carbonyl group that is inactivated both by KBH_4 and NH_2OH. This carbonyl group is thought to play a role in enzyme-substrate intermediate ormation. Presumably, a Schiff base involving the amino group of the substrate, glycine, and the carbonyl group of the enzyme protein B, can be a primary step of the reaction. An expected side reaction would be the hydrogen exchange between the methylene hydrogens of glycine and solvent. This assumption constitutes the theoretical basis for the tritium exchange assay for protein B activity. Neither selenoprotein A nor fraction C catalyze significant tritium exchange. Boiled protein B is completely inactive.

In addition to KBH_4 and NH_2OH, protein B is also inactivated by iodoacetamide. This suggests that protein B contains essential -SH groups. No organoselenium moiety has been detected in protein B.

c. Glycine Reductase Fraction C

A homogeneous preparation of fraction C has not yet been achieved, although it has been separated from the selenoprotein and protein B. The purification of this component is impeded by its instability after solubilization from the membrane bound complex. Thus, protein C is less well characterized.

The molecular weight, estimated by gel filtration on Sephacryl® S-200, is in excess of 250,000. Preliminary analyses by atomic absorption spectroscopy show that fraction C contains iron. Further characterization of the protein must await a solution to the instability problem.

B. Reduction of Proline

1. Chemical Reaction

The reduction of proline to δ-aminovalerate by cell suspensions of *C. sporogenes* was originally observed by Stickland in 1935.[13] It was not until 1956 that Stadtman isolated from *C. sticklandii* a crude enzyme preparation by ammonium sulfate precipitation which was referred to as proline reductase.[40] As in Equation 4, the enzyme catalyzes the reductive ring cleavage of proline in the presence of a number of dithiols, such as 1,3-dimercaptopropanol, 6,8-dimercaptooctanoate, 5,8-dimercaptooctanoate, and 1,4-dithiothreitol.[41]

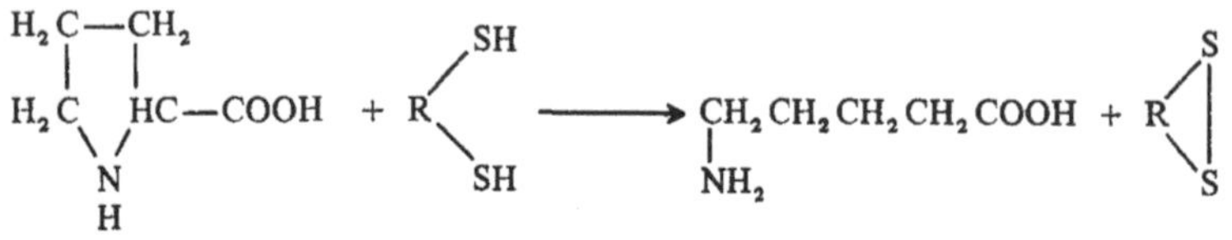

$$\text{proline structure} + R\diagup^{SH}_{SH} \longrightarrow CH_2CH_2CH_2CH_2COOH + R\diagup^{S}_{S}$$

$$\begin{array}{c} H_2C-CH_2 \\ | \quad\quad | \\ H_2C \quad HC-COOH \\ \diagdown N \diagup \\ | \\ H \end{array} + R\diagup^{SH}_{SH} \longrightarrow \begin{array}{c} CH_2CH_2CH_2CH_2COOH \\ | \\ NH_2 \end{array} + R\diagup^{S}_{S}$$

(4)

$$\begin{array}{c} \text{1,4-dithiothreitol} \\ \downarrow \end{array}$$

NADH $\longrightarrow$ flavoprotein $\longrightarrow$ Fe protein $\longrightarrow$ proline reductase $\longrightarrow$ D-proline

In the presence of the appropriate electron carrier proteins, NADH serves as the physiological electron donor for proline reduction. In contrast to the wide range of effective electron donors, the enzyme reduces the D-isomer of proline specifically. The L-isomer is reduced by crude preparations of proline reductase contaminated with an extremely active proline racemase. Azetidine-2-carboxylic acid, a proline analogue, and other amino acid analogues, such as DL-pipcolic acid, tetrahydropyrrole, and DL-ornithine will not substitue for D-proline as substrates. L-hydroxyproline is an inert substrate, but it is not clear whether the D-isomer is effective.

The stoichiometry of the proline reductase reaction was determined by direct chemical analyses of reactants and products. An equimolar ratio of reduced proline and oxidized -SH is established. The mechanism of reduction is not known, though it is established by radioisotope tracer studies that Δ'-pyrroline-2-carboxylic acid is not an intermediate of the reaction.[40] Hodgins and Abeles[42] showed that the reaction product, δ-aminovalerate contains one deuterium at the α-carbon position if the reaction is carried out in D_2O. A possible mechanism proposed by these investigators was based on the additional finding that proline reductase contains catalytically essential pyruvate groups (see the section entitled Proline Reductase). As illustrated in the following scheme (Figure 3), they postulated a Schiff base formation between the carbonyl group of pyruvate with the nitrogen of proline to make nitrogen a better leaving group, thus allowing a nucleophilic displacement at the α-carbon. Presumably, the nucleophile in this case would be the dithiol which exchanges the thiol hydrogen readily with D_2O.

An alternative mechanism which ascribes a significant function to pyruvate has been suggested by Hamilton.[43] As shown in Figure 4, this mechanism differs from that proposed by Hodgins and Abeles in that the dithiol adds to the minium ion that is the adduct of pyruvate and proline. In other respects, these proposed mechanisms share similar characteristics.

A distinct difference between the proline reduction and glycine reduction reactions is the lack of phosphate esterification coupled to proline reduction.[41] It thus appears that the metabolic role of proline reduction is to serve mainly as an electron sink. In view of the fact that both the glycine and proline systems share a common physiological reductant, namely NADH, it is not unlikely that competition for electron flow exists between the two systems. Preliminary indication of this occurrence in *C. sporogenes* is reported by Costilow, who showed that glycine reductase activity was depressed threefold in cultures grown with glycine and proline additions, as compared to cultures supplemented with glycine only.[23]

2. Assays of Proline Reductase
a. Ninhydrin Colorimetric Assay

Proline reductase activity is measured by determining the amount of substrate, proline, that is utilized in the presence of 1,4-dithiothreitol. Proline is quantitated by the acidic ninhydrin reagent of Chinard.[44] A limitation of this assay method is that subsaturating levels of proline must be used in order to detect colorimetrically a significant

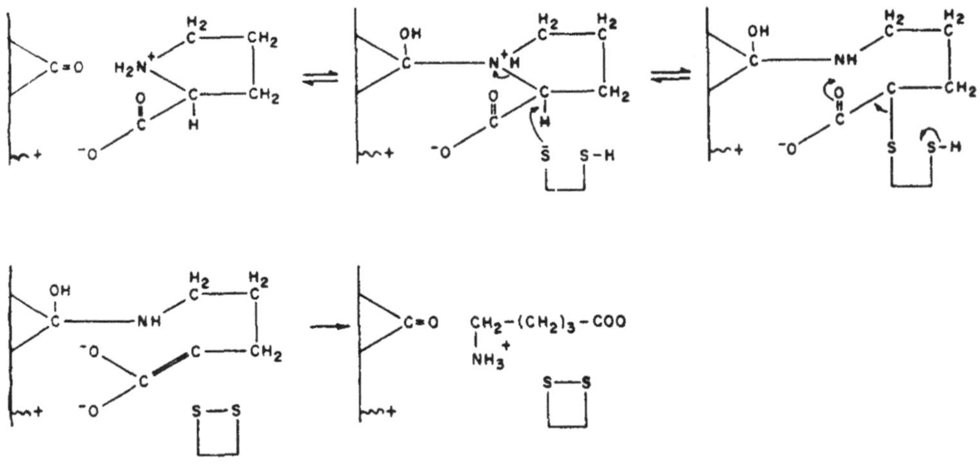

FIGURE 3. A postulated mechanism of action of proline reductase. (Reproduced from Hodgins, D. S. and Abeles, R. H., *Arch. Biochem. Biophys.*, 130, 274, 1969. With permission.)

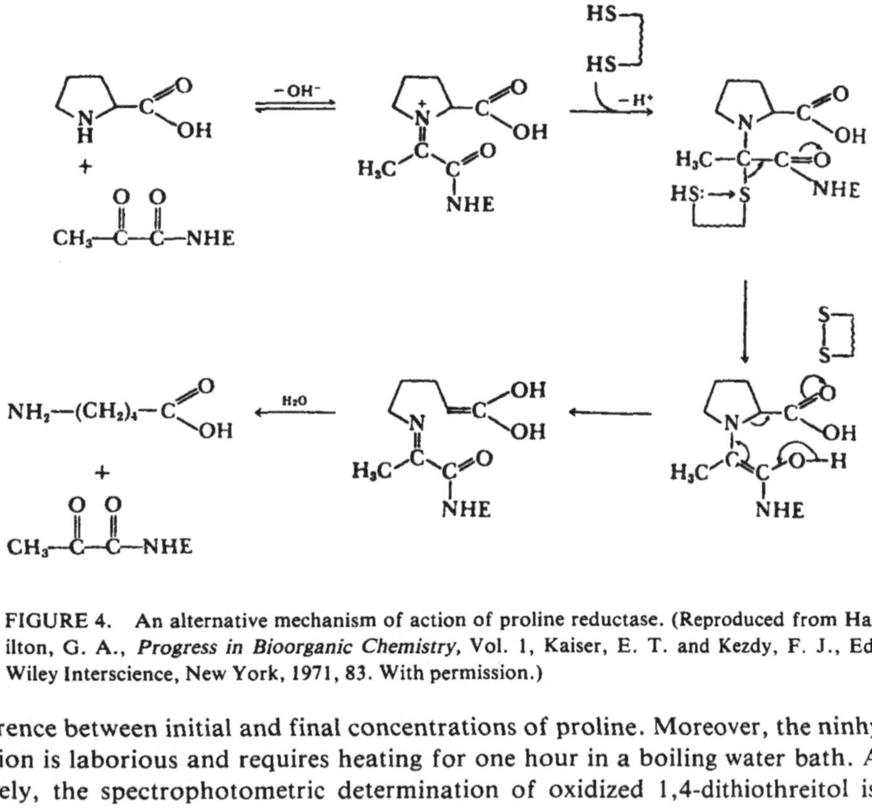

FIGURE 4. An alternative mechanism of action of proline reductase. (Reproduced from Hamilton, G. A., *Progress in Bioorganic Chemistry*, Vol. 1, Kaiser, E. T. and Kezdy, F. J., Eds., Wiley Interscience, New York, 1971, 83. With permission.)

difference between initial and final concentrations of proline. Moreover, the ninhydrin reaction is laborious and requires heating for one hour in a boiling water bath. Alternatively, the spectrophotometric determination of oxidized 1,4-dithiothreitol is less sensitive than the colorimetric quantitation of proline.

b. Fluorometric Reaction with o-Phthalaldehyde

A fluorogenic reagent, o-phthalaldehyde, was found to react specifically with primary amino acids.[45] It reacts with δ-aminovalerate, the product of the proline reductase reaction, to form a fluorescent adduct.[46] The water-soluble fluorescent adduct exhibits an excitation maximum at 340 nm and an emission maximum at 455 nm. This method is extremely sensitive and provides a direct measurement of enzymic product formation without interference from residual substrate.

3. Purification and Properties of the NADH-Linked Proline Reductase Complex
a. Proline Reductase

Following the initial isolation of a crude preparation of proline reductase by Stadt-man,[41] Hodgins and Abeles[42] further fractionated the extract from *C. sticklandii* by DEAE-cellulose and Sephadex® chromatography. However, final purification of the enzyme was achieved only recently.[47] The major problem which was not recognized earlier is the association of this enzyme with the cell membrane. Treatment with detergents, e.g., 1% sodium deoxycholate and 1.5% Triton® X-100, is necessary to release the enzyme from the membrane. By virtue of its hydrophobic nature, purification of proline reductase is facilitated by affinity chromatography with AH-Sepharose® 4B, a sepharose with covalently linked aminohexane spacers.

Using partially purified proline reductase, Hodgins and Abeles[42,48] demonstrated the presence of a catalytically essential pyruvoyl cofactor in the enzyme. This observation was subsequently confirmed and extended with the purified protein.[47,49] Inhibition of enzymic activity observed with a number of carbonyl reagents, such as borohydride, hydroxylamine, and phenylhydrazine, is due to derivitization of these essential pyruvoyl groups. Borohydride reduction of the protein results in the conversion of the covalently attached pyruvate to lactate. Indeed, much of the evidence for the identification of pyruvate is based on the isolation of [³H]lactate following NaB³H₄ reduction and acid hydrolysis. The labeled lactate from the protein was shown to cochromatograph with authentic lactate in a number of different solvent systems, and it migrated with the authentic compound when subjected to paper electrophoresis. Moreover, the phenacyl esters of the labeled product and of lactate were prepared and shown to be indistinguishable by a number of criteria. The evidence unequivocally shows that pyruvate is the carbonyl cofactor bound to the enzyme.

Structural studies of proline reductase revealed additional interesting features. The enzyme has a mol wt of 300,000. It is composed of ten identical subunits each with a C terminal glycine. In addition, there are ten pyruvate residues per mole of enzyme, and these residues are covalently linked to the peptide chains via amide bonds. No N terminal amino acid could be detected despite exhaustive investigation by various procedures, such as manual Edman degradation, aminopeptidase M digestion, and dansylation. A likely explanation may be that the N terminus is blocked by pyruvate. Evidence supporting this contention stems from studies on a 4600-dalton peptide released from proline reductase by very mild alkaline hydrolysis. Treatment with 0.1 *N* NaOH at 100°C for 10 min is sufficient to cleave this peptide from the protein. The peptide also shows a blocked N terminus, suggesting that it represents the N-terminal portion of the holoenzyme. The complementary large polypeptide formed by alkaline hydrolysis of the protein contains the same C terminus as the native enzyme, but its N terminus is glutamate. The C-terminal residue of the small peptide is serine. From these end terminal analyses, it is concluded that cleavage occurred between serine and glutamate. In view of the marked alkaline lability of the bond and the amino acid residues involved, it was suggested that an ester linkage may be present. However, as yet there is no direct chemical identification of this postulated ester bond. In view of the known occurrence of alkali labile peptide bonds in certain proteins, there is the alternative possibility that a particularly sensitive seryl-glutamate peptide bond is involved.

The presence of pyruvate poses many unanswered questions concerning its mode of incorporation into the polypeptide. Preliminary evidence indicates that the protein-bound pyruvate was formed by modification of a specific serine residue. However, the mechanism of conversion of serine to pyruvate in proline reductase is as yet unknown. Similarly in other pyruvate-containing enzymes, namely, histidine decarboxylase of *Lactobacillus* 30a,[50] S-adenosyl-L-methionine decarboxylase from *E. coli*,[51] *Saccharomyces cerevisiae*,[52] and rat liver[53] and phosphatidylserine decarboxylase from *E. coli*,[54] the precise mechanism of pyruvate incorporation is not elucidated.

Proline reductase exhibits no absorbance in the visible region, but in the UV region, the absorbances due to phenylalanine and to tyrosine are discernible. This observation is substantiated by amino acid analyses which show seven residues of tyrosine, eight residues phenylalanine, and no tryptophan. A total of 20 cysteic acid residues are found in the performic acid-oxidized enzyme. This result is in agreement with the 20 sulfhydryl groups titratable with Ellman's reagent in the reduced protein. Unlike the glycine reductase complex, no selenium has been detected in proline reductase.

b. Electron Transport Protein Components

The physiological reduction of proline coupled to NADH requires an enzyme complex consisting of a flavoprotein, an iron protein (Fe protein), and the proline reductase described in the previous section. Dithiothreitol reduces proline reductase directly in vitro, but the reducing equivalents from NADH must be transferred through the electron transport components to proline reductase. The carrier proteins are extremely labile after they are separated from the complex which is membrane bound. Proline reductase appears to be the most tightly bound to the membrane, and it is solubilized by detergent treatment. The carrier proteins can be separated from the complex on the basis of ionic strength. Extensive washing of a DEAE-cellulose column with low ionic strength buffer (20 to 50 mM phosphate buffer) results in the elution of membrane bound proline reductase. Only trace amounts of the carrier proteins are found in this fraction. The carrier proteins then are eluted simultaneously with 0.2 M phosphate buffer. The flavoprotein and the Fe protein are subsequently separated from each other by repeated chromatography on AH-Sepharose 4B and DEAE-cellulose, alternated with molecular sieve chromatography.

The procedures employed resulted in preparations of the flavoprotein that are greater than 90% pure. Preliminary characterization of this 57,000-dalton flavoprotein shows that it contains FAD and no detectable metals. It exhibits a fluorescence emission spectrum typical of flavins with a maximum at 525 nm. The excitation spectrum shows peaks at 378, 365, and a shoulder at 445 nm. The absorption spectrum shows a broad peak in the 420-nm region which is decreased by the addition of the substrate, NADH, or by dithionite.

The Fe protein contains 1 g·atom of iron per 250,000 daltons. Removal of iron by chelating agents, e.g., α,α'-dipyridyl, 1,10-phenanthroline, tiron (4,5 dihydroxy-*m*-benzenedisulfonic acid, disodium salt), and sodium diethyl dithiocarbamate, completely destroy the catalytic activity of the iron protein. As isolated, this protein exhibits no distinctive electronic absorption spectrum. However, upon reduction with dithionite, a new small peak appears at 405 nm. The relationship, if any, between this 405-nm peak and iron in the protein is not clear.

Although the characterizations of the flavoprotein and the Fe protein are preliminary, the evidence shown in Table 5 unequivocally demonstrates the essential roles of both proteins in coupling NADH oxidation to the reduction of D-proline by proline reductase. A probable pathway of proline reduction is as follows:

$$\text{NADH} \longrightarrow \text{flavoprotein} \longrightarrow \text{Fe protein} \longrightarrow \overset{\displaystyle \text{1,4-dithiothreitol}}{\underset{\displaystyle \downarrow}{\text{proline reductase}}} \longrightarrow \text{D-proline}$$

Table 5
COMPONENTS REQUIRED FOR RECONSTITUTION OF
NADH-LINKED D-PROLINE REDUCTASE SYSTEM

Enzyme component added	D-proline reduced (μmol)	
	NADH-linked	DTT^a-linked
Proline reductase, 12 μM	0	1.98
Flavoprotein, 10 μM	0	0
Fe protein, 2.5 μM	0	0
Proline reductase, 12 μM + flavoprotein, 10 μM	0	—
Proline reductase, 12 μM + Fe protein, 2.5 μM	0	—
Proline reductase, 12 μM + flavoprotein, 10 μM + Fe protein, 2.5 μM	0.88	—
Proline reductase, 12 μM + flavoprotein, 10 μM + Fe protein, 5 μM	1.11	—

^{a} DTT, 1,4-dithiothreitol.

IV. SUMMARY

In the 1930s, Stickland described the fermentation of pairs of amino acids by anaerobes, particularly the family Clostridiae. The reaction, which involves the coupled oxidation and reduction of the amino acids, constitutes an important energy-yielding process. Although the metabolic significance was recognized, the enzymes involved in this type of reaction have been studied only recently. The two enzyme systems that have been investigated are glycine reductase and proline reductase. Two of the three components of the glycine reductase complex (i.e., selenoprotein A and protein B) have been purified to homogeneity. Several unique features of this complex enzyme system have been elucidated. For example, the presence of a selenocysteine residue in the small, acidic selenoprotein A raises the possibilities of a role in the electron transport processes or in the phosphate transfer reaction. As yet, no phosphorylated protein intermediate has been identified, despite the unequivocal evidence that ADP esterification to yield ATP occurs concomitantly with glycine deamination. The report of a cysteine S-thiol phosphate in *E. coli* thioredoxin[39] is encouraging, for much of the difficulties in identifying a phosphorylated protein intermediate resides in its lability. Another aspect of the enzyme mechanism involves glycine deamination. The presence of a catalytically essential group in protein B suggests a mechanism involving Schiff base formation prior to deamination. Studies on the role of each of the protein components during catalysis will undoubtedly facilitate the elucidation of the overall enzyme mechanism. The purification of fraction C is therefore necessary.

All the protein components required for D-proline reduction have been purified. Thus, it is possible to study the electron transport processes associated with the reductive ring cleavage of D-proline. Moreover, with the reconstituted system consisting of homogenous protein components, it will be of interest to determine if there is ATP formation coupled to the electron flow between NADH and proline reductase. In view of the common electron donor, NADH, for both glycine reductase and proline reductase, it would be interesting to study whether these two pathways share common electron transport components.

ACKNOWLEDGMENT

The author thanks Thressa C. Stadtman for valuable suggestions in preparing this manuscript.

REFERENCES

1. **Quastel, J. H., Stephenson, M., and Whetham, M. D.,** Some reactions of resting bacteria in relation to anaerobic growth, *Biochem. J.,* 19, 304, 1925.
2. **Stickland, L. H.,** The chemical reactions by which *C. sporogenes* obtains its energy, *Biochem. J.,* 28, 1746, 1934.
3. **Woods, D. D.,** Further experiments on the coupled reactions between pairs of amino acids induced by *C. sporogenes, Biochem. J.,* 30, 1934, 1936.
4. **Kocholaty, W. and Hoogerheide, J. C.,** Dehydrogenation reactions by suspensions of *C. sporogenes, Biochem. J.,* 32, 437, 1938.
5. **Nisman, B.,** The Stickland reaction, *Bacteriol. Rev.,* 18, 16, 1954.
6. **Nisman, B., Raynaud, M., and Cohen, G. N.,** Extension of the Stickland reaction to several bacterial species, *Arch. Biochem.,* 16, 473, 1948.
7. **Woods, D. D. and Clifton, C. E.,** Hydrogen production and amino acid utilization by *C. tetanomorphum, Biochem. J.,* 31, 1774, 1937.
8. **Clifton, C. E.,** The utilization of amino acids and glucose by *C. botulinum, J. Bacteriol.,* 39, 485, 1940.
9. **Woods, D. D. and Trim, R.,** The metabolism of amino acids by *C. welchii, Biochem. J.,* 36, 501, 1942.
10. **Stadtman, T. C. and McClung, L. S.,** *Clostridium sticklandii,* nov. spec., *J. Bacteriol.,* 73, 218, 1957.
11. **Cardon, B. P. and Barker, H. A.,** Amino acid fermentations by *C. propionicum* and *Diplococcus glycinophilus, Arch. Biochem.,* 12, 165, 1947.
12. **Stickland, L. H.,** The oxidation of alanine by *C. sporogenes, Biochem. J.,* 29, 889, 1935.
13. **Stickland, L. H.,** The reduction of proline by *C. sporogenes, Biochem. J.,* 29, 288, 1935.
14. **Cohen-Bazier, G., Cohen, G. N., and Prevot, A. R.,** Nature et mode de formation des acides volatils dans les cultures de quelques bacteries anaerobies proteolytiques du groupe de *C. sporogenes.* Formation par reaction de Stickland des acides isobutyrique isovalerianique, et valerianique optiquement actif, *Ann. Inst. Pasteur, Paris,* 75, 291, 1948.
15. **Mamelak, R. and Quastel, J. H.,** Amino acid interactions in strict anaerobes (*C. sporogenes*), *Biochim. Biophys. Acta,* 12, 103, 1953.
16. **Stadtman, T. C.,** Proline and glycine reductases of Clostridium H.F., 3rd International Congress of Biochemistry, Brussels, 1955.
17. **Stadtman, T. C. and Elliot, P.,** A new ATP-forming reaction: the reductive deamination of glycine, *J. Am. Chem. Soc.,* 78, 2020, 1956.
18. **Stadtman, T. C., Elliot, P., and Tiemann, L.,** Phosphate esterification coupled with glycine reduction, *J. Biol. Chem.,* 231, 961, 1958.
19. **Barnard, G. and Akhtar, M.,** Stereochemistry of the glycine reductase of *Clostridium sticklandii, J. Chem. Soc. Chem. Commum.,* p. 980, 1975.
20. **Stadtman, T. C.,** Coupling of a DPNH generating system to glycine reduction, *Arch. Biochem. Biophys.,* 99, 36, 1962.
21. **Turner, D. C. and Stadtman, T. C.,** Purification of protein components of the clostridial glycine reductase system and characterization of protein A as a selenoprotein, *Arch. Biochem. Biophys.,* 154, 366, 1973.
22. **Andreesen, J. R. and Ljungdahl, L. G.,** Formate dehydrogenase of *Clostridium thermoaceticum.* Incorporation of selenium-75, and the effects of selenite, molybdate and tungstate on the synthesis of the enzyme, *J. Bacteriol.,* 116, 867, 1973.
23. **Costilow, R. N.,** Selenium requirement for the growth of *Clostridium sporogenes* with glycine as the oxidant in Stickland reaction system, *J. Bacteriol.,* 131, 366, 1977.
24. **Venugopalan, V. and Stadtman, T. C.,** Influence of growth conditions on glycine reductase activity of *C. sporogenes,* in preparation.
25. **Johnstone, R. M. and Quastel, J. H.,** Amino acid reductases, in *Methods in Enzymology,* Vol. 2 Colowick, S. P. and Kaplan, N. O., Eds., Academic Press, New York, 1955, 217.
26. **Cone, J. E., Martin del Rio, R., and Stadtman, T. C.,** Clostridial glycine reductase complex, purification and characterization of the selenoprotein component, *J. Biol. Chem.,* 252, 5337, 1977.
27. **Tanaka, H. and Stadtman, T. C.,** Selenium-dependent clostridial glycine reductase: purification and characterization of the two membrane associated protein components, *J. Biol. Chem.,* 254, 447, 1979.
28. **Stadtman, T. C. and Elliot, P.,** Metabolism of glycine by extracts of an amino acid fermenting *Clostridium* strain HF, *Bacteriol. Proc.,* p. 3, 1956.

29. **Stadtman, T. C.**, Phosphate esterification coupled to glycine reduction in enzyme preparations of *Clostridium sticklandii, Fed. Proc. Fed. Am. Soc. Exp. Biol.*, 16, 254, 1957.
30. **Sugiura, Y., Hojo, Y., Tamai, Y., and Tanaka, H.**, Selenium protection against mercury toxicity, binding of methylmercury by the selenohydryl-containing ligand, *J. Am. Chem. Soc.*, 98, 2339, 1976.
31. **Braun, V. and Schroeder, W. A.**, A reinvestigation of the hydrazinolytic procedure for the determination of C-terminal amino acids, *Arch. Biochem. Biophys.*, 118, 241, 1967.
32. **Cone, J. E., Martin del Rio, R., Davis, J. N., and Stadtman, T. C.**, Chemical characterization of the selenoprotein component of clostridial glycine reductase: identification of selenocysteine as the organoselenium moiety, *Proc. Natl. Acad. Sci. U.S.A.*, 73, 2659, 1976.
33. **Oh, S. H., Ganther, H. E., and Hoekstra, W. G.**, Selenium as a component of glutathione peroxidase isolated from ovine erythrocytes, *Biochemistry*, 13, 1825, 1974.
34. **Forstrom, J. W., Zakowski, J. J., and Tappel, A. L.**, Identification of the catalytic site of rat liver glutathione peroxidase as selenocysteine, *Biochemistry*, 17, 2639, 1978.
35. **Enoch, H. G. and Lester, R. L.**, The purification and properties for formate dehydrogenase and nitrate reductase from *Escherichia coli, J. Biol. Chem.*, 250, 6693, 1975.
36. **Jones, J. B., Dilworth, G. L., and Stadtman, T. C.**, Occurance of selenocysteine in the selenium-dependent formate dehydrogenase of *Methanococcus vannielii, Arch. Biochim. Biophys.*, 195, 255, 1979.
37. **Huber, R. E. and Criddle, R. S.**, Comparison of the chemical properties of selenocysteine and selenocystine with their sulfur analogs, *Arch. Biochem. Biophys.*, 122, 164, 1967.
38. **Günther, W. H. H.**, Methods in selenium chemistry. III. The reduction of diselenides with dithiothreitol, *J. Org. Chem.*, 32, 3931, 1967.
39. **Pigiet, V. and Conley, R. R.**, Isolation and characterization of phosphothioredoxin from *Escherichia coli, J. Biol. Chem.*, 253, 1910, 1978.
40. **Stadtman, T. C.**, Studies on the enzymic reduction of amino acids: a proline reductase of an amino acid-fermenting Clostridium, strain HF, *Biochem. J.*, 62, 614, 1956.
41. **Stadtman, T. C. and Elliot, P.**, Studies on the enzymic reduction of amino acids. II. Purification and properties of D-proline reductase and a proline racemase from *Clostridium sticklandii, J. Biol. Chem.*, 228, 983, 1957.
42. **Hodgins, D. S. and Abeles, R. H.**, Studies of the mechanism of action of D-proline reductase: the presence of covalently bound pyruvate and its role in the catalytic process, *Arch. Biochem. Biophys.*, 130, 274, 1969.
43. **Hamilton, G. A.**, The proton in biological redox reactions, in *Progress in Bioorganic Chemistry*, Vol. 1, Kaiser, E. T. and Kezdy, F. J., Eds., Wiley Interscience, New York, 1971, 83.
44. **Chinard, F. P.**, Photometric estimation of proline and ornithine, *J. Biol. Chem.*, 199, 91, 1952.
45. **Benson, J. R. and Hare, P. E.**, o-Phthalaldehyde: fluorogenic detection of primary amines in the picomole range. Comparison with fluorescamine and ninhydrin, *Proc. Natl. Acad. Sci. U.S.A.*, 72, 619, 1975.
46. **Seto, B.**, Proline reductase: a sensitive fluorometric assay with o-phthalaldehyde, *Anal. Biochem.*, 95, 44, 1979.
47. **Seto, B. and Stadtman, T. C.**, Purification and properties of proline reductase from *Clostridium sticklandii, J. Biol. Chem.*, 251, 2435, 1976.
48. **Hodgins, D. and Abeles, R. H.**, The presence of covalently bound pyruvate in D-proline reductase and its participation in the catalytic process, *J. Biol. Chem.*, 242, 5158, 1967.
49. **Seto, B.**, A pyruvate-containing peptide of proline reductase in *Clostridium sticklandii, J. Biol. Chem.*, 253, 4525, 1978.
50. **Riley, W. D. and Snell, E. E.**, Histidine decarboxylase of *Lactobacillus* 30a. IV. The presence of covalently bound pyruvate as the prosthetic group, *Biochemistry*, 7, 3520, 1968.
51. **Wickner, R. B., Tabor, C. W., and Tabor, H.**, Purification of adenosylmethionine decarboxylase from *Escherichia coli* W: evidence for covalently bound pyruvate, *J. Biol. Chem.*, 245, 2132, 1970.
52. **Cohn, M. S., Tabor, C. W., and Tabor, H.**, Identification of a pyruvoyl residue in S-adenosylmethionine decarboxylase from *Saccharomyces cerevisiae, J. Biol. Chem.*, 252, 8212, 1977.
53. **Demetriou, A. A., Cohn, M. S., Tabor, C. W., and Tabor, H.**, Identification of pyruvate in S-adenosylmethionine decarboxylase from rat liver, *J. Biol. Chem.*, 253, 1684, 1978.
54. **Satre, M. and Kennedy, E. P.**, Identification of bound pyruvate essential for the activity of phosphatidylserine decarboxylase of *Escherichia coli, J. Biol. Chem.*, 253, 479, 1978.
55. **Seto, B.**, Electron transport proteins associated with proline fermentation in *Clostridium sticklandii, Fed. Proc.*, 37, 1521, 1978.

Chapter 4

RESPIRATION WITH SULFATE AS ELECTRON ACCEPTOR

R. K. Thauer and W. Badziong

TABLE OF CONTENTS

I. INTRODUCTION

Two genera of strictly anaerobic bacteria, *Desulfovibrio* and *Desulfotomaculum*, are capable of utilizing sulfate (or reduced products thereof) as terminal electron acceptor for catabolic oxidation processes.[1,2] Sulfate is reduced to H_2S, the reaction being coupled with the synthesis of ATP from ADP and inorganic phosphate (P_i). The latter is most convincingly demonstrated by the finding that a few strains of *Desulfovibrio* can grow on H_2 plus sulfate as sole energy source.[3,4]

$$4H_2 + SO_4{}^{2-} + 2H^+ \longrightarrow H_2S + 4H_2O$$

$$\Delta G_o' = -37.7 \text{ kcal/mol}^a$$
$$(-158 \text{ kJ/mol})$$

(1)*

The oxidation of organic substrates by sulfate is generally incomplete in that it does not proceed beyond the level of acetate.[5,6] Recently, owever, a *Desulfotomaculum* species was isolated which can grow on acetate plus sulfate as sole energy source,[7] clearly establishing that oxidation of acetate to CO_2 with sulfate as electron acceptor is thermodynamically and mechanistically feasible.

$$\text{Acetate}^- + 3H^+ + SO_4{}^{2-} \longrightarrow H_2S + 2CO_2 + 2H_2O$$

$$\Delta G_o' = -15.05 \text{ kcal/mol}$$
$$(-63 \text{ kJ/mol})$$

(2)

The pathway of dissimilatory sulfate reduction** has not yet been completely resolved. That which is known can be outlined as follows:

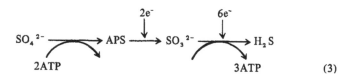

(3)

Sulfate is reduced to H_2S via adenylylsulfate (APS) and sulfite; 2 mol of ATP are required for the activation of sulfate to APS.[9] A total of 3 mol of ATP are generated during sulfite reduction to H_2S.[10] oupling of phosphorylation to sulfite reduction probably proceeds via a chemiosmotic mechanism.[5,11] There is still considerable controversy about how sulfite is reduced to H_2S. Also a consistent view of the electron carriers involved in dissimilatory sulfate reduction is lacking.

The physiology of sulfate-reducing bacteria has been reviewed by LeGall and Postgate;[12] the biochemistry of dissimilatory sulfate reduction has been reviewed by Siegel[13] and by Thauer et al.[5] We shall here focus attention on recent work in this selected area. Most of the work has been performed with *Desulfovibrio* sp. The article is thus mainly concerned with dissimilatory sulfate reduction in this genus.

---

* Except where indicated, $\Delta G_o'$ values were calculated from Gibbs-free energies of formation from the elements;[1] H_2, H_2S, and CO_2 are in the gaseous state, and all other substances are in aqueous solution.
** The catabolic reduction of sulfate is termed dissimilatory sulfate reduction or sulfate respiration.[8]

II. DISSIMILATORY SULFATE REDUCTION

During exponential growth of *Desulfovibrio vulgaris* (Marburg) on H_2 plus sulfate, H_2S is formed at a rate of 360 nmol/min/mg dry cells = 0.72 μmol/min/mg cell protein (Figure 1). Similar rates have been observed for other sulfate reducing bacteria. Enzymes involved in dissimilatory sulfate reduction should therefore have a specific activity in the order of 0.7 units per milligram cell protein (1 unit = 1 μmol/min).

A. Sulfate Activation

The initial step in dissimilatory sulfate reduction is the activation of sulfate with ATP to form adenosine 5'-phosphosulfate (APS).[14-16]

$$SO_4^{2-} + 2H^+ + ATP \rightleftharpoons APS + PP_i$$

$$\Delta G_o' = +11 \text{ kcal/mol}^{[17]}$$
$$(+46 \text{ kJ/mol}) \tag{4}$$

The reaction is catalyzed by a soluble ATP sulfurylase (EC 2.7.7.4; APT:sulfate adenyltransferase). The enzyme is present in sulfate-reducing bacteria in high activities (0.8 units per milligram cell protein) and has been purified from *Desulfovibrio desulfuricans* near to homogeneity.[18] Two equivalents of ATP are consumed in the ATP sulfurylase reaction. This inference is made on the premise that the pyrophosphate produced in the ATP sulfurylase reaction is completely hydrolyzed by inorganic pyrophosphatase (EC 3.6.1.1) present in high activities (19 units per mg cell protein) in the sulfate reducing bacteria.[19]

$$PP_i + H_2O \longrightarrow 2P_i$$

$$\Delta G_o' = -5.24 \text{ kcal/mol}^{[5]}$$
$$(-21.9 \text{ kJ/mol}) \tag{5}$$

The pyrophosphatase is subject to regulation by oxidation-reduction. The enzyme is inactivated in vivo upon addition of O_2 and can be reactivated in vivo and in vitro by treatment with reducing agents. Inactivation of pyrophosphatas prevents dissimilatory sulfate reduction as would be predicated from the thermodynamics of the ATP-sulfurylase reaction.[19]

Sulfate has to be activated before it can be reduced to sulfite, since the physiologically important reducing agents, e.g., H_2 (H^+/H_2; $E_o' = -414$ mV), formate (CO_2/formate; $E_o' = -432$ mV), or lactate (pyruvate/lactate; $E^{o'} = -190$ mV) are too positive in E_o' to reduce inorganic sulfate itself (SO_4^{2-}/HSO_3^-; $E_o' = -516$ mV.) The E_o' of the analogous reduction of APS is -60 mV. Reduction of APS is thus an exergonic reaction, whereas reduction of sulfate would be endergonic. For futher details on sulfate activation the reader is referred to the review by Siegel.[13]

B. APS Reduction

APS formed in the ATP sulfurylase reaction is reduced to sulfite:

$$APS + H_2 \rightleftharpoons HSO_3^- + AMP + H^+$$

$$\Delta G_o' = -16.4 \text{ kcal/mol}^{[20]}$$
$$(-68 \text{ kJ/mol}) \tag{6}$$

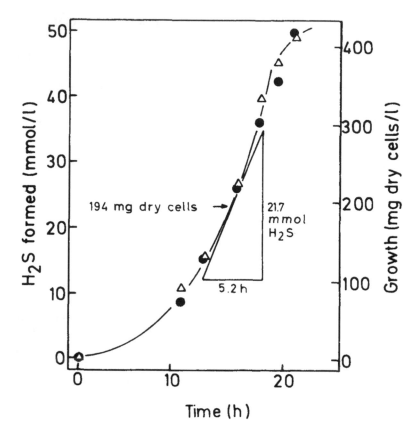

FIGURE 1. H_2S formation by *Desulfovibrio vulgaris* (Marburg) growing on H_2 plus sulfate as sole energy source. For conditions see Badziong and Thauer[10]. • = Growth; Δ = H_2S formed.

The reaction is catalyzed by APS reductase (EC 1.8.99.2), a cytoplasmic enzyme of mol wt 220,000 containing per molecule 1 FAD and 7 to 9 Fe/S.[21-23] The physiological electron donor (coenzyme) for the enzyme is not known. The relatively high redox potential of the APS/sulfite couple ($E_o' = -60$ mV) theoretically would allow the reduction of APS by various electron carriers, e.g., NADH ($E_o' = -320$ mV), cytochrome c_3 ($E_o' = -290$ mV), or menaquinone ($E_o' = -74$ mV). None of these, however, appear to be effective. Chemically reduced methyl viologen can serve as artificial an reductant. When tested with methyl viologen the specific activity of APS reductase in sulfate reducing bacteria is near 0.08 units per milligram cell protein.[21] The enzyme has been purified from *D. desulfuricans.*[22]

C. Sulfite Reduction

At present two mechanisms of sulfite reduction to H_2S seem possible. The first mechanism assumes that bisulfite is reduced to H_2S in a six-electron reduction without the formation of any free intermediates.[24]

$$HSO_3^- + 3H_2 + H^+ \longrightarrow H_2S + 3H_2O$$

$$\Delta G_o' = -42.4 \text{ kcal/mol}$$
$$(-177 \text{ kJ/mol})$$

(7)

The second mechanism assumes that bisulfite is reduced to H_2S via three two-electron reductions, with trithionate and thiosulfate as free intermediates (recycling sulfite pool mechanism).[25,26]

$$3SO_3{}^{2-} \xrightarrow{2e^-} S_3O_6{}^{2-} \xrightarrow[SO_3{}^{2-}]{2e^-} S_2O_3{}^{2-} \xrightarrow[SO_3{}^{2-}]{2e^-} H_2S \qquad (8)$$

$$3HSO_3{}^- + H_2 + H^+ \longrightarrow S_3O_6{}^{2-} + 3H_2O$$

$$\Delta G_0' = -11.1 \text{ kcal/mol}$$
$$(-46 \text{ kJ/mol}) \qquad (9)$$

$$S_3O_6{}^{2-} + H_2 \longrightarrow S_2O_3{}^{2-} + HSO_3{}^- + H^+$$

$$\Delta G_0' = -29.4 \text{ kcal/mol}$$
$$(-123 \text{ kJ/mol}) \qquad (10)$$

$$S_2O_3{}^{2-} + H_2 + H^+ \rightleftharpoons H_2S + HSO_3{}^-$$

$$\Delta G_0' = -1.9 \text{ kcal/mol}$$
$$(-8 \text{ kJ/mol}) \qquad (11)$$

The problem is that, dependent on the experimental conditions, either H_2S or trithionate is found as major or sole product(s) of sulfite reduction by purified bisulfite reductase from sulfate-reducing bacteria.[27-29] This subject has been discussed by Chambers and Trudinger[24] and by Siegel.[13] It is important to note that all the enzymes discussed below have specific activities that are too low to account for the observed rates of sulfite reduction to H_2S in vivo (Table 1).

Bisulfite reductase is a siroheme protein.[30-32] It is identical with desulfoviridin in *Desulfovibrio gigas*[33] with desulforubidin in *D. desulfuricans* strain Norway,[34] and with the CO-binding pigment P582 in *Desulfotomaculum* sp.[35,36] The enzyme catalyzes the reduction of bisulfite rather than sulfite with reduced methyl viologen. This has been deduced from the finding that the pH optimum of the reaction is near 6 which is considerably lower than the pK_2 of H_2SO_3 of 6.8.[33] Evidence has been published that reduced flavodoxin or reduced ferredoxin may be the physiological electron donor.[29,37-43] It should be noted, however, that the specific activity of the purified bisulfite reductase tested with flavodoxin is very low (<0.02 units per milligram purified enzyme, see Table 1) indicating that the physiological electron donor has not yet been elucidated.

At pH 6, low concentrations of reduced methyl viologen (1 mM), and high concentrations of bisulfite (5 mM), bisulfite reductase mediates the reduction of bisulfite to trithionate. At pH 7, high concentrations of reduced methyl viologen (10 mM) and low concentrations of bisulfite (1 mM), H_2S rather than trithionate is formed as major product.[27] The situation is further complicated by the recent finding that bisulfite is nonenzymatically reduced with reduced methyl viologen or reduced flavodoxin to dithionite.[44]

Table 1
ENZYMES POSSIBLY INVOLVED IN DISSIMILATORY SULFATE REDUCTION

Enzyme	Source	M_r	Chromophore	Specificity activity with methyl viologen as electron donor		Ref.
				Crude extract (Units/mg protein[a])	Purified enzyme (Units/mg protein[a])	
Bisulfite reductase I						
= Desulfoviridin	*Desulfovibrio gigas*	200,000	Siro-heme	0.146	0.63	33
			Siro-heme		0.1	27
	Desulfovibrio vulgaris	200,000	Siro-heme	0.04	0.18	25.43
					0.019[b]	
= Desulforubridin	*Desulfovibrio desulfuricans* Norway	226,000	Siro-heme	0.08	0.26	47
= P 582	*Desulfotomaculum nigrificans*	225,000	Siro-heme	0.16	0.41	34
		145,000	Siro-heme	—	0.002	35
			Siro-heme	0.006	0.07	36
			Siro-heme		0.125	52
Bisulfite-reductase II	*D. vulgaris*	50,000	Siro-heme		0.23	46
Sulfite-reductase	*D. vulgaris*	26,000	Siro-heme	0.07	0.9	47
Thiosulfate-forming enzyme	*D. vulgaris*	43,200	—		0.24	43
					0.17[c]	
Thiosulfate-reductase	*D. gigas*	220,000	—	0.2	83 (13.3)[c]	50
	D. vulgaris	16,300	—	—	—	49
	D. nigrificans	—	—	0.05	0.26	48

[a]   1 unit = 1 μmol substrate reduced per minute.
[b]   specific activity with flavodoxin as electron donor.
[c]   specific activity with cytochrome c_3 as electron donor.

Note: M_R = relative molecular weight.

$$2HSO_3^- + 2MV^- + 2H^+ \rightleftharpoons S_2O_4^{2-} + 2MV + 2H_2O$$

$$(12)$$

The formation of dithionite is thermodynamically favored at low pH and high concentrations of bisulfite. The rate of nonenzymatic dithionite formation is in the order of the rate of bisulfite reduction by purified bisulfite reductase.[44]

There is some evidence that the intracellular pH of sulfate reducing bacteria lies between 7 and 8.[45] Since, however, the intracellular concentration of sulfite and the physiological electron donor are not known, it is impossible to decide which products are formed by bisulfite reductase in vivo.

Besides bisulfite reductase, four other enzymes have been found which may be involved in sulfite reduction to H_2S in sulfate-reducing bacteria (Table 1):

1. Bisulfite reductase II[46]
2. Sulfite reductase[47]
3. Thiosulfate-forming enzyme[43]
4. Thiosulfate reductase[48-50]

Bisulfite reductase II is a siroheme protein which mediates the reduction of bisulfite to H_2S and thiosulfate, with reduced methyl viologen.[46]

$$3HSO_3^- + 5H_2 + H^+ \longrightarrow S_2O_3^{2-} + H_2S + 6H_2O$$

$$\Delta G_0' = -82.9 \text{ kcal/mol}$$
$$(-347 \text{ kJ/mol})$$

$$(13)$$

The physiological electron donor of the enzyme is not known. The pH optimum is near 6. The enzyme has been purified from *D. vulgaris* and has been shown to be distinct from bisulfite reductase I. The specific activity of the purified enzyme is below 0.25 units per milligram protein (Table 1).

Sulfite reductase is a siroheme protein which catalyzes six-electron reduction of sulfite to H_2S, with reduced methyl viologen.[47] The physiological electron donor is not known. The enzyme has been purified from *D. vulgaris* and has been shown to be distinct from bisulfite reductase. The specific activity of the purified enzyme is 0.9 units per milligram protein (Table 1). Sulfite reductase has been assumed to be involved in assimilatory sulfate reduction rather than in dissimilatory sulfate reduction.[47]

Thiosulfate-forming enzyme mediates the reduction of trithionate to H_2S and sulfite, with reduced methyl viologen[43]

$$S_3O_6^{2-} + HSO_3^- + H_2 \longrightarrow S_2O_3^{2-} + 2HSO_3^- + H^+$$

$$\Delta G_0' = -29.4 \text{ kcal/mol}$$
$$(-123 \text{ kJ/mol})$$

$$(14)$$

The pH optimum is near 6. The formation of thiosulfate by the enzyme is dependent on the simultaneous presence of bisulfite. The enzyme has been purified from *D. vulgaris.* The specific activity of the purified enzyme is below 0.25 units per milligram protein, when tested with either cytochrome c_3 or reduced methyl viologen as electron donor (Table 1).

Thiosulfate reductase mediates the reduction of thiosulfate to H_2S and bisulfite with reduced methyl viologen.

$$S_2O_3{}^{2-} + H_2 + H^+ \rightleftharpoons H_2S + HSO_3{}^-$$

$$\Delta G_0{}' = -1.9 \text{ kcal/mol}$$
$$(-8 \text{ kJ/mol})$$

(15)

Cytochrome cc_3 appears to be the physiological electron donor (see below). The pH optimum is near 7.5. The enzyme has been purified to homogeneity from *Desulfotomaculum nigrificans*[48] *D. vulgaris*[49] and *D. gigas.*[50] For specific activities of the enzymes, see Table 1.

The presence of the thiosulfate-forming enzyme and of thiosulfate reductase in sulfate-reducing bacteria has been taken as evidence that dissimilatory sulfate reduction proceeds via trithionate and thiosulfate.[29] This interpretation neglects the fact that growing sulfate-reducing bacteria reduce sulfate to H_2S at a specific rate of near 700 nmol/min/mg cell protein, whereas the specific activity of the purified thiosulfate-forming enzyme is only 240 nmol/min/mg protein.[43] Sulfate-reducing bacteria are capable of utilizing trithionate and thiosulfate as terminal electron acceptors for growth.[51] It is therefore conceivable that both the thiosulfate-forming enzyme and thiosulfate reductase have a function only when the bacteria are growing in the presence of trithionate or thiosulfate.

D. H₂, Formate, and Lactate Oxidation

Hydrogen, formate, and lactate are the substrates commonly oxidized by sulfate reducing bacteria.[1,2,12] The enzymes involved are a periplasmic hydrogenase,[53] a periplasmic formate dehydrogenase,[54,55] a membrane-bound lacate dehydrogenase,[56,57] and a cytoplasmic pyruvate ferredoxin oxidoreductase[58,59] (Table 2). Hydrogenase and formate dehydrogenase have been purified. Hydrogenase catalyzes the reversible reduction of cytochrome c_3 with H_2;[60,61] formate dehydrogenase mediates the reduction of cytochrome c_3[55] or cytochrome c_{553}[54] with formate. The electron acceptor for lactate dehydrogenase is not known; it is not NAD or NADP. Pyruvat ferredoxin oxidoreductase catalyzes the oxidation of pyruvate to acetyl-CoA and CO_2, with either ferredoxin or flavodoxin as electron acceptor.[58,59]

Desulfotomaculum acetoxidans grows on acetate and sulfae as sole energy source; acetate is oxidized to CO_2. The enzymes and reactions involved in acetate oxidation have not been elucidated.[7]

E. Electron and Hydrogen Carriers

A consistent view of the electron and hydrogen carriers involved in dissimilatory sulfate reduction is lacking. *Desulfovibrio sp.* contain c-type cytochromes, b-type cytochromes, menaquinone, ferredoxin, flavodoxin, rubredoxin, desulforedoxin, and pyridine nucleotides. In *Desulfotomaculum sp.*, c-type cytochromes appear to be lacking.[2] A possible function of the carriers can be deduced from their redox potential (Table 3). This is, however, somewhat awkward, since the redox potentials of some of the carriers and electron acceptors is known only with considerable uncertainty.

c-Type cytochromes: *Desulfovibrio* contain cytochrome c_3, cytochrome cc_3, and cytochrome c_{553}. Cytochrome c_3 is a four-heme-containing cytochrome of mol wt 13,000.[73,74,79-88] It is located in the periplasmic space[53,89] and functions as the coenzyme of the periplasmic hydrogenase.[60,61] Cytochrome c_3 has also been reported to be the coenzyme of thiosulfate reductase[50,87,90] and of the thiosulfate-forming enzyme[43] which are cytoplasmic enzymes. A location of cytochrome c_3 in both the periplasm and cytoplasm is, however, unlikely because of the manner of synthesis of proteins which have to be exported to the periplasm. Cytochrome cc_3 is an eight-heme-containing

Table 2
DEHYDROGENASE INVOLVED IN DISSIMILATORY SULFATE REDUCTION

Enzyme	Source	M_r	Chromophore	Ref.
Hydrogenase	Desulfovibrio vulgaris	89,000	7—9 Fe/7—9 S	62
		52,000	11—13 Fe/11—13 S	63
		60,000	3—4 Fe/3—4 S	64
		45,000		65
		<100,000		66,67
	Desulfovibrio gigas	89,500	11—13 Fe/11—13 S	68,61
	Desulfovibrio desulfuricans	60,000		60
Formate-dehydrogenase	D. vulgaris	—	—	54,55
Lactate-dehydrogenase	D. gigas	—	—	56,57
Pyruvate: ferredoxin oxidoreductase	Desulfovibrio desulfuricans	—	—	58,59

Note: M_r = relative molecular weight.

Table 3
REDOX POTENTIALS OF ELECTRON DONORS, ELECTRON ACCEPTORS, AND CARRIERS INVOLVED OR POSSIBLY INVOLVED IN DISSIMILATORY SULFATE REDUCTION

Redox component	E_0' (mV)	Ref.
Pyruvate/acetate + CO_2	−659	[a]
Acetaldehyde/acetate	−582	[a]
SO_2^- /HSO_3^-	−516	[a]
Pyruvate/acetyl-CoA + CO_2	−500	5
SO_3^- /$S_2O_4^{2-}$	−471	69
Ferredoxin I ox/red[b]	−450	70—72
Methyl viologen ox/red	−440	69
CO_2/formate	−432	[a]
Flavodoxin ox/red ($E_{0,1}'$)	−400	39
$S_2O_3^{2-}$ /HSO_3^- + H_2S	−372	[a]
Benzyl viologen ox/red	−360	69
$NADP^+$/$NADPH$ + H^+	−324	69
NAD^+/$NADH$ + H^+	−320	69
Ferredoxin ox/red	−310	38
Cytochrome c_3 ox/red	−290	73,74
S^o/H_2S	−240	[a]
SO_4^{2-} /H_2S	−209	[a]
Acetylaldehyde/ethanol	−197	[a]
Pyruvate/lactate	−190	[a]
HSO_3^- /$S_3O_6^{2-}$	−173	[a]
Ferredoxin II ox/red[b]	−130	70—72
HSO_3^- /H_2S	−106	[a]
Cytochrome c_{553} ox/red	−100	12
Menaquinone ox/red	−74	75,76
APS/SO_3^{2-} + AMP	−60	20
Rubredoxin ox/red	−60	77
Desulforedoxin ox/red	−35	78
$S_4O_6^{2-}$ /$S_2O_3^{2-}$	+ 24	[a]
$S_3O_6^{2-}$ /$S_2O_3^{2-}$ + HSO_3^-	+ 225	[a]

[a] Calculated from Gibbs-free energies of formation from the elements;[5] H_2, H_2S, and CO_2 are in the gaseous state, and all other substances are in aqueous solution.

[b] Ferredoxin = monomeric iron sulfur protein of mol wt 6,000; ferredoxin I = trimeric ferredoxin of mol wt 18,000; ferredoxin II = tetrameric ferredoxin of mol wt. 24,000.

protein of mol wt 26,000 which is associated with the cytoplasmic membrane.[90-92] Amino acid compositions show that cytochrome cc_3 is not a dimer of cytochrome c_3.[12,93] Cytochrome cc_3 has been reported to be involved in thiosulfate reduction.[50,90] Thiosulfate reductase of *D. vulgaris* has the highest activity, when both cytochrome cc_3 and c_3 are present as donors. Cytochrome c_{553} is a one-heme-containing periplasmic cytochrome of mol wt 6500[54] (9100).[93] The cytochrome shows sequence homologies with mitochondrial cytochrome *c*.[11] The cytochrome has been reported to function as coenzyme of formate dehydrogenase,[54] although the specificity of the enzyme for cytochrome c_{553} remains to be demonsrated.

b-Type cytochromes are detected in the membrane fraction of both *Desulfovibrio*[94] and *Desulfotomaculum*[2,7] spp. They appear to be functionally associated with the membrane fumarate reductase present in high quantities in many slfate-reducing bacteria.[5,94,95] *b*-Cytochromes have, however, also been found in sulfate-reducing bacteria not capable of utilizing fumarate as terminal electron acceptor and not containing a catabolic fumarate reductase,[45] indicating that *b*-cytochromes also may have a function in dissimilatory sulfate reduction.

Menaquionone-6 is found in all sulfate-reducing bacteria so far investigated.[95-97] As for cytochrome *b*, it has been suggested that menaquionone is functionally associated with the membrane-bound fumarate reductase.[95] Menaquinone could, however, be demonstrated in *D. vulgaris* (Marburg), a strain which can grow on H_2 plus sulfate as sole energy source and which is unable to use fumarate as electron acceptor.[45] On the basis of thermodynamic considerations, Wagner et al. proposed that menaquionone could be involved in APS recution and trithionate reduction rather than thiosulfate reduction.[76] The validity of this proposal is dependent upon whether or not trithionate and thiosulfate are physiological intermediates in dissimilatory sulfate reduction.

Ferredoxins from sulfate-reducing bacteria are iron-sulfur prteins of mol wt 6000, containing four nonheme iron and four acid-labile sulfur per mole.[38,87,98] Trimers and tetramers can be isolated with mol wt 18,000 and 24,000, respectively.[70-72] The iron-sulfur protein has been shown to be the coenzyme of pyruvate ferredoxin oxidoreductase.[58] It has been discussed as a possible electron donor for bisulfite reductase.[37,38,47] The specificity of this enzyme for ferredoxin remains to be demonstrated.

Flavodoxin is a FMN flavoprotein of mol wt 16,000.[99-103] It is a low-potential one-electron carrier operating between the semiquinone and fully reduced state.[39] The flavoprotein can substitute for ferredoxin in all reactions.[39,58] When growing on media high in iron, *Desulfovibrio spp.* contain both flavodoxin and ferredoxin. In some species, synthesis of flavodoxin is greatly stimulated by iron deficiency at the expense of ferredoxin, while in others, this is not so and there might be a functional difference.[11] *D. gigas* appears to be an example of the first type of behavior,[91] while in *D. vulgaris*, flavodoxin seems to predominate.[103]

Rubredoxin is an iron protein of mol wt 6000 containing 1 mol of iron.[77,87,104-108] It has been purified from *D. gigas*,[105,106] *D. vulgaris*,[107,108] and *D. desulfuricans*.[87] Cells of *D. vulgaris* contain a NADH rubredoxin oxidoreductase.[109] Neither the function of this enzyme nor of rubredoxin is known.

Desulforedoxin is an iron protein of mol wt 7900, containing 2 mol of iron per mole protein, but no acid-labile sulfur.[78,100] It can be reduced by dithionate, indicating that it is an electron carrier.[78,110] The function of this newly found redox component remains to be elucidated.

Pyridine nucleotides do not appear to be involved in dissimilatory sulfate reduction. None of the dehydrogenases or reductases are NAD or NADP dependent. A NADH dehydrogenase capable of linking the reduction of inorganic sulfur compounds to the oxidation of NADH has not been found.

F. Topography

Hydrogenase,[53,62,63,68] cytochrome c_3,[53,89] and cytochrome c_{553}[54] have been shown to be located in the periplasmic space of *Desulfovibrio spp.* grown on sulfate plus lactate or sulfate plus H_2[45] (Figure 2). Both proteins can be washed off the cells after protoplast formation or are liberated by osmotic shock. The percentage washed off or liberated varies from species to species; a considerable part always remains membrane bound, indicating that hydrogenase and cytochrome c_3 are peripheric membrane proteins.

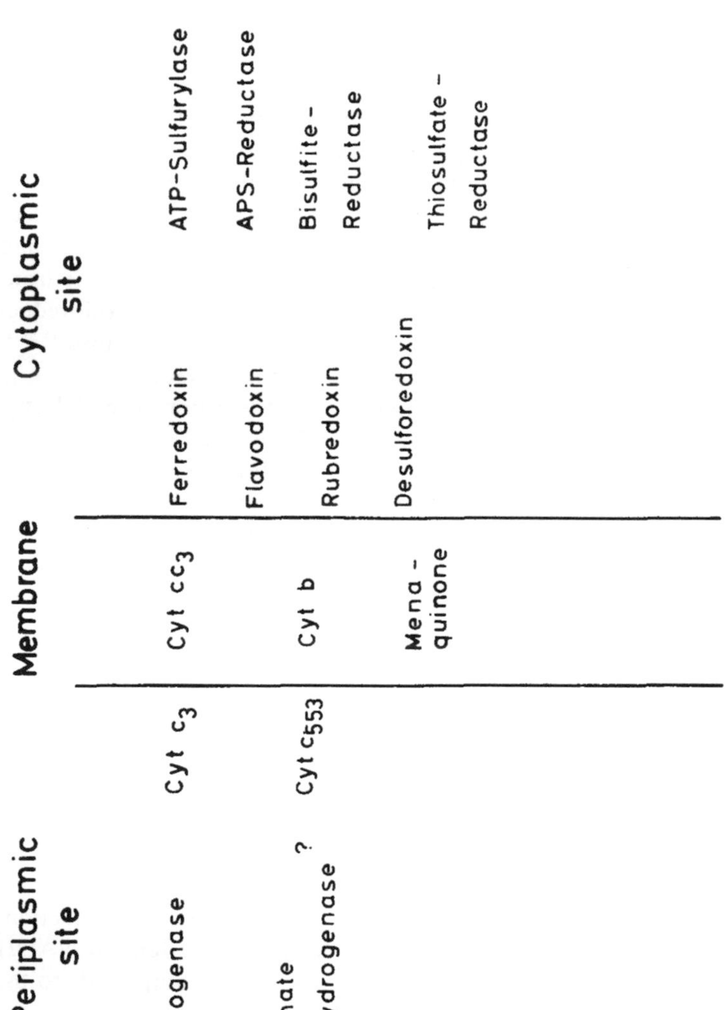

Periplasmic site	Membrane	Cytoplasmic site	
Hydrogenase	Cyt c_3	Cyt cc_3	Ferredoxin	ATP-Sulfurylase
			Flavodoxin	APS-Reductase
Formate Dehydrogenase ?	Cyt c_{553}	Cyt b	Rubredoxin	Bisulfite - Reductase
		Mena - quinone	Desulforedoxin	Thiosulfate - Reductase

FIGURE 2.   Location of enzymes and redox carriers in *Desulfovibrio* spp.[45]

The location of c-type cytochromes on the periplasmic site of the membrane is of special interst. Cytochrome c_2 from *Rhodopseudomonas spheroides* and *R. capsulata*,[111] cytochrome c from *Micrococcus denitrificans*,[112] and cytochrome c_{552} from *Escherichia coli*[113] are all located in the periplasmic space of the respective bacteria. Racker et al.[114] showed that inmitochondria, cytochrome c is outside the inner membrane. The location of the c-type cytochromes mentioned above on only one side of the membrane may be taken to indicate that the function of these cytochromes are very similar.[5]

The particulate fraction of *Desulfovibrio spp.* contains cytochrome cc_3,[90-92] b-type cytochromes,[45,94] and menaquinone-6.[45,95-97] *Desulfovibrio spp.* appear to be devoid of intracytoplasmic membranes;[4] therefore these electroncarriers must be located in the cytoplasmic membrane.

ATP sulfurylase, APS reductase, bisulfite reductase, thiosulfate reductase, ferredoxin, and flavodoxin are soluble cytoplasmic proteins. APS reductase, bisulfite reductase, and thiosulfate reductase are released from protoplasts only after lysis.[45,50,53] The location of ATP- sulfurylase, ferredoxin and flavodoxin is deduced from the fact that ATP (substrate of ATP sulfurylase) and pyruvate ferredoxin oxidoreductase (enzyme connected with ferredoxin and flavodoxin) are cytoplasmic components.[53]

III. PHOSPHORYLATION COUPLED TO DISSIMILATORY SULFATE REDUCTION

A. Stoichiometry of Coupling
1. In Vivo Evidence

The first evidence that dissimilatory sulfate reduction is coupled to phosphorylation was supplied by Peck.[115] He showed that cell suspensions of *D. vulgaris* reduce sulfate with H_2 to H_2S in a completely mineral salts medium. Since two ATPs are required for the activation of sulfate to APS, he concluded that the reduction of APS to H_2S must be coupled with the formation of at least 2 mol of ATP; these are required for the activation of sulfate to APS. This interpretation was substantiated by demonstrating that sulfate reduction by cell suspensions was inhibited by 2,4-dinitrophenol and methyl viologen, whereas sulfite reduction was not.[115]

Recently the amount of ATP formed during sulfate respiration was estimated from growth yields corrected for energy of maintenance (Y^{max}).[10] $Y_{SO_4^{2-}}$ and $Y^{max}_{S_2O_3^{2-}}$ were determined as a function of the growth rate (μ) for *D. vulgaris* (Marburg) growing in H_2 plus sulfate and H_2 plus thiosulfate. Double reciprocal plots of Y vs. μ were linear (Figure 3). From the intercepts on the ordinate, a $Y^{max}_{SO_4^{2-}}$ of 12.2 g/mol and a $Y^{max}_{S_2O_3^{2-}}$ of 33.5 g/mol were obtained. On the basis of known biochemical pathways, a Y^{max}_{ATP} of 12.5 g/mol was calculated. A comparison of $Y^{max}_{SO_4^{2-}}$ with Y^{max}_{ATP} indicates that during growth of *D. vulgaris* (Marburg) on H_2 plus sulfate, approximately 1 mol of ATP is formed per mole of sulfate reduced to H_2S.[10]

$$SO_4^{2-} \xrightarrow[\quad 1ATP \quad]{8e-} H_2S \qquad (16)$$

$Y^{max}_{S_2O_3^{2-}}$ was 2.7 times higher than $Y^{max}_{SO_4^{2-}}$, suggesting that during growth on H_2 plus thiosulfate, nearly 3 mol of ATP are formed per mole of $S_2O_3^{2-}$ reduced to H_2S. The reduction of thiosulfate to H_2S proceeds via sulfite which is generated in a reaction

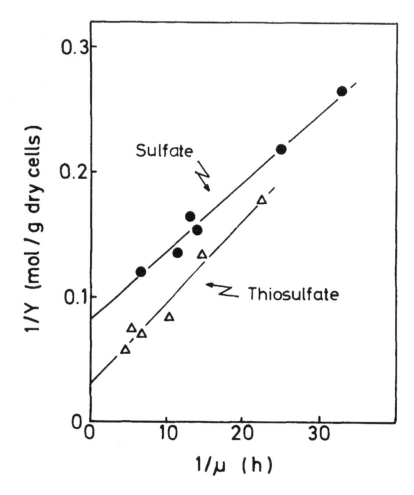

FIGURE 3.   Double reciprocal plot of the molar growth yield (Y) vs. the specific growth rate (μ).[10] *Desulfovibrio vulgaris* (Marburg) was grown on H_2 plus sulfate and H_2 plus thiosulfate as sole energy source.

neither requiring nor forming ATP.[10] It was therefore concluded that the reduction of sulfite to H_2S is coupled with the synthesis of 3 mol of ATP.[10]

$$S_2O_3{}^{2-} \rightleftharpoons \overset{\overset{\displaystyle 2e-}{\downarrow}}{} SO_3{}^{2-} \underset{3ATP}{\overset{\overset{\displaystyle 6e-}{\downarrow}}{\longrightarrow}} H_2S \qquad (17)$$

This data support the present view that 2 mol ATP have to be expended to prime the sulfate prior to its reduction.

$$SO_4{}^{2-} \underset{2ATP}{\overset{\overset{\displaystyle 2e-}{\downarrow}}{\longrightarrow}} SO_3{}^{2-} \underset{3ATP}{\overset{\overset{\displaystyle 6e-}{\downarrow}}{\longrightarrow}} H_2S \qquad (18)$$

Other growth yield data are available for sulfate-reducing bacteria.[116-120] They have, however, been obtained with nonexpotentially growing cultures in which the pH was not kept constant. Since both the pH and the growth rate have been shown to significantly affect the growth yield of sulfate reducing bacteria,[10] these data cannot be used to predict ATP gains.[11]

2. In Vitro Evidence

Cell-free preparations of *D. gigas* grown on lactate plus sulfate were shown by Peck to catalyze the reduction of sulfite to sulfide, with H_2 as electron donor with the concomitant esterification of phosphate.[121] A soluble protein fraction and membrane particles were required for both electron transfer and phosphorylation. Phosphorylation was uncoupled by 2,4-dinitrophenol, pentachlorophenol, 2-*n*-heptyl- 4-hydroxyquinoline-*N*-oxide, and gramicidin. Oligomycin and antimycin were without effect. The observed P/2e ratios were 0.05 to 0.2 (see Barton et al.).[56]

B. Mechanism of Coupling

Growth-yield studies indicate that in sulfate-reducing bacteria, the reduction of sulfate to H_2S with H_2 is coupled with the net synthesis of 1 mol of ATP, whereas the reduction of sulfite to H_2S is linked with the net formation of 3 mol ATP.[10] Coupling probably proceeds via the chemiosmotic mechanism. This is indicated by the finding that coupling is sensitive to uncouplers of electron transport phosphorylation[115,121] and that sulfate-reducing bacteria contain an active membrane-bound ATPase.[122] In Figure 4, a chemiosmotic model is presented which accounts for the observed stoichiometries and for the topography of the proteins involved. In the model, sulfite is assumed to be reduced to H_2S via six-electron reduction.

Hydrogenase is located on the periplasmic site of the cytoplasmic membrane; the enzymes involved in sulfate reduction to H_2S are located in the cytoplasm (Figure 3). The reduction of sulfate with H_2 to H_2S must therefore be a transmembrane redox process. The model assumes that APS reductase is linked to hydrogen via hydrogen carriers (electroneutral transport of electrons through the membrane), whereas bisulfite reductase is interconnected with hydrogenase via electron carriers (electrogenic electron transport through the membrane). Thus, only the reduction of sulfite to H_2S and not the reduction of APS to sulfite is associated with the generation of a pH gradient (inside alkaline) and of a membrane potential (inside negative). Only 6 protons are formed in the periplasmic space per 4 H_2 oxidized. The six protons are utilized to drive the synthesis of ATP. If, as in mitochondria and aerobic bacteria, two protons are required to poise the ATPase in direction of ATP synthesis,[123] then 3 mol of ATP are formed per mole of sulfate reduced to H_2S; from the 3 mol, 2 mol are consumed in the activation of sulfate.

During growth of the bacteria on H_2 plus sulfate, two protons are consumed per mole sulfate reduced to H_2S, and therefore, the medium becomes alkaline.

$$2H^+ + SO_4^{2-} + 4H_2 \longrightarrow H_2S + 4H_2O \qquad (19)$$

This is accounted for in the model by the uptake of sulfate into the cell by a sulfate^{2-}/2H$^+$ symport. This could also be a sulfate^{2-}/2Na$^+$ symport, since sodium extrusion is accomplished in procaryotes via an electroneutral H$^+$/Na$^+$ antiport.[5] H_2S is a weak acid, 50% of it being undissociated at pH 7. H_2S most likely leaves the cell in the undissociated form via free diffusion (Figure 4).

Recently a chemiosmotic model for sulfate respiration has been proposed by Wood.[11] The model is based on the assumption that dissimilatory sulfate reduction is

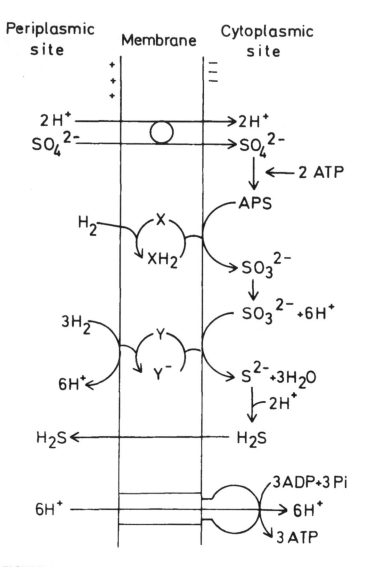

FIGURE 4.   A chemiosmotic model for coupling of phosphorylation to dissimilatory sulfate reduction. X = hydrogen carrier; Y = electron carrier.[45]

coupled with the net synthesis of 2 mol rather than 1 mol of ATP, an assumption for which experimental evidence is lacking.

IV. ACKNOWLEDGMENT

This work was supported by a grant from the Deutsche Forschungsgemeinschaft and by the Fonds der Chemischen Industrie. We thank N. P. Wood for reading the manuscript and for making many helpful suggestions.

REFERENCES*

1. **Postgate, J. R. and Campbell, L. L.**, Classification of *Desulfovibrio* species, the nonsporulating sulfate reducing bacteria, *Bacteriol. Rev.*, 30, 732, 1966.

1a. **Postgate, J.R.**, in *The Sulphate-Reducing Bacteria*, Cambridge University Press, Cambridge, 1979.

2. **Campbell, L. L. and Postgate, J. R.**, Classification of the spore-forming sulfate reducing bacteria, *Bacteriol. Rev.*, 29, 359, 1965.

3. **Sorokin, Y. J.**, Role of Carbon dioxide and acetate in biosynthesis by sulphate-reducing bacteria, *Nature (London)*, 210, 551, 1966.

4. **Badziong, W., Thauer, R. K., and Zeikus, J. G.**, Isolation and characterization of *Desulfovibrio* growing on hyrogen plus sulfate as the sole energy source, *Arch. Microbiol.*, 116, 41, 1978.

5. **Thauer, R. K., Jungermann, K., and Decker, K.**, Energy conservation in chemotrophic anaerobic bacteria, *Bacteriol. Rev.*, 41, 100, 1977.

6. **Decker, K., Jungermann, K., and Thauer, R. K.**, Energy production in anaerobic organism, *Angew. Chem. Int. Ed. Engl.*, 9, 138, 1970.

7. **Widdel, F. and Pfennig, N.**, A new anaerobic, sporing, acetate-oxidzing sulfate-reducing bacterium, *Desulfotomaculum* (emend.) *acetoxidans*, *Arch. Microbiol.*, 112, 119, 1977.

8. **Postgate, J. R.**, Sulfate reduction by bacteria, *Annu. Rev. Microbiol.*, 13, 505, 1959.

9. **Peck, H. D.**, Sulfation linked to ATP cleavage, in *The Enzymes*, Vol. 10, Boyer, P. D., Ed., Academic Press, New York, 1974, 651.

10. **Badziong, W. and Thauer, R. K.**, Growth yields and growth rates of *Desulfovibrio vulgaris* (Marburg) growing on hydrogen plus sulfate and hydrogen plus thiosulfate as the sole energy sources, *Arch. Microbiol.*, 117, 209, 1978.

11. **Wood, P. M.**, A chemiosmotic model for sulphate respiration, *FEBS Lett.*, 95, 12, 1978.

12. **LeGall, J. and Postgate, J. R..**, The physiology of sulfate-reducing bacteria, in *Advances in Microbial Physiology*, Vol. 10, Rose, A. H. and Tempest, D. W., Eds., Academic Press, London, 1973, 81.

13. **Siegel, L. M.**, Biochemistry of the sulfur cycle, in *Metabolic Pathways*, Vol. 7, Greenberg, D. M., Ed., Academic Press, New York, 1975, 217.

14. **Ishimoto, M. and Fujimoto, D.**, Adenosine-5'-phosphosulfonate as an intermediate in the reduction of sulfate by a sulfate-reducing bacterium, *Proc. Jpn. Acad. Sci.*, 35, 243, 1959.

15. **Peck, H. D.**, The ATP-dependent reduction of sulfate with hydrogen in extracts of *Desulfovibrio desulfuricans*, *Proc. Natl. Acad. Sci., U.S.A.* 45, 701, 1959.

16. **Peck, H. D., Jr.**, The role of adenosine-5'-phosphosulfate in the reduction of sulfate to sulfite by *Desulfovibrio desulfuricans*, *J. Biol. Chem.*, 237, 198, 1962.

17. **Robbins, P. W. and Lipmann, F.**, Enzymatic synthesis of adenosine-5'-phosphosulfate, *J. Biol. Chem.*, 233, 686, 1958.

18. **Akagi, J. M. and Campbell, L. L.**, Studies on thermophilic sulfate-reducing bacteria. III. Adenosine triphosphate sulfurylase of *Clostridium nigrificans* and *Desulfovibrio desulfuricans*, *J. Bacteriol.*, 84, 1194, 1962.

19. **Ware, D. A. and Postgate, J. R.**, Physiological and chemical properties of a reductant-activated inorganic pyrophosphatase form *Desulfovibrio desulfuricans*, *J. Gen. Microbiol.* 67, 145, 1971.

20. **Egami, F. J., Ischimoto, M., and Tanigishi, S.**, The electron transfer from cytochromes to terminal electron acceptor in nitrate respriration and sulfate respiration, in *Haematin Enzymes*, Falk, J. E., Lemberg, R., and Morton, R. K., Eds., Pergamon Press, Oxford, 1961, 392.

21. **Ishimoto, M. and Fujimoto, D.**, Biochemical studies on sulfate-reducing bacteria. X. Adenosine-5'-phosphosulfate reductase, *J. Biochem. (Tokyo)*, 50, 299, 1961.

22. **Peck, H. D., Deacon, T. E., and Davidson, J. T.**, Studies on adenosine 5'-phosphosulfate reductase from *Desulfovibrio desulfuricans* and *Thiobacillus thioparus*. I. The assay and purification, *Biochim. Biophys. Acta*, 96, 429, 1965.

23. **Michaels, G. B., Davidson, J. T., and Peck, H. D.**, Studies on the mechanism of adenylylsulfate reductase from the sulfate-reducing bacterium, *Desulfovibrio vulgaris*, in *Flavins and Flavoproteins*, Kamin, H., Ed., University Park Press, Baltimore, 1971, 555.

24. **Chambers, L. A. and Trudinger, P. A.**, Are thiosulfate and trithionate intermediates in dissimilatory sulfate reduction?, *J. Bacteriol.*, 123, 36, 1975.

25. **Kobayashi, K., Takahashi, and Ishimoto, M.**, Biochemical studies of sulfate-reducing bacteria. XI. Purification and some properties of sulfite reductase, desulfoviridin, *J. Biochem. (Tokyo)*, 72, 879, 1972.

---

* The review of the literature for this article was concluded in December, 1978.

26. **Kobayashi, K., Seki, Y., and Ishimoto, M.,** Biochemical studies on sulfate-reducing bacteria. XIII. Sulfite reductase from *Desulfovibrio vulgaris* — mechanism of trithionate, thiosulfate, and sulfide formation and enzymic properties, *J. Biochem. (Tokyo),* 75, 519, 1974.

27. **Jones, H. E. and Skyring, G. W.,** Effect of enzymatic assay conditions in sulfite reduction catalyzed by desulfoviridin from *Desulfovibrio gigas, Biochim. Biophys. Acta,* 377, 52, 1975.

28. **Drake, H. L. and Akagi, J. M.,** Product analysis of bisulfite reductase activity isolated from *Desulfovibrio gigas, J. Bacteriol.,* 126, 733, 1976.

29. **Drake, H. L. and Akagi, J. M.,** Bisulfite reductase of *Desulfovibrio vulgaris,* explanation for product formation, *J. Bacteriol.,* 132, 139, 1977.

29a. **Drake, H. L., Akagi, J. M.,** Dissimilatory reduction of bisulfite by *Desulfovibrio vulgaris, J. Bacteriol.,* 136, 916, 1978.

30. **Murphy, M. J. and Siegel, L. M.,** Siroheme and sirohydrochlorin: the basis for a new type of porphyrin-related prosthetic group common to both assimilatory and dissimilatory sulfite reductases, *J. Biol. Chem.,* 248, 6911, 1973.

31. **Murphy, M. J., Siegel, L. M., Tove, S. R., and Kamin, H.,** Siroheme: a new prosthetic group participating in six-electron reduction reactions catalyzed by both sulfite and nitrite reductases, *Proc. Natl. Acad. Sci., U.S.A.,* 71, 612, 1974.

32. **Murphy, M. J., Siegel, L. M., and Kamin, H.,** An iron tetrahydroporphyrin prosthetic group common to both assimilatory and dissimilatory sulfite reductases, *Biochem. Biophys. Res. Commun.,* 54, 82, 1973.

33. **Lee, J. P. and Peck, H. D., Jr.,** Purification of the enzyme reducing bisulfite to trithionate from *Desulfovibrio gigas* and its identification as desulfoviridin, *Biochem. Biophys. Res. Commun.,* 45, 583, 1971.

34. **Lee, J. P., Yi, C., LeGall, J., and Peck, H. D., Jr.,** Isolation of a new pigment, desulforubidin, from *Desulfovibrio desulfuricans* (Norway Strain) and its role in sulfite reduction, *J. Bacteriol.,* 115, 453, 1973.

35. **Trudinger, P. A.,** Carbon monoxide-reacting pigment from *Desulfotomaculum nigrificans* and its possible relevance to sulfite reduction, *J. Bacteriol.,* 104, 158, 1970.

36. **Akagi, J.M. and Adams, V.,** Isolation of bisulfite reductase activity from *Desulfotomaculum nigrificans,* and its identification as the carbon monoxide-binding pigment P582, *J. Bacteriol.,* 116, 392, 1973.

37. **Akagi, J. M.,** The participation of a ferredoxin of *Clostridium nigrificans* in sulfite reduction, *Biochem. Biophys. Res. Commun.,* 21, 72, 1965.

38. **LeGall, J. and Dragoni, N.,** Dependence of sulfite reduction on a crystallized ferredoxin from *Desulfovibrio gigas, Biochem. Biophys. Res. Commun.,* 23, 145, 1966.

39. **LeGall, J. and Hatchikian, E. C.,** Purification et proprietes d'une flavoproteine intervenant dans la reduction du sulfite par *Desulfovibrio gigas, C. R. Acad. Sci. D,* 264, 2580, 1967.

40. **Moura, J. J. G., Xavier, A. V., Cookson, D. J., Moore, G. R., Williams, R. J. P., Bruschi, M., and LeGall, J.,** Redox states of cytochrome c_3 in the absence and presence of ferredoxin, *FEBS Lett.,* 81, 275, 1977.

41. **Dickerson, R. E. and Timkovich, R.,** Cytochromes *c* in *The Enzymes,* Vol. 11, Boyer, P. D., Ed., Academic press, New York, 1975, 397.

42. **Xavier, A. V. and Moura, J. J. G.,** NMR studies of electron carrier proteins from sulphate-reducing bacteria, *Biochimie,* 60, 327, 1978.

43. **Drake, H. L. and Akagi, J. M.,** Characterization of a novel thiosulfate-forming enzyme isolated from *Desulfovibrio vulgaris, J. Bacteriol.,* 132, 132, 1977.

44. **Mayhew, S. G., Abels, R., and Platenkamp, R.,** The production of dithionite and SO^-_2 by chemical reaction of (bi) sulphite with methyl viologen semiquionone, *Biochem. Biophys. Res. Commun.,* 77, 1397, 1977.

45. **Badziong, W., Ditter, B., and Thaur, R. K.,** Acetate and carbon dioxide assimilation by *Desulfovibrio vulgaris* (Marburg), growing on hydrogen and sulfate as sole energy source, *Arch. Microbiol.,* 123, 301, 1979.

45a. **Badziong, W. and Thaur, R. K.,** Vectorial electron transport in *Desulfovibrio vulgaris* (Marburg) growing on hydrogen plus sulfate as sole energy source, *Arch. Microbiol.,* 1980, in press.

46. **Drake, H. L. and Akagi, J. M.,** Purification of a unique bisulfite-reducing enzyme from *Desulfovibrio vulgaris, Biochem. Biophys. Res. Commun.,* 71, 1214, 1976.

47. **Lee, J. P., LeGall, J., and Peck, H. D., Jr.,** Isolation of assimilatory- and dissimilatory-type sulfite reductases from *Desulfovibrio vulgaris, J. Bacteriol.,* 115, 529, 1973.

48. **Nakatsukasa, W. and Akagi, J. M.,** Thiosulfate reductase isolated from *Desulfotomaculum nigrificans, J. Bacteriol.,* 98, 429, 1969.

49. **Haschke, R. H. and Campbell, L. L.,** Thiosulfate reductase of *Desulfovibrio vulgaris, J. Bacteriol.,* 106, 603, 1971.

50. **Hatchikian, E. C.**, Purification and properties of thiosulfate reductase from *Desulfovibrio gigas*, *Arch. Microbiol.*, 105, 249, 1975.

51. **Postgate, J. R.**, The reduction of sulphur compounds by *Desulfovibrio desulphuricans*, *J. Gen. Microbiol.*, 5, 725, 1951.

52. **Akagi, J. M., Chan, M., and Adams, V.**, Observation on the bisulfite reductase (P 582) isolated from *Desulfotomaculum nigrificans*, *J. Bacteriol.*, 120, 240, 1974.

53. **Bell., G. R., LeGall, J., and Peck H. D., Jr.**, Evidence for the periplasmic location of hydrogenase in *Desulfovibrio gigas*, *J. Bacteriol.*, 120, 994, 1974.

54. **Yagi, T.**, Formate: cytochrome oxidoreductase of *Desulfovibrio vulgaris*, *J. Biochem. (Tokyo)*, 66, 473, 1969.

55. **Riederer-Henderson, M. A. and Peck, H. D., Jr.**, Formic dehydrogenase of *Desulfovibrio gigas*, *Bacteriol. Proc.*, 134, 70, 1970.

56. **Barton, L. L., LeGall, J., and Peck, H. D., Jr.**, Phosphorylation coupled to oxidation of hydrogen with fumarate in extracts of the sulfate reducing bacterium, *Desulfovibrio gigas*, *Biochem. Biophys. Res. Commun.*, 41, 1036, 1970.

57. **Barton, L. L. and Peck, H. D.**, Phosphorylation coupled to electron transfer between lactate and fumarate in cell-free extracts of the sulfate reducing anaerobe, *Desulfovibrio gigas*, *Bacteriol. Proc.*, 155, 1971.

58. **Akagi, J. M.**, Electron carriers for the phosphoroclastic reaction of *Desulfovibrio desulfuricans*, *J. Biol. Chem.*, 242, 2478, 1967.

59. **Suh, B. and Akagi, J. M.**, Pyruvate-carbon dioxide exchange reaction of *Desulfovibrio desulfuricans*, *J. Bacteriol.*, 91, 2281, 1966.

60. **Yagi, T., Honya, M., and Tamiya, N.**, Purification and properties of hydrogenases of different origins, *Biochim. Biophys. Acta.*, 153, 699, 1968.

61. **Bell, G. R., Lee, J. -P., Peck, H. D., Jr., and LeGall, J.**, Reactivity of *Desulfovibrio gigas* hydrogenase toward artificial and natural electron donors or acceptors, *Biochimie*, 60, 315, 1978.

62. **Yagi, T., Kimura, K., Daidoji, H., Sakai, F., Tamura, S., and Inokuchi, H.**, Properties of purified hydrogenase from the particulate fraction of *Desulfovibrio vulgaris*, Miyazaki *J. Biochem. (Tokyo)*, 79, 661, 1976.

63. **van der Westen, H. M., Mayhew, S. G., and Veeger, C.**, Separation of hydrogenase from intact cells of *Desulfovibrio vulgaris*: purification and properties, *FEBS Lett.* 86, 122, 1978.

64. **LeGall, J., Dervartanian, D. V., Spilker, E., Lee, J. P., and Peck, H. D., Jr.**, Evidence for the involvement on non-heme iron in the active site of hydrogenase from *Desulfovibrio vulgaris*, *Biochim. Biophys. Acta*, 234, 525, 1971.

65. **Haschke, R. and Campbell, L. L.**, Purification and properties of hydrogenase from Desulfovibrio vulgaris, *J. Bacteriol.*, 105, 249, 1971.

66. **Krasna, A. I., Riklis, E., and Rittenberg, D.**, The purification and properties of the hydrogenase of *Desulfovibrio desulfuricans*, *J. Biol. Chem.*, 235, 2717, 1960.

67. **Yagi, T.**, Solubilization, purification and properties of particulate hydrogenase from *Desulfovibrio vulgaris*, *J. Biochem. (Tokyo)*, 68, 649, 1970.

68. **Hatchikian, E. C., Bruschi, M., and LeGall, J.**, Characterization of the periplasmic hydrogenase from *Desulfovibrio gigas*, *Biochem. Biophys. Res. Commun.*, 82, 451, 1978.

69. **Loach, P. A.**, Oxidation-reduction potentials, absorbance bands and molar absorbance of compounds used in biochemical studies, in *Handbook of Biochemistry*, Sober, H. A., Ed., CRC Press, Cleveland, 1970, J-33.

70. **Bruschi, M., Hatchikian, E. C., LeGall, J., Moura, J. J. G., and Xavier, A. V.**, Purification, characterization and biological activity of three forms of ferredoxin from the sulfate-reducing bacterium *Desulfovibrio gigas*, *Biochim. Biophys. Acta*, 449, 275, 1976.

71. **Cammack, R., Rao, K. K., Hall, D. O., Moura, J. J. G., Xavier, A. V., Bruschi, M., LeGall, J., Deville, A., and Gayda, J. P.**, Spectroscopic studies of the oxidation-reduction properties of the three forms of ferredoxin from *Desulfovibrio gigas*, *Biochim. Biophys. Acta*, 490, 311, 1977.

72. **Moura, J. J. G., Xavier, A. V., Bruschi, M., LeGall, J.**, NMR characterization of three forms of ferredoxin from *Desulphovibrio gigas*, a sulfate reducer, *Biochim. Biophys. Acta*, 459, 278, 1977.

73. **Yagi, T. and Maruyama, K.**, Purification and properties of cytochrome c_3 of *Desulfovibrio vulgaris*, Miyazaki *Biochim. Biophys. Acta*, 243, 214, 1971.

74. **Dervartanian, D. V., Xavier, A. V., and LeGall, L.**, EPR determination of the oxidation-reduction potentials of the hemes in cytochrome c_3 from *Desulfovibrio vulgaris*, *Biochimie*, 60, 321, 1978.

75. **Schnorf, U.**, Der Einfluβ von Substituenten auf Redoxpotential und Wuchsstoffeigenschaften von Chinonen, *Eidg. Tech. Hochsch. (Zürich)*, Diss.Nr. 3871, 1966.

76. **Wagner, G. C., Kassner, R. J., and Kamen, M. D.**, Redox potentials of certain vitamins K: implications for a role in sulfite reduction by obligately anaerobic bacteria, *Proc. Natl. Acad. Sci. U.S.A.*, 71, 253, 1974.

77. **Newman, D. J. and Postgate, J. R.**, Rubredoxin from a nitrogen-fixing variety of *Desulfovibrio desulfuricans*, *Eur. J. Biochem.*, 7, 45, 1968.

78. **Moura, I., Xavier, A. V., Cammack, R., Bruschi, M., and LeGall, J.**, A comperative spectroscopic study of two non-heme iron proteins laking labile sulphide from *Desulfovibrio gigas Biochim. Biophys. Acta*, 533, 156, 1978.

79. **Postgate, J. R.**, Cytochrome c_3 and desulphoviridin; pigments of the anaerobe *Desulphovibrio desulphuricans*, *J. Gen. Microbiol.*, 14, 545, 1956.

80. **Ambler, R. P., Bruschi, M., and LeGall., J.**, The structure of cytochrome c'_3 from *Desulfovibrio gigas*, (NCIB 9332), *FEBS Lett.* 5, 115, 1969.

81. **Drucker, H., Campbell, L. L., and Woody, R. W.**, Optical rotatory properties of the cytochromes c_3 from three species of *Desulfovibrio*, *Biochemistry*, 9, 1519, 1970.

82. **Drucker, H., Trousil, E. B., Campbell, L. L., Barlow, E., and Margoliash, E.**, Amino acid composition, heme content, and molecular weight of cytochrome c_3 of *Desulfovibrio desulfuricans* and *Desulfovibrio vulgaris*, *Biochemistry*, 9, 1515, 1970.

83. **Ambler, R. P., Bruschi, M., and LeGall, J.**, The amino acid sequence of cytochrome c_3 from *Desulfovibrio desulfuricans* (strain El Agheila Z, NCIB 8380), *FEBS Lett.* 18, 347, 1971.

84. **Meyer, T. E., Bartsch, R. G., and Kamen, M. D.**, Cytochrome c_3, a class of electron transfer heme proteins found in both phototrophic and sulfate reducing bacteria, *Biochim. Biophys. Acta*, 245, 453, 1971.

85. **Dobson, C. M., Hoyle, N. J., Geraldes, C. F., Wright, P. E., Williams, R. J. P., Bruschi, M. and LeGall, J.**, An outline structure of cytochrome c_3 and a consideration of its properties, *Nature (London)*, 249, 425, 1974.

86. **Frey, M., Haser, R., Pierrot, M., Bruschi, M., and LeGall, J.**, Preliminary crystallographic study on cytochrome c_3 of *Desulfovibrio desulfuricans* (Strain Norway), *J. Mol. Biol.*, 104, 741, 1976.

87. **Bruschi, M., Hatchikian, C. E., Golovleva, L. A., and LeGall, J.**, Purification and characterization of cytochrome c_3, ferredoxin, and rubredoxin isolated from *Desulfovibrio desulfuricans*, Norway, *J. Bacteriol.*, 129, 30, 1977.

88. **Dervartanian, D. V. and LeGall, J.**, Studies on the reaction of imidazol with cytochrome c_3 from *Desulfovibrio vulgaris*, *Biochim. Biophys. Acta*, 502, 458, 1978.

89. **LeGall, J., Mazza, G., and Dragoni, N.**, Le cytochrome c_3 des *Desulfovibrio gigas*, *Biochim. Biophys. Acta*, 99, 385, 1965.

90. **Bruschi, M., LeGall, J., Hatchikian, C. E., and Dubourdieu, M.**, Cristallisation et propriétés d'un cytochrome intervenant dans la reduction du thiosulfate par *Desulfovibrio gigas*, *Bull. Soc. Fr. Physiol. Veg.*, 15, 381, 1969.

91. **Hatchikian, E. C., LeGall, J., Bruschi, M., and Dubourdieu, M.**, Regulation of the reduction of sulfite and thiosulfate by ferredoxin, flavodoxin and cytochrome cc'_3 in extracts of the sulfate reducer *Desulfovibrio gigas*, *Biochim. Biophys. Acta*, 258, 701, 1972.

92. **Ambler, R. P., Bruschi, M., and LeGall., J.**, Biochimie comparee des cytochromes des bacteries sulfato-reductrices, in *Recent Advances in Microbiology*, 10th Int. Cong. Microbiology, Mexico, Perez-Niravete, A. and Pelasz, D. Ed., 1971, 25.

93. **Bruschi, M. and LeGall, J.**, C-type cytochromes of *Desulfovibrio vulgaris*, Amino acid composition and end groups of cytochrome c_{553}, *Biochim. Biophys. Res. Commun.*, 38, 607, 1970.

94. **Hatchikian, E. C. and LeGall, J.**, Evidence for the presence of a *b*-type cytochrome in the sulfate reducing bacterium *Desulfovibrio gigas* and its role in the reduction of fumarate by molecular hydrogen, *Biochim. Biophys. Acta*, 267, 479, 1972.

95. **Hatchikian, E. C.**, On the role of menaquinone-6 in the electron transport of hydrogen: fumarate reductase system in the strict anaerobe *Desulfovibrio gigas*, *J. Gen. Microbiol.*, 81, 261, 1974.

96. **Weber, M. M., Matschiner, J. T., and Peck, H. D.**, Menaquinone-6, in the strict anaerobes *Desulfovibrio vulgaris* and *Desulfovibrio gigas*, *Biochim. Biophys, Res. Commun.*, 38, 197, 1970.

97. **Maroc, J., Azerad, R., Kamen, M. D., and LeGall, J.**, Menaquinone (MK-6) in the sulfate-reducing obligate anaerobe, *Desulfovibrio*, *Biochim. Biophys. Acta*, 197, 87, 1970.

98. **Zubieta, J. A., Mason, R., and Postgate, J. R.**, A four iron ferredoxin from *Desulfovibrio desulfuricans*, *Biochim. J.*, 133, 851, 1973.

99. **Dubourdieu, M. and LeGall, J.**, Chemical study of two flavodoxin extracted from sulfate reducing bacteria, *Biochim. Biophys. Res. Commun.*, 38, 965, 1970.

100. **Watenpaugh, K. D., Sieker, L. C., Jensen, L. H., LeGall, J., and Dubourdieu, M.**, Structure of the oxidized form of a flavodoxin at 2.5 Å resolution: resolution of the phase ambiquity by anomalous scattering, *Proc. Natl. Acad, Sci. U.S.A.* 69, 3185, 1972.

101. **Watenpaugh, K. D., Sieker, L. C., and Jensen, L. H.**, The binding of riboflavin-5′-phosphate in a flavoprotein: flavodoxin at 2.0 Å resolution, *Proc. Natl. Acad. Sci. U.S.A.* 70, 3857, 1973.

102. **Watenpaugh, K. D., Sieker, L. C., and Jensen, L. H.**, A crystallographic structural study of the oxidation states of *Desulfovibrio vulgaris*, flavodoxin, in *Flavins and Flavoproteins*, Singer, T. P., Ed., Elsevier, Amsterdam, 1976, 405.

103. **Dubourdieu, M. and Fox, J. L.**, Amino acid sequence of *Desulfovibrio vulgaris* flavodoxin, *J. Biol. Chem*, 252, 1453, 1977.

104. **Bruschi, M. and LeGall, J.**, Purification et propriétés d'une rubredoxine isolee a partir de *Desulfovibrio vulgaris* (souche NCIB 8303), *Biochim. Biophys. Acta*, 263, 279, 1971.

105. **Bruschi, M.**, The amino acid sequence of rubredoxin from the sulfate-reducing bacterium, *Desulfovibrio gigas*, *Biochem. Biophys. Res. Commun.*, 70, 615, 1976.

106. **Pierrot, M., Haser, R., Frey, M., Bruschi, M., LeGall, J., Sieker, L. C., and Jensen, L. H.**, Some comparisons between two crystallized anaerobic bacterial rubredoxin from *Desulfovibrio gigas* and *D. vulgaris*, *J. Mol. Biol.*, 107, 179, 1976.

107. **Bruschi, M.**, Non-heme iron proteins: the amino acid sequence of rubredoxin from *Desulfovibrio vulgaris*, *Biochim. Biophys. Acta*, 434, 4, 1976.

108. **Adman, E. T., Sieker, L. C., Jensen, L. H., Bruschi, M., and LeGall, J.**, A structural model of rubredoxin from *Desulfovibrio vulgaris*, at 2Å resolution, *J. Mol. Biol.*, 112, 113, 1977.

109. **LeGall, J.**, Purification Partielle et Etude de la NAD: Rubredoxine oxydo-reductase de *D. gigas*, *Ann. Inst. Pasteur, Paris*, 114, 109, 1968.

110. **Moura, I., Bruschi, M., LeGall, J., Moura, J. J. G., and Xavier, A. V.**, Isolation and characterization of desulforedoxin, a new type of non-heme iron protein from *Desulfovibrio gigas*, *Biochem. Biophys. Res. Commun.*, 75, 1037, 1977.

111. **Prince, R. C., Baccarini-Melandri, A., Hauska, G. A., Melandri, B. A., and Crofts, A. R.**, Assymmetry of an energy transducing membrane. The location of cytochrome c₂ in *Rhodopseudomonas spheroides* and *Rhodopseudomonas capsulata*, *Biochim. Biophys. Acta*, 387, 212, 1975.

112. **Scholes, P. B., McLain, G., and Smith, L.**, Purification and properties of a c-type cytochrome from *Micrococcus denitrificans*, *Biochemistry*, 10, 2072, 1971.

113. **Fujita, T. and Sato, R.**, Studies on soluble cytochromes in *Enterobacteriaceae*. III. Localization of cytochrome c-552 in the surface layer of cells, *J. Biochem. (Tokyo)*, 60, 568, 1966.

114. **Racker, E., Burstein, C., Loyter, A., and Christiansen, R. O.**, The sidedness of the inner mitochondrial membrane, in *Electron Transport and Energy Conservation*, Tager, J. M., Papa, S., Quagliariello, E., and Slater, E. C., Eds., Adriatica Editrice, Bari, 1970, 235.

115. **Peck, H. D., Jr.**, Evidence for oxidative phosphorylation during the reduction of sulfate with hydrogen by *Desulfovibrio desulfuricans*, *J. Biol. Chem.*, 235, 2734, 1960.

116. **Senez, J. C.**, Some considerations on the energetics of bacterial growth, *Bacteriol. Rev.*, 26, 95, 1962.

117. **Vosjan, J. H.**, ATP generation by electron transport in *Desulfovibrio desulfuricans*, *Antonie van Leeuwenhoek J. Microbiol. Serol.*, 36, 584, 1970.

118. **Khosrovi, B., MacPherson, R., and Miller, J. D. A.**, Some observations on growth and hydrogen uptake by *Desulfovibrio vulgaris*, *Arch. Microbiol.*, 80, 324, 1971.

119. **Vosjan, J. H.**, Respiration and fermentation of the sulphate-reducing bacterium *Desulfovibrio desulfuricans* in a continuous culture, *Plant Soil*, 43, 141, 1975.

120. **Magee, E. L., Jr., Ensley, B. D., and Barton, L. L.**, An assessment of growth yields and energy coupling in *Desulfovibrio*, *Arch. Microbiol.*, 117, 21, 1978.

121. **Peck, H. D., Jr.**, Phosphorylation coupled with electron transfer in extracts of the sulfate reducing bacterium, *Desulfovibrio gigas*, *Biochem. Biophys. Res. Commun.*, 22, 112, 1966.

122. **Guarraia, L. J., and Peck, H. D., Jr.**, Dinitrophenol-stimulated adenosine triphosphase activity in extracts of *Desulfovibrio gigas*, *J. Bacteriol.*, 106, 890, 1971.

123. **Papa, S.**, Proton translocation reactions in the respiratory chains, *Biochim. Biophys. Acta*, 456, 39, 1976.

Chapter 5

RESPIRATION IN THE AMMONIA-OXIDIZING CHEMOAUTOTROPHIC BACTERIA

J. W. Drozd

TABLE OF CONTENTS

I. INTRODUCTION

In this review, I will restrict myself to a discussion of respiration and ammonia-oxidation in the ammonia-oxidizing chemoautotrophic bacteria typified by *Nitrosomonas* sp. I will not elaborate on ammonia oxidation or nitrification by heterotrophic bacteria or fungi because they do not use ammonia oxidation as the sole source of energy, and many of the reactions are not oxidations; the reader is referred to other research papers[1-8] for a discussion of these organisms. I will, however, discuss ammonia oxidation by the methane-oxidizing bacteria because these bacteria have several close similarities[9,10] to the ammonia-oxidizing chemoautotrophic bacteria and may be able to derive energy from ammonia oxidation.

It is as well to point out at the start of this review that although compared to *Escherichia coli* relatively little work has been done on the physiology and biochemistry of the ammonia-oxidizing bacteria, these bacteria are in fact of very great importance in many microbial processes. For example, in effluent treatment processes where in the overall process of nitrification which is the conversion of organic nitrogen into nitrate, the rate-limiting step may often be the oxidation of ammonia to nitrite.[11,12] This oxidation is thought to be mainly catalyzed by chemoautotrophic ammonia-oxidizing bacteria of the family Nitrobacteraceae, genera *Nitrosomonas, Nitrosospira, Nitrosococcus*, and *Nitrosolobus*. Indeed if one regards ammonia-oxidation as an essential part of the biological nitrogen cycle (Figure 1), it is surprising that in comparison with the vast amount of work on biological nitrogen fixation, so little work has been done on biological ammonia oxidation. An understanding of ammonia oxidation might permit rates of nitrification in effluent treatment processes to be increased, especially where there is a low carbon-to-nitrogen ratio, or conversely to inhibit ammonia-oxidation in soils after the application of an ammonia-based nitrogen fertilizer. The latter is possible by the addition of specific inhibitors of ammonia-oxidation, e.g., 2-chloro-6-(trichloromethyl)pyridine.[13-15] The prevention of ammonia oxidation to nitrate in soils can be advantageous because ammonia is less readily leached from soils than the anionic nitrate ion and is thus available to the plant for a longer period of time; also with ammonia, the plant does not have to use reductant to reduce nitrate to ammonia.[16] Furthermore, nitrates and nitrites leached from soils can give rise to intense eutrophication in bodies of fresh water into which the run-off water flows.

II. SUBSTRATE

Many excellent recent reviews have dealt with various aspects of the biochemistry and physiology of ammonia oxidation by chemautotrophic bacteria of the family Nitrobacteraceae[11,17-26] and in this review, a short account will be presented of some pertinent literature together with some recent findings. The oxidation of ammonia to nitrite can be written as:

$$NH_4^+ + 1.5O_2 \rightarrow NO_2^- + H_2O + 2H^+ \qquad (1)$$

The substrate is written as the ammonium ion, although measurements of the maximum rate of ammonia utilization and the K_m of whole cells for ammonia oxidation at various pH values[27,28] suggests that ammonia is the actual substrate, and there is little reaction with the ammonium ion. These results are not simply due to the presence of a specific ammonium ion transport mechanism which limits the maximum rate of flux of ammonium ion across the cytoplasmic membrane such that the enzyme(s) responsible for ammonia oxidation are unsaturated with substrate, but a very rapid diffusion

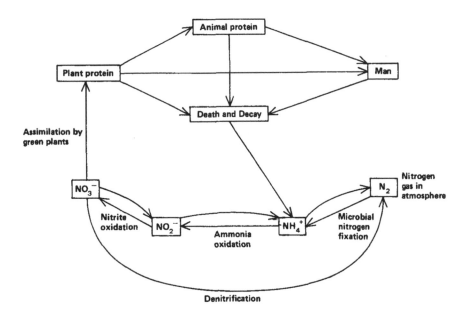

FIGURE 1. Schematic representation of the biological nitrogen cycle.

of free ammonia across the cytoplasmic membrane to the enzyme(s) at higher pH values, because there is also a decrease in the K_m of ammonia-dependent oxygen uptake with an increase in pH in a crude extract of *Nitrosomonas europaea*[27] which is commensurate with free ammonia as the substrate for the enzyme(s). The K_m value for ammonia is between 18 and 56 μM in both whole cells and cell-free extracts (the pKa value for the $NH_4^+ = NH_3 + H^+$ equilibrium is pH 9.25 at 30°C). Caution should be exercised in the interpretation of these experiments because:

1. Oxygen uptake is only an indirect measure of ammonia oxidation.
2. The substrate ammonia is itself an inhibitor of its own oxidation at high concentrations.
3. Nitrite, the final product of ammonia oxidation, is also an inhibitor of ammonia oxidation.

For these reasons, results which are obtained by the indirect method of oxygen uptake measurements over a short period of time[27,28] when the time of inhibition is limited and the cells are nongrowing may not always correspond to results obtained with growing cultures over a longer period of time, although Laudelout et al.[29,30] obtained good agreement between predictions from the short-term oxygen uptake experiments and a mathematical model with batch culture experiments with pure and mixed cultures.[30] Ammonia is also the species which possibly crosses the cytoplasmic membrane because the rapid decrease in pH value during batch culture growth of ammonia oxidizers can possibly be partly explained by the liberation of a proton on the outside of the membrane if the cells can in fact transport the NH_4^+ species as NH_3. This leaves the question of pH control within the cell undecided because a rapid influx of ammonia into the cell would cause a rapid rise in pH which would have to be controlled. It might be mentioned that in chromatophores of photosynthetic systems exposed to ammonium ions, ammonia crosses the membrane and reionizes on the inside which causes an uncoupling of photophosphorylation by mediating the collapse of the light-induced proton[31] gradient (positive inside). The phenomenon of acidification during ammonium ion uptake is not restricted to ammonia-oxidizing bacteria; in nearly all heterotrophic bacteria, there is a decrease in pH when ammonium ions are taken up which could be consistent with the transport of the ammonia species, the transport of ammonium ion with a hydroxyl ion, or the transport of the ammonium ion, with the subsequent extrusion of a proton or uptake of a hydroxyl ion. We require more knowledge of ammonium ion transport in bacteria in general to be able to answer these questions, and it is a pity that so little work has been done on ammonium transport[32-35] compared with the vast literature on the transport of the carbon substrate(s) in bacteria. Undoubtedly one reason for this is the technical difficulty of working with the system and lack of a large number of substrate analogues, although methylamine has been used.[35] Interestingly, Cole[36] recently suggested that the extensive intracytoplasmic membrane system[37] often seen in ammonia oxidizers (Figure 2) may be a result of the utilization of a gaseous substrate (ammonia) which might require a peripheral membrane phase where the enzymes of ammonia oxidation may reside. This hypothesis is attractive, but there is no direct evidence for this, and other explanations can be put forward, for example, that the membrane network is associated with so-called energy-dependent reverse electron flow. Alternatively the low ATP yield during substrate oxidation and high ATP demand for growth[38] means that a high respiration rate is needed to support growth, and a high respiration rate possibly means a large amount of cytochromes, flavoproteins, membrane-bound dehydrogenases, etc. which may result in an extended membrane network. I will return to these points later.

III. GROWTH OF THE ORGANISMS

Most workers have grown *Nitrosomonas* in batch culture in a basic mineral salts medium, but there are only a few reports on growth in continuous culture.[39] Early workers used a medium in which a carbonate precipitate was present, but a precipitate-free medium was developed by Engel and Alexander,[40] and the medium was further improved by Loveless and Painter.[41] There is no "secret" about the growth of *Nitrosomonas*, although some publications leave the reader with this impression. The points to remember are:

1. The culture has a slow doubling time, with a minimum of about 8 hr. This corresponds to a specific growth rate of 0.087/hr. Loveless and Painter[41] have indexed much of the published growth rate data.
2. The organisms exhibit a very low yield (grams of cells formed per mole of ammonia used) on ammonia, i.e., they show a very high rate of ammonia oxidation for a given growth rate. Most people have used a measure of nitrite formed as an estimation of yield, and one can assume that approximately 1 mol of ammonia gives rise to 1 mol of nitrite. Naturally there will be some ammonia incorporated into biomass, but this is negligible compared to the amount of ammonia oxidized to nitrite. There is, however, the untested possibility that gaseous oxides of nitrogen might be formed (see later). Loveless and Painter[41] obtained yields in pH controlled batch culture of between 0.42 and 1.4 g cells per mole of nitrite produced. The high value may have been an error due to mineral precipitates. I have obtained yields[101] for similar pH-controlled batch cultures of *N. europaea* of between 0.6 and 0.8 g cells formed per mole of nitrite produced, with a maximum specific growth rate of 0.02 to 0.03/hr. Note that the specific growth rate (μ/hr) is related to the culture doubling time (t_d, hr) by the relationship:

$$\mu = \frac{\ln 2}{t_d} \qquad\qquad (2)$$

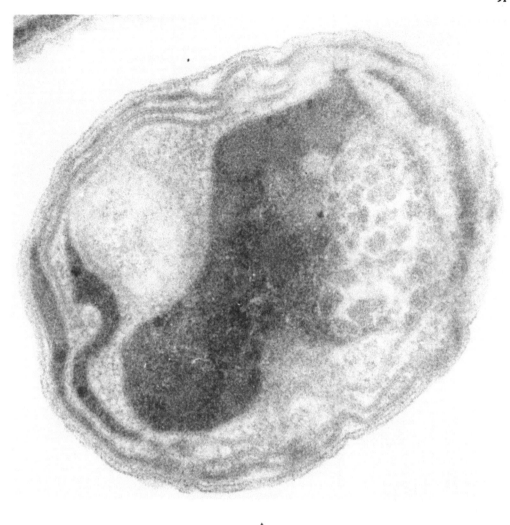

A

FIGURE 2. Electron micrographs of thin sections of (A) autotrophically grown batch culture cells of *Nitrosomonas europaea* and (B) *Thiobacillus neapolitanus* C grown with sodium thiosulphate. Sections stained with uranyl acetate and lead citrate. The bar represents 0.5 μ. The samples were prepared and the pictures taken by Mr. R. Pillinger and Mrs. S. A. POTTER.

The *in situ* rate of ammonia oxidation (or more properly the rate of nitrite formation) is then given by:

$$q = \frac{\mu}{Y} \tag{3}$$

where q is the rate of nitrite formation (mol/g/hr), and Y is the observed yield (grams of cells produced per mole of ammonia oxidized or nitrite formed). From the above data, rates of ammonia oxidation can be calculated as being approximately between 30 and 200 mmol of nitrite produced per g/hr, with most values in the range of 30 to 60 mmol nitrite produced per g/hr. Naturally one complication in these calculations would be the rapid oxidation of ammonia by nongrowing cells, i.e., when substrate oxidation and possible energy generation is "uncoupled" from growth. By analogy, to the growth of heterotrophic bacteria

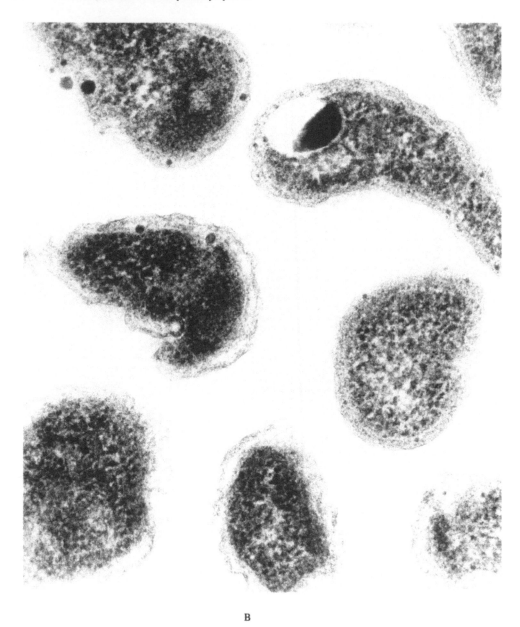

B

such a situation should be minimal under energy-limitation in continuous culture,[42] but such growth data is very scanty for ammonia-oxidizing bacteria. It should also be noted that, unlike the situation in heterotrophic bacteria, in autotrophic bacteria the carbon and energy sources are supplied quite separately. Thus, one can obtain an energy (NH_4^+)-limited, carbon sufficient (excess CO_2) culture or vice-versa. The production of any carbon overflow metabolites[42] should be minimal under CO_2 limitation. Skinner and Walker[39] grew *N. europaea* in batch and continuous culture on an improved medium, but it is not always clear as to what was the growth limiting nutrient, and unfortunately the nitrite concentration in the culture was not measured. In early experiments, oxygen was the growth-limiting nutrient, but for higher cell densities, it is not always clear if the ammonia concentration in the influent medium (up to 0.12 *M*) limited the growth rate or not. Nevertheless at a maximum dilution rate of approximately 0.03/hr (the culture washed out at higher dilution rates), the yield on ammonia can be calculated to be approximately 0.7 g dry weight mole ammonia^{-1}; this gives an in situ rate of ammonia oxidation of 43 mmol/g/hr. The maximum observed growth rate in batch culture was 0.086/hr, but more often maximum values of 0.04 to 0.06/hr were seen. What is clear is that much better growth yield data are required for the growth of *Nitrosomonas* in continuous culture. Nevertheless the above data emphasize the low observed growth yield on ammonia. Unfortunately no data are available on the so-called maintenance ammonia consumption[43] which, if high, could cause the observed yields at the low growth rates to be considerably lower than the theoretical yield in the absence of any maintenance substrate consumption. If a very high ammonia concentration is initially used in batch culture, then growth inhibition is observed.[11,41] The only way of overcoming this is to use ammonia pH control on batch culture or to use ammonia limitation in chemostat culture.

3. *Nitrite, the product of ammonia oxidation, is inhibitory to growth of Nitrosomonas.* In early studies, Meyerhof[44] noted a 36% inhibition of oxygen uptake in *Nitrosomonas* by an added final concentration of 100 mM nitrite, and complete inhibition by 300 mM nitrite, and Painter[11] gives more data on this and related work in his review. Laudelout et al.[30] carried out an elegant analysis of the effect of pH on a mixed culture of *Nitrosomonas* and *Nitrobacter* and fitted the data to a mathematical model. However, they did not look at the effect of pH on nitrite inhibition of ammonia oxidation, although they noted that ammonia was itself more inhibitory than an equivalent ammonium ion concentration. The final concentration of nitrite found in batch culture can be very high; Loveless and Painter[41] found a final value of up to 179 mM nitrite. Drozd et al.[102] obtained a final maximum nitrite concentration of up to 0.5 M. Interestingly when this high concentration of nitrite was added to the culture before inoculation, no growth was seen which suggested that growth ceased because of nitrite accumulation. More work needs to be done on the kinetics of nitrite inhibition of ammonia oxidation and respiration in *Nitrosomonas* and to identify the precise site(s) and possible pH dependence of nitrite inhibition. The latter point is of interest because the undissociated weak acid could act as an uncoupler of oxidative phosphorylation. In heterotrophic bacteria, a final nitrite concentration of 10 mM causes an inhibition of growth in *Paracoccus denitrificans*.[45]

IV. CYTOCHROMES

Reduced minus oxidized difference spectra of intact cells and membrane preparations of *N. europaea* suggest the presence of cytochromes of the *b, c, o,* and *a* types,[23,46] as well as a cytochrome P$_{450}$[47] or a P$_{460}$ system,[48] as identified by their carbon monoxide binding spectra. Interestingly there are no reports o any cytochrome oxidases of the *d* type, and there are no isolates which lack cytochrome(s) *c*. The presence of an oxidase of the *d* type might not be consistent with the presence of a sole site of energy transduction at the cytochrome *c* oxidase region of the respiratory chain[49] (see later). Unfortunately relatively little work[50-53] has been done on the biochemistry of the reported cytochrome P$_{450}$ or P$_{460}$ systems, so it is only speculation[25] as to whether such systems are really involved in the initial hydroxylation of ammonia to produce hydroxylamine. There are several cytochrome *c* species,[52,53] one of which has quite a high E$_0'$ value[54] and reacts with CO in a manner analogous to the high potential CO-binding cytochrome *c* identified in methylotrophic bacteria.[55] The content of cytochromes of the *c* type in *Nitrosomonas* cells is probably very high, and it is possibly this which gives the cells their characteristic red-orange color, although this has yet to be unambiguously proved. There is also a lack of kinetic data for the identification of the cytochrome oxidase(s) present. For more detailed accounts of the cytochromes, the reader is referred to the excellent recent reviews of Suzuki[23] and Aleem.[25]

V. AMMONIA OXIDATION

It is probably fair to say that there is a large degree of uncertainty as to the exact mechanism of ammonia oxidation.[23,25] The detailed experimental evidence for an oxygenase type of reaction catalyzed by a cytochrome P.460 or P.450 type of complex is extremely poor. The system is usually written as:

$$NH_3 + O_2 + XH_2 \rightarrow NH_2OH + X + H_2O \qquad (4)$$

but the evidence is quite circumstantial with the reaction above catalyzed by a hypothetical ammonia hydroxylase and XH_2, a reduced cytochrome (P_{450}, P_{460}, or *c*?). Rees and Nason[51] found a small incorporation of $^{18}O_2$ from $^{18}O_2$ into nitrite during ammonia oxidation by *Nitrosomonas* cells which suggested a direct incorporation of some oxygen into the substrate at some stage in its oxidation, but not necessarily at the site of a proposed ammonia oxygenase. It should be possible to test this further by the recovery of labeled hydroxylamine in a manner analogous to the demonstration that methanol was the direct product of methane oxidation in methylotrophic bacteria.[56] Hooper[57] noted that the addition of hydroxylamine eliminated the initial lag in ammonia oxidation by resting cells. This could be explained if hydroxylamine oxidation gave rise to the reduced species of X in Equation 4, i.e., XH_2 which is necessary for the initial hydroxylation of ammonia.

What is known is that the oxidation of ammonia to hydroxylamine is unlikely to generate much energy and the reaction may be endergonic,[21,24,25,26] e.g., Aleem[21] calculated that for the oxidation of ammonia to hydroxylamine, the $\Delta G'_o$ value (free energy change) was $+3.85$ Kcal/mol, or as Schlegel[24] has written it, the couple NH_4^+ / NH_2OH has a high potential (calculated by Aleem)[21] with an E'_o of $+0.899$ V. Although such calculations give an indication of the situation, thy should be used with some caution. Often considerable assumptions are made about th state and concentration of reactants and products within the bacterial cell, and values obtained for a standard free energy change are not the same as the actual free energy change when the activities of reactants and products may be far removed from equilibrium concentrations. The reader should study the paper of Burton and Krebs[58] to see an example where the above points were taken into consideration. Many of the theoretical studies with *Nitrosomonas* involve unstated assumptions on the acivities and states of reactants and products, whereas Burton and Krebs outline all the problems involved in such calculations. Aleem[21] and Kelly[26] do, however, state that their calculations are based on a pH of 7.0 and a concentration of reactants except the hydrogen ions in the standard state. As stated earlier, such a calculation gives a value of $\Delta G'_o$ of $+3.85$ Kcal mol for the reaction in Equation 4. It would be extremely interesting to know the activities of the reactant and products especially if the enzyme is membrane bound; such measurements might give a very different value to $\Delta G'$.

What is known about the ammonia oxidation reaction is that it is inhibited[59] by a wide range of metal-binding agents, such as allylthiourea, uncouplers of oxidative phosphorylation, electron acceptors such as phenazine methosulphte, and carbon monoxide. The data suggest the presence of a metal such as copper which is chelated by metal-binding agents and a carbon monoxide-binding factor, e.g., a haemoprotein. The electron acceptors possibly oxidize the intermediate electron donor XH_2, and in fact the electron acceptors stimulate hydroxylamine oxidation. The effect of uncouplers of oxidative phosphorylation is harder to explain; they would tend to collapse a membrane potential (see later) and decrease the rate of ATP formation. This might suggest that a membrane potential was directly involved in ammonia oxidation or indirectly by driving the reduction of X to XH_2, e.g., via reverse electron flow driven by a protonmotive force, where XH_2 is the propsed source of reductant for the hypothetical ''ammonia oxygenase'' enzyme complex. If ammonia oxidation was somehow tightly coupled to CO_2 fixation, then an uncoupler would prevent $NAD(P)^+$ reduction by reverse electron flow (see later) and thus inhibit ammonia oxidation.

If there is a monooxygenase type of reaction, then in accord with Equation 4, this must be taken into account in oxygen uptake calculations to determine the amount of oxygen consumed during electron transport-linked energy generation. For example, although only a few studies have been made to check if the stoicheiometry in Equation 1 holds in practice with growing cultures (and none of these studies involved accurate steady-state chemostat experiments), most of the batch culture studies indicated that the actual rations observed were fairly close to the theoretical value. Wezernak and Gannon[60] in batch cultures obtained a ratio of 1.4 mol of oxygen used per mole of ammonia oxidized to nitrite; from their highest yield of cells on ammonium of 1.37 g mol ammonium^{-1} and with a growth rate at 0.03/hr, this would give an in situ oxygen uptake rate of 31 mmol oxygen per g/hr and a maximum observed yield on oxygen of 0.98 g cells mole oxygen^{-1}. This yield on oxygen represents oxygen consumed in the hypothetical ammonia mono-oxygenase reaction and in the cytochrome oxidase reaction. It is probably only the uptake of oxygen in the cytochrome oxidase reaction which is directly associated with ATP production via respiratory chain electron transport and coupled oxidative phosphorylation. If we assume that overall half a mole of oxygen is consumed in the ammonia hydroxylase reaction, then we can find the yield on oxygen for energy coupled, i.e., cytochrome oxidase-mediated oxygen uptake from the relationship at a given growth rate, μ (per hr) of:

$$q_{O_2} \text{ (energy)} = q_{O_2} \text{ (total)} - \tfrac{1}{2}q_{NH_4^+} \tag{5}$$

or

$$\frac{1}{Y_{O_2}} \text{ (energy)} = \frac{1}{Y_{O_2}} \text{ (total)} - \frac{0.5}{Y_{NH_4^+}} \tag{6}$$

For the above typical data, this gives a Y_{O_2} (energy) of 1.52 g mol when the YO_2 (total) is 0.98 g/mol, and the Y_{NH_4} is 1.37 g/mol. It is to be stressed that these are all observed yields, with no correction for so-called maintenance oxygen or amonia consumption, and it must be realized that because of the slow growth rates of these bacteria, the maintenance substrate consumption might constitute a large part of the observed oxygen and ammonia uptake rates. There is also the consideration that in calculations based on ammonia disappearance, there will be the direct incorporation of some of the ammonia nitrogen into biomass, and it will not all be oxidized.

Although little is known about the ammonia hydroxylase system in *Nitrosomonas* sp., considerably more is known about a similar enzyme, the methane monooxygenase enzyme complex in methane-oxidizing bacteria.[9,10] This enzyme is a nonspecific hydroxylase which, as well as hydroxylating methane to methanol,[56,61] will convert carbon monoxide into carbon dioxide,[62] ammonia into hydroxylamine,[62-64] ethylene into ethylene oxide, propylene into propylene oxide,[65] benzene into phenol,[65] toluene into benzyl alcohol and cresol, and cyclohexane into cyclohexanol.[65] There are two types of hydroxylase system; one is found in the serine pathway methylotroph *Methylosinus trichosporium* OB 3b and is dependent on a reduced high-potential carbon monoxide-binding cytochrome *c* as a source of electrons:[66]

$$CH_4 + O_2 + c \text{ (reduced)} + 2H^+ \rightarrow$$

$$CH_3OH + H_2O + c \text{ (oxidized)} \tag{7}$$

while the other system[65] found in the ribulose monophosphate pathway methane oxidizer *Methylococcus capsulatus* is NADH dependent:

$$CH_4 + O_2 + NADH + H^+ \rightarrow CH_3OH + H_2O + NAD^+ \tag{8}$$

In *M. trichosporium* OB 3b, the reduced cytochrome *c* is generated at the level of methanol dehydrogenase and can probably also be generated at the level of formaldehyde and formate dehydrogenases (Figure 3), while in *M. capsulatus*, the NADH is generated at the level of formaldehyde and formate dehydrogenases (Figure 3). Obviously the oxidation of ammonia to hydroxylamine in methane-oxidizing bacteria is very similar to the reactions in the Nitrobacteraceae, and recently Suzuki et al.[67] noted that methane and methanol could competitively inhibit ammonia oxidation in intact cells and extracts of *N. europaea*, although these substrates are not themselves oxidized by *N. europaea*,[28] and we (Drozd et al.)[102] have not detected any methanol formed from methane by intact cells of *N. europaea* in the presence or absence of ammonia, although such cells did catalyze a slow oxygen-dependent production of propylene oxide from propylene (Figure 4), phenol from benzene, and cyclohexanol from cyclohexane. Obviously these nonspecific hydroxylation reactions are very similar to those catalyzed by the methane-oxidizing bacteria and suggest a strong similarity between these hydroxylase (oxygenase) systems. Before this analogy is taken too far, it should be remembered that:

1. *Nitrosomonas* does not appear to actually hydroxylate methane or oxidize methanol[28,67] or to grow with methane as the source of energy.
2. Biochemically the route of CO_2 fixation in *Nitrosomonas* (Calvin cycle) is more similar to the ribulose monophosphate pathway of formaldehyde incorporation,[9,10] especially as the enzyme ribulose diphosphate carboxylase has been found in *M. capsulatus*.[68] This would imply a NADH-dependent[65,69] hydroxylase in *Nitrosomonas*, whereas because of the scarcity of NADH (see later) and the improbability that hydroxylamine oxidation can be directly coupled to NAD^+ reduction, one might expect a cytochrome-linked hydroxylase as found in the serine pathway organism *M. trichosporium*[66]
3. The maximum in situ rate of ammonia oxidation in *Methylococcus* NCIB 11083 grown in continuous culture on methane[64] was only 0.62 mmol g^{-1} h^{-1}; this is considerably less than the maximum observed rate of ammonia oxidation in *Nitrosomonas* of 40 to 50 mmol/g/hr.
4. *Methylococcus* cannot grow with ammonia as the source of energy and fix CO_2.[104]

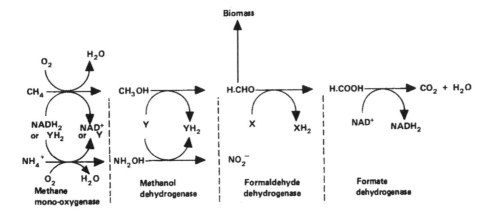

FIGURE 3. Schematic representation of the pathway of substrate oxidation in methane-oxidizing bacteria. In *Methylosinus trichosporium* OB 3b, the reductant for the methane monooxygenase is a reduced form of carbon-monoxide binding cytochrome c (YH$_2$) generated at the level of methanol and possibly formaldehyde dehydrogenase.[66] In *Methylococcus* sp., the methane monooxygenase is NADH$_2$-dependent[65,69], with NADH$_2$ generated at the level of formate and possibly formaldehyde dehydrogenase.[61] The probable pathway of ammonia oxidation to nitrite is also shown.

If the monooxygenase cannot conserve energy then all the reductant generated by hydroxylamine oxidation (probably reduced cytochrome c) will be recycled to the monooxygenase, with no net energy or NADP(H) generation. Even if grown with the toxic hydroxylamine there would be a requirement for energy-dependent reverse electron flow to reduce NAD(P)$^+$. For these reasons, any similarities between these organisms may be due to convergent evolution.

Despite our lack of knowledge as to many of the basic details of the ammonia-hydroxylase (oxygenase) system in the Nitrobacteraceae, it is clear that if it is an oxygenase, then the reductant, XH$_2$ in Equation 4, must be generated either directly or indirectly from hydroxylamine oxidation. As the redox couple (E$_o'$ value) of the hydroxylamine/nitrite couple is 66 mV,[21] there can only be the direct reduction of electron transfer components by hydroxylamine with a redox couple value which is more positive than this.

VI. HYDROXYLAMINE OXIDATION

There is good evidence that hydroxylamine oxidation is fairly closely linked to cytochrome reduction and subsequent energy generation. The argument is really as to where electrons from hydroxylamine enter the electron transport chain. Aleem and Lees[46] indicated that a hydroxylamine cytochrome c reductase might be involved in hydroxylamine oxidation, with the possible involvement of a flavin and cytochrome b. Nicholas and Jones[70] prepared extracts of *Nitrosomonas* that would oxidize hydroxylamine to nitrite, provided a suitable electron acceptor such as cytochrome c or phenazine methosulphate was added. This was probably the first report of a cell-free system capable of oxidizing hydroxylamine, and the interesting observation was made that phosphate inhibited the reaction and was competitive with cytochrome c. Earlier, Hofman and Lees[71] found that washed cell suspensions of *Nitrosomonas* could oxidize hydroxylamine to nitrite, and Lees[17] proposed the metabolism of ammonia to nitrite proceeded via hydroxylamine and an unknown intermediate, probably nitroxyl (NOH). The enzymic reduction of cytochrome c by hydroxylamine was inhibited by

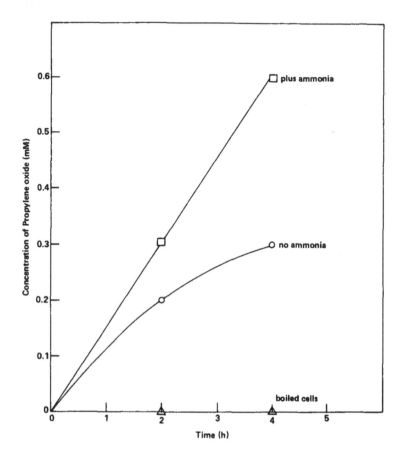

FIGURE 4. Formation of propylene oxide from propylene by washed, intact cells of *Nitrosomonas europaea* at approximately 5 mg/ml in nitrogen-free medium with and without the addition of 50 mM NH$_4$Cl. As a control, a cell sample was boiled for 10 min. The reaction was at 30°C in a 250-ml flask with 50:50, air to propylene gas mixture.

mepacrine, but FAD reversed this inhibition which suggested that a flavin was involved.[46,72] The enzymic system termed hydroxylamine cytochrome *c* reductase seems to be involved only in the oxidation of NH$_2$OH to the level of the unstable intermediate, nitroxyl (NOH). Although some of the above evidence suggested a flavin was involved, it could not be detected in a purified enzyme which contained only *b*- and *c*-type cytochromes,[52,53,73] but no flavin. Naturally with an enzyme preparation, an electron acceptor is required, e.g., phenazine, methosulphate, dichlorophenolindolphenol, or cytochrome *c*.[74] Experimentally less than 50% of the hydroxylamine oxidized may appear as nitrite,[70,72] and the measured ratios of 2 mol of electron acceptor reduced per mole hydroxylamine oxidized[72,73] suggested that a compound at the oxidation level of nitroxyl (NOH) was the intermediate product produced by dehydrogenation of hydroxylamine. Nitric oxide has also been identified as a product of hydroxylamine oxidation and is thus also a possible precursor to nitrite,[75-77] although it could in crude enzyme preparations also arise from the reduction of nitrite by a nitrite reductase activity which is closely associated with the hydroxylamine oxidase activity during purification.[78,79] Although nitrite is always thought of as the final product of ammonia oxidation by *Nitrosomonas*, nitrate can also be found. Anderson[75] found some nitrate produced from hydroxylamine in extracts of *Nitrosomonas* with oxygen and methylene blue. Recently Hooper et al.[74] found that although nitrate was not produced in the oxidation of ammonia or hydroxylamine by intact cells of *Nitrosomonas*, a partially purified hydroxylamine oxidoreductase enzyme system catalyzed the aerobic oxidation of hydroxylamine to nitrite and nitrate. The reaction was most rapid in the presence of the artificial electron acceptor PMS when 60 to 80% of the hydroxylamine utilized was recovered as nitrite and nitrate in roughly equimolar quantities. Hydrogen peroxide was also formed in the reaction. Because nitrate was not formed by intact cells, it was suggested that the formation of nitrate in the crude enzyme preparation was a reaction not normally found in intact cells. In the enzyme preparation, nitrate was not formed from nitrite, and it was suggested that the precursor of nitrate was an *N*-oxide of oxidation state between hydroxylamine and nitrite. Diethyldithiocarbamate inhibited nitrate, but not nitrite formation, and was itself oxidized to *bis*(diethyldithiocarbamoyl)disulfide. It was suggested that diethyldithiocarbamate reduced a reactive form of oxygen and thus prevented the oxidation of the *N*-oxide to nitrate or reduced a nitrogenous precursor of nitrate to a compound which was subsequently oxidized to nitrite. In fresh intact cells given hydroxylamine,[75] there is 1 mol of nitrite produced per mole of hydroxylamine added:

$$NH_2OH + O_2 \rightarrow NO_2^- + H^+ + H_2O \qquad (9)$$

In cells which had been stored for 3 to 4 weeks, only 60% of the hydroxylamine added was recovered as nitrite, and it was suggested that nitrous oxide was found. Intact cells did not oxidize hydroxylamine anaerobically, but when a low concentration of the artificial electron acceptor methylene blue was added, a mixture of nitrous and nitric oxides was formed equivalent to the hydroxylamine added.

The above account presents a somewhat confusing picture of the mechanism(s) of hydroxylamine oxidation to nitrite. This is partially accounted for by the difficulties encountered in preparing reproducible cell-free extracts, the relatively small amount of work done on the system, and the problem of isolating any stable intermediates between hydroxylamine and nitrite. In a recent review, Suzuki,[23] on the basis of published data and unpublished work from his laboratory, suggested an oxidation scheme: $NH_4^+ \rightarrow NH_2OH \rightarrow (NOH) \rightarrow NO_2^-$ or, under certain conditions, $2(NOH) \rightarrow N_2O + H_2O$. He enlarged this scheme to the following overall mechanism:

$$NH_3 + O_2 + AH_2 \rightarrow NH_2OH + A + H_2O \qquad (10)$$

$$NH_2OH + 2 \text{ cytochrome } c(Fe^{+++}) \rightarrow$$
$$(NOH) + 2H^+ + 2 \text{ cytochrome } c(Fe^{++}) \qquad (11)$$

$$2 \text{ cytochrome } c(Fe^{++}) + \tfrac{1}{2}O_2 + 2H^+ \rightarrow$$
$$2 \text{ cytochrome } c(Fe^{+++}) + H_2O \qquad (12)$$

$$(NOH) + A + H_2O \rightarrow NO_2^- + AH_2 + H^+ \qquad (13)$$

$$\text{Sum: } NH_3 + 1.5 \ O_2 \rightarrow NO_2^- + H_2O + H^+ \qquad (14)$$

This is an attractive scheme, but many aspects remain to be proven, e.g., the nature of the hypothetical ammonia hydroxylase in Reaction 10 or the proof of nitroxyl (NOH) as an intermediate and the nature of the hypothetical nitroxyl dehydrogenation scheme in Reaction 13. Aleem[25] suggested a cytochrome P_{460}-type ammonia hydroxylase and nitroxyl oxidation scheme:

$$(NOH) \xrightarrow{\quad \frac{1}{2}O_2 \quad} HNO_2 \qquad\qquad (15)$$

This system poses a problem in the uptake of half a mole of oxygen per mole of substrate in the reactions, whereas in the system of Suzuki, 1 mol of oxygen is taken up per mole of ammonia hydroxylated in Reaction 10. To explain the oxidation of hydroxylamine in the absence of ammonia, it is necessary to postulate nitroxyl oxidation (Reaction 13) without coupling to Reaction 10 which in fact acts like a terminal electron acceptor for AH_2 oxidation. Possibly the AH_2 species is oxidized via cytochrome c and cytochrome oxidase with concomitant energy generation. Alternatively the hypothetical nitroxyl oxidase (Reaction 13) could be analogous to nitrite reductase. In the steady-state oxidation of ammonia, all the reactions are highly coupled and must proceed at the same rate. Suzuki[23] points out that artificial electron acceptors, such as PMS or methylene blue, may inhibit ammonia oxidation by causing a complete oxidation of cytochrome c or AH_2 (which may be closely related to cytochrome c). The E_o' value of the NH_2OH/NO_2^- system is $+0.066$ V,[21] so it is highly unlikely that the oxidation of hydroxylamine to nitrite could be coupled to NADH formation (E_o' of -0.32 V) without energy-dependent reverse electron flow (see later) which makes it unlikely that AH_2 can be equated with $NADH_2$. Because of the complexities of the *Nitrosomonas* system, it is useful to consider ammonia oxidation in the methane oxidizing bacteria.

I have earlier shown the mechanisms of methane oxidation in the bacteria which carry out these reactions (Equations 7 and 8), and it is probable that ammonia is oxidized to hydroxylamine in an analogous manner. The methane monooxygenase reaction probably does not conserve energy as ATP, i.e., the transfer of electrons from $NADH_2$ or reduced cytochrome c to the monooxygenase enzyme complex is not linked to ATP formation, unlike such a transfer of electrons down the cytochrome chain to oxygen. The methane oxidizing bacteria also oxidize hydroxylamine,[63,64] with 1 mol of oxygen taken up per mole of hydroxylamine oxidized[64] and a high (50%) aerobic steady-state reduction of the cytochrome(s) c in intact cells.[64] The most thorough study of this system is in *M. capsulatus* by Dalton.[63] The ammonia-oxidizing activity was dependent on O_2 and $NADH_2$ and was inhibited by methane and specific inhibitors of the methane mono-oxygenase system, e.g., acetylene and 8-hydroxyquinoline, but surprisingly neither methane nor ammonia-oxidizing activity was inhibited by metal chelating agents, such as thiourea or $\alpha\alpha'$-dipyridyl. The evidence suggested that ammonia was oxidized by the methane monooxygenase enzyme system. Hydroxylamine was oxidized to nitrite in a reaction system that could be coupled to phenazine methosulphate, but not to cytochrome c; this reaction did not compete with methanol oxidation, and a semipurified methanol dehydrogenase would not oxidize hydroxylamine. The K_m for ammonia in the extracts was very high at 87 mM and unlike the system in *Nitrosomonas*, the K_m increased with an increase in pH above pH 7, and ammonia oxidation was completely inactive at pH 9. In intact cells, the K_m was also high, 31 mM at pH 7, and increased with an increase in pH. In *Methylococcus* NCIB 11083, the situation is somewhat different,[64] and a measurement of ammonia-dependent oxygen uptake gives a much lower K_m for ammonia oxidation in intact cells (Figure 5) than that found in *Methylococcus capsulatus* Bath. Also (see Drozd e. al.)[102] hydroxylamine oxidation is partially competitive with methanol oxidation in intact cells. Dual wavelength spectroscopy indicated that ammonia oxidation gave no aerobic steady-state cytochrome c

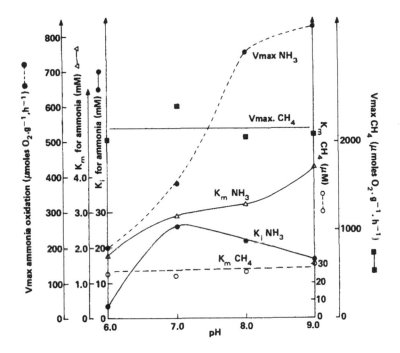

FIGURE 5. The Michaelis constant, K_m and maximum velocity, V_{max}, of methane and ammonium chloride dependent oxygen uptake by washed, intact cells of *Methylococcus* NCIB 11083 in a nitrogen-free growth medium. The reaction[63] was at 45°C in a Rank Oxygen Electrode (Rank Bros., Bottisham, Cambridge, U.K.), with a final cell density of 2 mg dry weight per ml to 5 mg dry weight per ml. The reaction was run for 5 min before the addition of substrate. At each pH tested, the ammonium chloride was a competitive inhibitor of methane-dependent oxygen uptake,[63,64] and the inhibitor constants, Ki values, are shown for the inhibition of methane oxidation by ammonia.

(or *b* or NAD$^+$) reduction in intact cells (surprisingly methane also did not), but hydroxylamine gave a 50% steady-state reduction of cytochrome *c*. Obviously the availability of semipurified methane monooxygenase system[65,66,69] will enable detailed studies to be done on this system, but care should be used before the results are extrapolated to the *Nitrosomonas* system.

VII. ENERGY TRANSDUCTION

Energy generation is thought to occur at the levels o hydroxylamine and possibly nitroxyl (NOH) oxidation, although if nitroxyl oxidation occurs according to Equation 13 with AH$_2$ used in the hydroxylation of ammonia, it is unlikely that this would generate energy, as no system is known to this reviewer in which electron transfer to an oxygenase system, i.e., the reoxidation of AH$_2$ in Reaction 13 by Reaction 10, is coupled to energy generation. This would mean that the only step in energy coupling would be at the level of hydroxylamine oxidation to nitroxyl. Naturally when hydroxylamine and not ammonia is oxidized, then AH$_2$ could be oxidized via the respiratory chain and energy could be generated. On this basis, *Nitrosomonas* sp. should show a higher molar growth yield on hydroxylamine than on ammonia, but there are no reports of growth on the rather unstable and toxic hydroxylamine; perhaps the organism could be grown in hydroxylamine limited in continuous culture.

Because of the possible involvement of a flavin and cytochrome *b* in hydroxylamine cytochrome *c* reductase activity[46] and an E'_0 value of $+66$ mV for the NH_2OH/NO_2^- couple, Aleem[21] suggested that up to two sites of oxidative phosphorylation might be associated with hydroxylamine oxidation. Phosphorylating membrane vesicles only exhibited P/O ratios of 0.2 when coupled to hydroxylamine oxidation,[80] but this is consistent with the low efficiency of oxidative phosphorylation often exhibited by many bacterial membrane preparations. The chemiosmotic theory of oxidative phosphorylation proposed by Mitchell[81] postulates that the respiratory chain is spatially oriented across the energy-conserving membrane(s) into proton-translocating loops, each loop consisting of a hydrogen carrier and an electron carrier. Two protons are translocated outwards per electron pair transferred down the respiratory chain, and this proton gradient which consists of a pH (acid outside) and an electrical (positive outside) gradient can drive ATP synthesis via a reversible membrane-bound proton-translocating ATPase such that 1 g mole of ATP is synthesised per 2 g equivalent of H^+ translocated. The H^+/O ratios associated with hydroxylamine oxidation[28] of approximately 2 would be indicative of a P/O ratio of 1, and a suitable proton-translocating scheme (Figure 6) can be constructed which would comprise of hydroxylamine cytochrome *c* reductase on the outer or inner side of the membrane, cytochrome *c* on the outside, and the cytochrome oxidase(s) on the inner side of the membrane. Unfortunately no probes have been used to determine on which side of the membrane the hydroxylamine cytochrome *c* reductase is situated, but as hydroxylamine is generated inside the membrane, it might be logical for the hydroxylamine to be oxidized on the inside of the membrane. Perhaps the system spans across the membrane such that hydroxylamine is oxidized on the inner membrane surface, but the protons (Equation 11) are released on the outside. The electrons are then carried by cytochrome *c*. In the presence of the uncoupler carbonyl cyanide *m*-chlorophenylhydrazone, which renders the membrane proton permeable by catalyzing proton equilibration across the membrane(s), only a very minute alkalinization of the external medium was seen when oxygen was added to the system.[28] This would indicate only a small net pH change across the membrane; however, if two protons are produced on the outside of the membrane and two protons are consumed on the inside of the membrane per electron pair in the cytochrome oxidase reaction (Figure 6), then a larger initial net inward movement of protons would not be expected. However, this would depend on how fast the protons equilibrated across the membrane in the presence of the uncoupler. Such experimentation should be carried out to determine the topology of the various electron carriers, hydrogen carriers and dehydrogenases.

The presence of this one "site" of energy coupling would uggest that in the oxidation of 1 mol of ammonia to hydroxylamine, only 1 mol of ATP is produced. With a maximum observed yield on ammonia of between approximately 0.8 and 1.4 g mol, this would give an observed Y_{ATP} value of between 0.8 and 1.4 g dry weight cells produced per mole of ATP generated. This is an oversimplification because:

1.   Some of the ammonia nitrogen is incorporated into biomass and is not oxidized; an estimate of nitrite produced should indicate the amount of ammonia oxidized (In practice this error is negligible.)
2.   Some of the electrons from reduced cytochrome *c* are used to reduce $NAD(P)^+$ and are not used for energy generation (see later). This will mean that less than 1 mol of ATP is produced per mole of hydroxylamine oxidized (Figure 7).

If ammonia or ammonium ion had to be transported into the cell by an energy-dependent uptake process (see earlier) in which 1 mol of ATP or its equivalent was used per mole of substrate transported, then the organisms could not grow because a

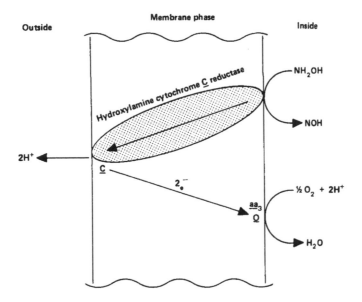

FIGURE 6. A possible schematic representation of proton-translocating Loop 3 in the membrane(s) of *Nitrosomonas europaea*. There is as yet little evidence[28] for the orientation of the electron and hydrogen carriers as shown.

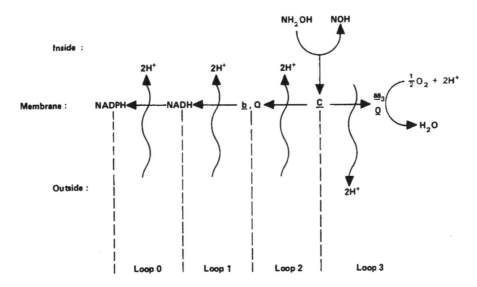

FIGURE 7. A schematic representation of energy-dependent reverse electron flow in *Nitrosomonas europaea*. The protonmotive force generated by Loop 3 would be used to drive Loops 1 and 2 and possibly loop 0 in the reverse direction, i.e., these loops would translocate protons inwards.[87,88,90,91,100] Naturally some of the protonmotive force would also be used for ATP synthesis via the membrane-bound proton-translocating ATPase complex. It is assumed that 1 g mol of ATP is synthesized per 2 g equivalent H^+ translocated.

maximum of only 1 mol of ATP is generated per mole of ammonia oxidized. It is important to find out:

1. How ammonia is transported and the energy requirement or gain of any transport process (For example, if the ammonium ion enters as ammonia and a proton is

left behind, this would contribute to the protonmotive force and contribute to the energy gain from substrate oxidation.)

2.   If nitroxyl oxidation is coupled to energy generation.
3.   If electrons can be transferred to an oxygenase with coupling to energy generation, i.e., formation of a transmembrane proton gradient.

With regard to (1), it is tempting to speculate that the failure of the organism to grow well at more acid pH values, as well as being due to the very toxic nature of nitrous acid compared to the nitrite anion,[82] could be due to the zero energy yield from ammonium oxidation if a large amount of energy was involved in ammonium uptake. A factor not yet discussed is the mechanism of nitrite transport down a concentration gradient out of the cell. At neutral or alkaline pH values, nitrite will be present as the anion (pKa of 3.4), and the transport of an anion out of the cell would tend to collapse any transmembrane protonmotive force, positive outside. Approximately, 1 mol of nitrite will be transported out of the cell per mole of ammonia oxidized. There is no evidence for any specific nitrite transport mechanism; even in *Nitrobacter*, the K_m for nitrite oxidation is 1.6 mM[82] at pH 7.7 which may suggest the lack of a transport system for nitrite. Cobley[83,84] also observed a high K_m for nitrite oxidation in *Nitrobacter*. Obviously more work needs to be done on nitrite transport out of the cell, as well as on ammonium transport into the cell, and one could speculate on the presence of an NH_4^+/NO_2^- antiport system, but there is no evidence for this, although it is an attractive hypothesis as it would maintain electroneutrality and the transport of nitrite out of the cell down a concentration gradient could be linked to ammonium transport into the cell up a concentration gradient.

VIII. NADH PRODUCTION

A special feature of the growth of *Nitrosomonas* is that they must derive all their reducing power from ammonia oxidation to achieve reductive carbon dioxide assimilation. The direct reduction of pyridine nucleotides by ammonia or hydroxylamine oxidation is very unfavorable,[85] the E_o' value of the $NAD^+/NADH$ couple is −320 mV, and the E_o' value of the NH_2OH/NO_2^- couple is +66 mV. The mechanism of pyridine reduction by hydroxylamine would be analogous to the ATP-dependent reduction of NAD^+ in mitochondria by succinate, as demonstrated by Chance.[86] In a classic paper, Aleem[87] demonstrated that cell-free preparations from *Nitrosomonas* would catalyze an active ATP-dependent $NAD(P)^+$ reduction by succinate, hydroxylamine, or ferrocytochrome *c*. This reaction was stimulated by oligomycin and inhibited by flavoprotein inhibitors and uncouplers of oxidative phosphorylation. On the basis of these studies, Aleem[21] postulated that energy coupling at "Site" 3 is used for ATP generation, while "Sites" 2 and 1 are used for ATP-driven reverse-electron flow to NAD^+. These results can be re-evaluated in terms of the chemiosmotic theory which postulates that all the transmembrane oxidation-reduction loops are fully reversible. Thus, some of the protonmotive force generated by Loop 3 could be used to drive Loops 1 and 2 (and possibly loop 0, if there is an energy-dependent transhydrogenase)[88] in the reverse direction. It is important that the preparation used by Aleem[87] was a cell extract which was subject only to low-speed centrifugation and thus probably contained a large amount of membrane vesicles. There was a 5-min. lag in the ATP-dependent reduction of $NAD(P)^+$; this would be consistent with ATP hydrolysis occurring with the build-up of a transmembrane protonmotive force. The reversal of Loops 1, 2, and possibly 0 and the reduction of $NAD(P)^+$ by an applied pH gradient has yet to be demonstrated in *Nitrosomonas*; the occurrence of H^+/O ratios of up to 4.4 with endogenous substrates in unstarved cells[28] suggests that there can be the generation of a protonmotive force possibly by the oxidation of a large intracellular quinone pool, this implies that at least Loop 2 is present. Whether there will be some sort of reversed protonmotive Q cycle[89] remains to be elucidated. Poole and Haddock[90] have observed an energy-linked reduction of NAD^+ in membranes from various *E. coli* K-12 strains and Shahak et al.[91] have observed reversed electron flow in isolated chloroplasts by the imposition of an artificial transmembrane protonmotive force. *Nitrosomonas* would be an ideal tool in which to probe the functioning of reverse electron flow and in which to study the regulation of the respiration rate.[92] It is probable that the rate of respiration will be governed by the back pressure of the protonmotive force on the rate of electron flow through Loop 3, i.e., hydroxylamine-cytochrome *c* reductase, cytochrome *c*, and the cytochrome oxidase(s). The utilization of the protonmotive force via reactions, such as ATP synthesis, reverse electron flow (which in the chemiosmotic model can proceed without any direct participation of ATP), and substrate transport will in large depend on the growth rate of the organisms. A high rate of electron flow through Loop 3 will be required because of the large demand for ATP and NAD(P)H in the chemo-autotrophic bacteria. Kelly[26] has done detailed calculations of energy coupling in the chemoautotroph *Thiobacillus denitrificans*. Similar calculations can be done for *Nitrosomonas* sp. From an observed approximate growth yield of 0.8 g dry weight cells per mole ammonia oxidized and from the observation that the dried biomass is 44% carbon, it can be easily calculated that 0.35 g carbon, i.e., 0.03 mol of CO_2 are fixed per mole of ammonia oxidized or 33 mol of ammonia are oxidized per mole of CO_2 fixed. If all this CO_2 is fixed via the Calvin cycle in which 3ATP and 2NADH are required (Equation 16) to fix 1 mol of CO_2 to the level of fructose (CH_2O), then the reductant requirement (4H) per mole of CO_2 could be met by the oxidation of 1.5 mol of ammonia (six electrons per mole ammonia oxidized), with 31.5 mol of ammonia oxidized (95%) for energy requirements:

$$6CO_2 + 18ATP + 12NAD (P) H + 12H^+ \rightarrow$$

$$\text{fructose-6-P} + 18ADP + 17Pi + 12NAD (P)^+ \qquad (16)$$

The reader is referred to the review of Kelly[26] for a full discussion of these points. The situation in *T. denitrificans* grown on thiosulphate is complicated by the production of ATP via substrate level phosphorylation.[26] In *Thiobacillus ferroxidans* grown on iron,[93] there is no substrate level phosphorylation, but only oxidative phosphorylation with a likely P/O ratio of 1.[26] The rue growth yield in the absence of maintenance is 1.33 g dry weight of cells per mole iron oxidized at first siht corresponds to 2.66 g of cells produced per mole of ATP generated. However, as Kelly[26] has discussed, there is the use of electrons from iron oxidation for $NAD(P)^+$ reduction. Better growth yield data is required for *Nitrosomonas* to enable the systems to be compared. For an account of much of the early work on reversed electron flow in chemoautotrophs and kinetic and steady-state measurements of cytochrome oxidation and reduction, the reader is referred to the review by Kiesow.[18]

IX. CONCLUSIONS

The study of respiration in *Nitrosomonas* offers a fascinating field of study for the microbial physiologist or biochemist because it is a largely unresearched area, and there are many aspects of the system for which better data is required or for which there is as yet no data. There is also the added interest that an oxygenase enzyme system is probably involved in the initial hydroxylation of ammonia. The organisms should not be regarded as some rather obscure objects of study; they are very important in the natural environment where they form an essential part of the biological nitrogen cycle, and their activities are especially important in such processes as effluent treatment and the nitrogen cycle in the soil. The latter is of special relevance to the use of ammonia-based fertilizers.

Cultures of *N. europaea* exhibit a very rapid rate of ammonia and oxygen uptake, e.g., at a specific growth rate of 0.02/hr, with a yield on ammonia of 0.8 g/mol, the q (ammonia) will be 25 mmol/g/hr, the q (total oxygen) will be 37 mmol/g/hr, and the specific rate of nitrite formation (and its transport out of the cell) will also be approximately 25 mmol/g/hr. Interestingly this indicates that the culture would consume one third of its own weight of ammonia per hour. This high respiratory activity, c.f., heterotrophic bacteria may be one reason why the organism produces a complex internal membrane system (Figure 2). This is speculation, and absolute measurements of the total membrane and respiratory chain components are required. Other possible roles of the membrane system are that it is in some way connected with the ammonia hydroxylation enzyme complex or with energy-dependent reverse electron flow. The former possibility would seem most likely because there is no obvious extensive internal membrane system in thin sections of the chemoautotroph *Thiobacillus neapolitans* C (Figure 2) in which there is also an energy-dependent reverse electron flow,[26] but the methane-oxidizing bacteria do have an *extensive internal membrane network*, whereas the methanol-oxidizing bacteria do not.[61] The only major difference between the latter two groups of organisms is that the methane-oxidizing bacteria hve a methane monooxygenase enzyme complex; in terms of electron transport linked, i.e., cytochrome oxidation-mediated oxygen uptake rates, the organisms will exhibit very similar rates at a given growth rate. Another alternative as suggested by Cole[36] is that there is a correlation between gas metabolism other than oxygen and the membrane network, but hydrogen-oxidizing bacteria do not have any such obvious extensive internal membrane network.[24]

The reason for the very high rates of substrate metabolism and oxygen uptake is that the organisms must expend a large amount of energy and reductant to synthesize cellular material from CO_2, but there is only a very limited amount of energy available from substrate oxidation. The oxidation of the substrate does not directly lead to the NAD(P)H required for cell synthesis, but energy and electrons must also be expended to reduce $NAD(P)^+$ to NAD(P)H. There is also the possibility that ammonia transport into the cell and nitrite transport out of the cell could consume a large amount of energy, especially the transport of the ammonium ion at acid pH values, although it is possible that ammonium transport could indirectly contribute to the protonmotive force if only ammonia is actually transported with a proton left outside the cytoplasmic membrane. Very little is known about these transport processes, and it would be highly interesting to grow *Nitrosomonas* ammonia limited in chemostat culture and measure the apparent K_s of the culture for ammonia, where the K_s refers to the extracellular ammonia concentration which will support half the maximum specific growth rate. Under such conditions of growth, the organisms might produce a high-affinity energy-dependent ammonium transport system. As I have argued, if such a system were to consume one ATP or equivalent per mole of ammonium transported, then growth would be impossible if intracellular ammonia oxidation produced an absolute maximum of only 1 mol of ATP per mole of ammonia oxidized. This calculation ignores the reductant, i.e., electron demand for $NAD(P)^+$ reduction. Such continuous culture studies might also enable accurate estimates of growth yield and maintenance substrate consumption to be made and attempts could be made to utilize hydroxylamine as the growth substrate under conditions of hydroxylamine limitation.

In the text and above discussion, a large amount of stress has been laid on the amount of energy generated from substrate oxidation. However, in a very interesting analysis, Anthony[94] has highlighted that the growth yield of most autotrophic and certain methylotrophic bacteria is reductant rather than energy limited. This is in contrast to the situation in most heterotrophic bacteria where the growth yield is limited by the ATP supply which in turn is governed by the P/O ratio. In the methylotrophs, some NADH is generated directly from substrate oxidation, but in *Nitrosomonas* all the NAD(P)H is generated by energy-dependent reverse electron flow; in this situation, an increase in the P/O ratio might increase the cell yield because an increase in P/O ratio with a constant of substrate oxidation would allow a greater rate of NAD^+ reduction with an unchanged rate of ATP generation. In terms of the energy-dependent reverse electron flow, the oxido-reduction loops should be quite reversible; thus, provided NADH dehydrogenase is present, and it probably is,[9] it should be possible for NADH to be oxidized to give a apparent P/O ratio of 3. The organisms must have a mechanism for preventing NAD(P)H oxidation during periods when cell biosynthesis is arrested. Presumably during balanced growth, the intracellular NAD(P)H pool is very low, and it may be that the K_m of the NADH dehydrogenase for NADH is much higher than the pool level of NADH or that there is some regulation of dehydrogenase activity as in certain heterotrophic bacteria[95] which prevents the oxidation of NAD(P)H. The condition of growth limitation by oxygen is probably the most likely situation in *Nitrosomonas* in many natural environments and certainly inside sewage flocs or biological films. The physical conditions within films or flocs are very complex. There will be gradients of pH and dissolved oxygen, ammonium, and nitrate/nitrate concentration. These factors coupled to a lack of accurate data for the diffusion coefficients of substrates in such environments will make kinetic analysis difficult. It is also probable that much of the metabolic activity exhibited by *Nitrosomonas* in the natural environment is performed by nongrowing cells,[96] i.e., substrate oxidation will occur but growth of the cells cannot occur because of a nutrient limitation or starvation, where the nutrient is not ammonia or oxygen. Under these conditions, substrate oxidation may be "uncoupled" from cell growth and the NAD(P)H pool may increase in concentration until eventually the concentration of NAD(P)H is so high that reverse electron flow and substrate oxidation are inhibited. In this context, it is interesting that in our laboratory washed cell suspensions of *Nitrosomonas* never oxidized ammonia so rapidly as growing cells. This suggests that the rate of ammonia oxidation may be well regulated or closely linked to cell growth and to metabolic demands.

The nature of energy transduction in *Nitrosomonas* makes it essential that proton-translocating Loop 3 is functional. This is absolutely dependent on the presence of a high-potential cytochrome *c* in a manner exactly analogous to the requirement for this cytochrome in the functioning of Loop 3 in heterotrophic bacteria.[88] The high concentration of *c*-type cytochrome(s) in *Nitrosomonas* may be required to support a high rate of electron flux through Loop 3. The reader is referred to several recent articles[97-99] for a discussion of the possible functioning of Loop 3 in mitochondria.

The nature of the reductant for the ammonia hydroxylase is still uncertain; circumstantial evidence suggests a cytochrome P_{450} or P_{460} system, but there is no absolutely conclusive evidence for this. It would seem unlikely that the organisms would use NAD(P)H because this is only indirectly produced from substrate oxidation in an energy-dependent process, and its use would be an enormous drain on the supply of reducing power. It is more likely that some electron carrier is used which can be directly reduced by substrate oxidation.

This review is not a complete synopsis of all the literature, and I am aware of obvious omissions, but I hope that I have highlighted some areas where I think that there is a large amount of uncertainty and where reliable data are required. Perhaps some readers might be stimulated to take an interest in these very fascinating and economically important groups of bacteria; if so then I feel that my imperfect presentation will have had some positive effect.

X. ADDENDUM

The similarity between the methane-oxidizing bacteria and the Nitrobacteraceae is further strengthened by the observation that acetylene inhibits ammonia oxidation in *N. europaea* which is analogous to the inhibition in the methane-oxidizing bacteria of methane oxidation by acetylene.[103]

REFERENCES

1. **Hirsch, P., Overrein, L., and Alexander, M.**, Formation of nitrite and nitrate by *Actinomycetes* and fungi. *J. Bacteriol.*, 82, 442, 1961.
2. **Marshall, K. C. and Alexander, M.**, Nitrification by *Aspergillus. flavus, J. Bacteriol.*, 83, 572, 1962.
3. **Gunner, H. B.**, Nitrification by *Arthrobacter globiformis, Nature (London)*, 197, 1127, 1963.
4. **Aleem, M. I. H., Lees, H., and Lyric, R.**, Ammonium oxidation by cell-free extracts of *Aspergillus wentii, Can. J. Biochem.*, 42, 989, 1964.
5. **Doxtader, K. G. and Alexander, M.**, Nitrification by growing and replacement cultures of *Aspergillus, Can. J. Microbiol.*, 12, 807, 1966.
6. **Molina, J. A. E. and Alexander, M.**, Formation of Nitrate from 3-nitropropionate by *Aspergillus flavus, J. Bacteriol.*, 105, 489, 1971.
7. **Verstraete, W. and Alexander, M.**, Heterotrophic nitrification by *Arthrobacter* sp., *J. Bacteriol.*, 110, 955, 1972.
8. **Verstraete, W. and Alexander, M.**, Mechanisms of nitrification by *Arthrobacter* sp., *J. Bacteriol.*, 110, 962, 1972.
9. **Whittenbury, R. and Kelly, D. P.**, Autotrophy, a conceptual phoenix, *Symp. Soc. Gen. Microbiol.*, 27, 121, 1977.
10. **Quayle, J. R. and Ferenci, T.**, Evolutionary aspects of autotrophy, *Microbiol. Revs.*, 42, 251, 1978.
11. **Painter, H. A.**, A review of literature on inorganic nitrogen metabolism in microorganisms, *Water Res.*, 4, 393, 1970.
12. **Wong-Chong, G. M. and Loehr, R. C.**, The kinetics of microbial nitrification, *Water Res.* 9, 1099, 1975.
13. **Campbell, N. E. R. and Aleem, M. I. H.**, The effect of 2-chloro,6-(trichloromethyl)pyridine on the chemo-autotrophic metabolism of nitrifying bacteria. I. Ammonia and hydroxylamine oxidation by *Nitrosomonas, Antonie van Leeuwenhoek J. Microbiol. Serol.*, 31, 124, 1965.
14. **Chen, R. L., Keeney, D. R., and Konrad, J. G.**, Nitrification in sediments of selected Wisconsin Lakes, *J. Environ. Qual.*, 1, 151, 1972.
15. **Webb, K. L. and Wiebe, W. J.**, Nitrification on a coral reef, *Can. J. Microbiol.*, 21, 1427, 1975.
16. **Moore, P. D.**, Conservation of nitrogen in climax ecosystems, *Nature (London)*, 254, 184, 1975.
17. **Lees, H.**, Energy metabolism in the chemolithotrophic bacteria, *Ann. Rev. Microbiol.*, 14, 83, 1960.
18. **Kiesow, L.**, Energy-linked reactions in chemoautotrophic organisms, *Curr. Top. Bionerg.*, 2, 195, 1967.
19. **Peck, H. D.**, Energy-coupling mechanisms in chemolithotrophic bacteria, *Ann. Rev. Microbiol.*, 21, 489, 1968.
20. **Wallace, W. and Nicholas, D. J. D.**, The biochemistry of nitrifying micro-organisms, *Biol. Rev., Cambridge Philos. Soc.*, 44, 359, 1969.
21. **Aleem, M. I. H.**, Oxidation of inorganic nitrogen compounds, *Ann. Rev. Plant Physiol.*, 21, 67, 1970.
22. **Kelly, D. P.**, Autotrophy: concepts of lithotrophic bacteria and their organic metabolism, *Ann. Rev. Microbiol.*, 25, 177, 1971.

23. Suzuki, I., Mechanisms of inorganic oxidation and energy coupling, *Ann. Rev. Microbiol.*, 28, 85, 1974.
24. Schlegel, H. G., Mechanisms of chemo-autotrophy, in *Marine Ecology*, Vol. 2 (Part 1), Kinne, O., Ed., John Wiley & Sons, London, 1975, 9.
25. Aleem, M. I. H., Coupling of energy with electron transfer reactions in chemolithotrophic bacteria, *Sym. Soc. Gen. Microbiol.*, 27, 351, 1977.
26. Kelly, D. P., Bioenergetics of chemolithotrophic bacteria, in *Companion to Microbiology*, Bull, A. T. and Meadow, P. M., Eds., Academic Press, London, 1978, 363.
27. Suzuki, I., Dular, U., and Kwok, S. C., Ammonia or ammonium ion as substrate for oxidation by *Nitrosomonas europaea* cells and extracts, *J. Bacteriol.*, 120, 556, 1974.
28. Drozd, J. W., Energy coupling and respiration in Nitrosomonas europaea, *Arch. Microbiol.*, 110, 257, 1976.
29. Laudelout, H., Lambert, R., Friapiat, J. L., and Pham, M. L., Effet de la température sur la vitesse d'oxydation de l'ammonium en nitrate par des cultures mixtes de nitrifiants, *Ann. Microbiol. (Inst. Pasteur)*, 125, 75, 1974.
30. Laudelout, H., Lambert, R., and Pham, M. L., Influence du pH et de la pression partielle d'oxygene sur la nitrification, *Ann. Microbiol. (Inst. Pasteur)*, 127, 367, 1976.
31. Henderson, P. J. F., Ion transport by energy conserving biological membranes, *Ann. Rev. Microbiol.*, 25, 393, 1971.
32. Kleiner, D., Ammonium uptake by nitrogen fixing bacteria. 1. Azotobacter vinelandii, *Arch. Microbiol.*, 104, 163, 1975.
33. Kleiner, D., Ammonium uptake and metabolism by nitrogen fixing bacteria, *Arch. Microbiol.*, 111, 85, 1976.
34. Bergter, F., Schumann, H., and Koburger, M., Abhängigkeit der spezifischen Wachstumsrate von der Ammoniumkonzentration bei *Escherichia coli* ML 30, *Z. Allg. Mikrobiol.*, 17, 183, 1977.
35. Stevenson, R. and Silver, S., Methylammonium uptake by *Escherichia coli*: evidence for a bacterial NH_4^+ transport system, *Biochem. Biophys. Res. Commun.*, 75, 1133, 1977.
36. Cole, J. A., Microbial gas metabolism, *Adv. Microb. Physiol.*, 14, 1, 1976.
37. Murray, R. G. E. and Watson, S. W., Structure of *Nitrosocystis oceanus* and comparison with *Nitrosomonas* and *Nitrobacter*, *J. Bacteriol.*, 89, 1594, 1965.
38. Stouthamer, A. H., Energetic aspects of the growth of micro-organisms, *Symp. Soc. Gen. Microbiol.*, 27, 285, 1977.
39. Skinner, F. A. and Walker, N., Growth of *Nitrosomonas* in batch and continuous culture. *Arch. Microbiol.*, 38, 339, 1961.
40. Engel, M. S. and Alexander, M., Growth and autotrophic metabolism of *Nitrosomonas europaea*, *J. Bacteriol.*, 76, 217, 1958.
41. Loveless, J. E. and Painter, H. A., The influence of metal ion concentration and pH value on the growth of a Nitrosomonas strain isolated from activated sludge, *J. Gen. Microbiol.*, 52, 1, 1968.
42. Neijssel, O. M. and Tempest, D. W., Bioenergetic aspects of aerobic growth of *Klebsiella aerogenes* NCTC 418 in carbon-limited and carbon-sufficient chemostat cultures, *Arch. Microbiol.*, 107, 215, 1976.
43. Pirt, S. J., The maintenance energy of bacteria in growing cultures, *Proc. R. Soc. London Ser. B.*, 163, 224, 1965.
44. Meyerhof, O., Untersuchungen uber den Atmungsvorgang nitrifizierenden Bakterien. I. Die Atmung des Nitratbildners, *Pflugers Arch. Gesamte Physiol. Menschen Tiere*, 164, 353, 1916.
45. Van Verseveld, H. W., Meijer, E. M., and Stouthamer, A. H., Energy conservation during nitrate respiration in *Paracoccus denitrificans*. *Arch. Microbiol.*, 112, 17, 1977.
46. Aleem, M. I. H. and Lees, H., Autotrophic enzymic systems. I. Electron transport systems concerned with hydroxylamine oxidation in *Nitrosomonas*, *Can. J. Biochem. Physiol.*, 41, 763, 1963.
47. Rees, M. K. and Nason, A., A P.450 like cytochrome and a soluble terminal oxidase identified as cytochrome *o* from *Nitrosomonas europaea*, *Biochem. Biophys. Res. Commun.*, 21, 248, 1965.
48. Erickson, R. H. and Hooper, A. B., Preliminary characterisation of a variant CO-binding heme protein from *Nitrosomonas*, *Biochim. Biophys. Acta*, 275, 231, 1972.
49. Jones, C. W., Aerobic respiratory systems in bacteria, *Symp. Soc. Gen. Microbiol.*, 27, 23, 1977.
50. Rees, M. and Nason, A., A P450-like cytochrome and a soluble terminal oxidase identified as cytochrome *o* from *Nitrosomonas europaea*, *Biochem. Biophys. Res. Commun.*, 32, 301, 1965.
51. Rees, M. K. and Nason, A., Incorporation of atmospheric oxygen into nitrite formed during ammonia oxidation by *Nitrosomonas europaea*, *Biochim. Biophys. Acta*, 113, 398, 1966.
52. Rees, M. K., Studies of the hydroxylamine metabolism of *Nitrosomonas europaea*, I. Purification of hydroxylamine oxidase. *Biochemistry*, 7, 353, 1968.

53. **Rees, M. K.**, Studies of the hydroxylamine metabolism of *Nitrosomonas europaea*. II. Molecular properties of the electron transport particle, hydroxylamine oxidase, *Biochemistry*, 7, 366, 1968.

54. **Tronson, D. A., Ritchie, G. A. F., and Nicholas, D. J. D.**, Purification of the *c*-type cytochromes from *Nitrosomonas europaea*, *Biochim. Biophys. Acta*, 310, 331, 1973.

55. **Tonge, G. M., Knowles, C. J., Harrison, D. E. F., and Higgins, I. J.**, Metabolism of one carbon compounds: cytochromes of methane- and methanol-utilizing bacteria, *FEBS Lett.*, 44, 106, 1974.

56. **Higgins, I. J. and Quayle, J. R.**, Oxygenation of methane by methane-grown *Pseudomonas methanica* and *Methanomonas methanooxidans*, *Biochem. J.*, 118, 201, 1970.

57. **Hooper, A. B.**, Lag phase of ammonia oxidation by resting cells of *Nitrosomonas europaea*, *J. Bacteriol.*, 97, 968, 1969.

58. **Burton, K. and Krebs, H. A.**, The free energy changes associated with the individual steps on the tricarboxylic acid cycle, glycolysis and alcoholic fermentation and with the hydrolysis of the pyrophosphate groups of adenosinetriphosphate, *Biochem. J.*, 54, 94, 1953.

59. **Hooper, A. B. and Terry, K. R.**, Specific inhibitors of ammonia oxidation in *Nitrosomonas*, *J. Bacteriol.*, 115, 480, 1973.

60. **Wezernak, C. T. and Gannon, J. J.**, Oxygen-nitrogen relationships in autotrophic nitrification, *Appl. Microbiol.*, 15, 1211, 1967.

61. **Quayle, J. R.**, The metabolism of one-carbon compounds by microorganisms, *Ann. Rev. Microbiol.*, 7, 119, 1972.

62. **Ferenci, T., Strom, T., and Quayle, J. R.**, Oxidation of carbon monoxide and methane by *Pseudomonas methanica*, *J. Gen. Microbiol.*, 91, 79, 1975.

63. **Dalton, H.**, Ammonia oxidation by the methane oxidising bacterium *Methylococcus capsulatus* strain Bath, *Arch. Microbiol.*, 114, 273, 1977.

64. **Drozd, J. W., Godley, A., and Bailey, M. L.**, Ammonia oxidation by methane-oxidizing bacteria, *Proc. Soc. Gen. Microbiol.*, 5, 66, 1978.

65. **Colby, J., Stirling, D. I., and Dalton, H.**, The soluble methane mono-oxygenase of *Methylococcus capsulatus* (Bath), *Biochem. J.*, 165, 395, 1977.

66. **Tonge, G. M., Harrison, D. E. F., and Higgins, I. J.**, Purification and properties of the methane mono-oxygenase enzyme system from *Methylosinus trichosporium* OB 3b, *Biochem. J.*, 161, 333, 1977.

67. **Suzuki, I., Kwok, S. C., and Dular, U.**, Competitive inhibition of ammonia oxidation in *Nitrosomonas europaea* by methane, carbon monoxide or methanol, *FEBS Lett.*, 72, 117, 1976.

68. **Taylor, S.**, Evidence for the presence of ribulose 1,5-biophosphate carboxylase and phosphoribokinase in *Methylococcus capsulatus* (Bath), *FEMS Microbiol. Lett.*, 2, 305, 1977.

69. **Colby, J. and Dalton, H.**, Resolution of the methane mono-oxygenase of *Methylococcus capsulatus* (Bath) into three components, *Biochem. J.*, 171, 461, 1978.

70. **Nicholas, D. J. D. and Jones, O. T. G.**, Oxidation of hydroxylamine in cell-free extracts of *Nitrosomonas europaea*, *Nature (London)*, 185, 512, 1960.

71. **Hofman, T. and Lees, H.**, The biochemistry of nitrifying organisms. 4. The respiration and intermediate metabolism of *Nitrosomonas*, *Biochem. J.*, 54, 579, 1953.

72. **Falcone, A. B., Shug, A. L., and Nicholas, D. J. D.**, Oxidation of hydroxylamine by particles from *Nitrosomonas*. *Biochem. Biophys. Res. Commun.*, 9, 126, 1962.

73. **Hooper, A. B. and Nason, A.**, Characterisation of hydroxylamine cytochrome *c* reductase from the chemoautotroph *Nitrosomonas europaea* and *Nitrosocystis oceanus*, *J. Biol. Chem.*, 240, 4044, 1965.

74. **Hooper, A. B., Terry, K. R., and Maxwell, P. C.**, Hydroxylamine oxidoreductase of *Nitrosomonas*. Oxidation of diethyldithiocarbamate concomitant with stimulation of nitrite synthesis, *Biochim. Biophys. Acta*, 462, 141, 1977.

75. **Anderson, J. H.**, The metabolism of hydroxylamine to nitrite by *Nitrosomonas*, *Biochem. J.*, 91, 8, 1964.

76. **Anderson, J. H.**, Studies on the oxidation of ammonia by *Nitrosomonas*, *Biochem. J.*, 95, 688, 1965.

77. **Ritchie, G. A. F. and Nicholas, D. J. D.**, Identification of the sources of nitrous oxide produced by oxidative and reductive processes in *Nitrosomonas europaea*, *Biochem. J.*, 126, 1181, 1972.

78. **Hooper, A. B.**, A nitrite-reducing enzyme from *Nitrosomonas europaea*. Preliminary characterization with hydroxylamine as electron donor, *Biochim. Biophys. Acta*, 162, 49, 1968.

79. **Ritchie, G. A. F. and Nicholas, D. J. D.**, The partial characterisation of purified nitrite reductase and hydroxylamine oxidase from *Nitrosomonas europaea*, *Biochem. J.*, 138, 471, 1974.

80. **Ramaiah, A. and Nicholas, D. J. D.**, The synthesis of ATP and the incorporation of ^{32}P by cell-free preparations from *Nitrosomonas europaea*, *Biochim. Biophys. Acta*, 86, 459, 1964.

81. **Mitchell, P.**, Chemiosmotic coupling in oxidative and photosynthetic phosphorylation, *Biol. Rev. Cambridge Philos. Soc.*, 41, 445, 1966.

82. **Boon, B. and Laudelout, H.,** Kinetics of Nitrite oxidation by *Nitrobacter winogradskyi, Biochem. J.,* 85, 440, 1862.

83. **Cobley, J. G.,** Energy-conserving reactions in phosphorylating electron transport particles from *Nitrobacter winogradskyi, Biochem. J.,* 156, 481, 1976.

84. **Cobley, J. G.,** Reduction of cytochromes by nitrite in Electron-transport particles from *Nitrobacter winogradskyi, Biochem. J.,* 156, 493, 1976.

85. **Gibbs, M. and Schiff, J. A.,** Chemosynthesis: the energy relations of chemoautotrophic organisms, in *Plant Physiology,* Steward, F. C., Ed., Academic Press, New York, 1960, 279.

86. **Chance, B.,** The interaction of energy and electron transfer reactions in mitochondria. II. General properties of adenosine triphosphate linked oxidation of cytochrome and reduction of pyridine nucleotides. *J. Biol. Chem.,* 236, 1544, 1961.

87. **Aleem, M. I. H.,** Generation of reducing power in chemosynthesis. II. Energy-linked reduction of pyridine nucleotides in the chemoautotroph *Nitrosomonas europaea, Biochim. Biophys. Acta,* 113, 216, 1966.

88. **Jones, C. W., Brice, J. M., Downs, A. J., and Drozd, J. W.,** Bacterial respiration driven proton translocation and its relationship to respiratory chain composition, *Eur. J. Biochem.,* 52, 265, 1975.

89. **Mitchell, P.,** The proton-motive Q cycle: a general formulation, *FEBS Lett.,* 59, 137, 1975.

90. **Poole, R. K. and Haddock, B. A.,** Energy-linked reduction of nicotinamide adenine dinucleotide in membranes derived from normal and various respiratory-deficient mutant strains of *Escherichia coli* K 12, *Biochem. J.,* 144, 77, 1974.

91. **Shahak, Y., Hardt, H., and Avron, M.,** Acid-base driven reverse electron flow in isolated chloroplasts, *FEBS Lett.,* 54, 151, 1975.

92. **Harrison, D. E. F.,** The regulation of respiration rate in growing bacteria, *Adv. Microb. Physiol.,* 14, 243, 1976.

93. **Jones, C. A. and Kelly, D. P.,** Energetics of *Thiobacillus ferrooxidans* grown on ferrous iron in the chemostat, *Proc. Soc. Gen. Microbiol.,* 4, 73, 1977.

94. **Anthony, C.,** The prediction of growth yields in methylotrophs, *J. Gen. Microbiol.,* 104, 91, 1978.

95. **Ackrell, B. A. C., Erickson, S. K., and Jones, C. W.,** The respiratory chain NADPH dehydrogenase of *Azotobacter vinelandii, Eur. J. Biochem.,* 26, 22, 1972.

96. **Laudelout, H., Simonart, P-C, and van Droogenbroeck, R.,** Calorimetric measurement of free energy utilisation by *Nitrosomonas* and *Nitrobacter, Arch. Microbiol.,* 63, 256, 1968.

97. **Petersen, Lars. Chr.,** On the Mechanism of the cytochrome *c* oxidase reaction, *Eur. J. Biochem.,* 85, 339, 1978.

98. **Wikstrom, M. K. F. and Saari, H. T.,** The mechanism of energy conservation and transduction by mitochrondrial cytochrome *c* oxidase, *Biochim. Biophys. Acta,* 462, 347, 1977.

99. **Wikstrom, M. and Krab, K.,** Cytochrome *c* oxidase is a proton pump, *FEBS Lett.,* 91, 8, 1978.

100. **Mitchell, P.,** Membranes of cells and organelles, *Symp. Soc. Gen. Microbiol.,* 20, 121, 1970.

101. **Drozd, J. W.,** unpublished data.

102. **Drozd, J. W., Godley, A. R., and Bailey, M. C.,** unpublished observation.

103. **Hynes, R. K. and Knowles, R.,** Inhibition by acetylene of ammonia oxidation in *Nitrosomonas europaea, FEMS Microbiol. Lett.,* 4, 319, 1978.

104. **Drozd, J. W., Bailey, M. L., and Godley, A. R.,** unpublished data.

Chapter 6

RESPIRATION IN CHEMOAUTOTROPHS OXIDIZING SULFUR COMPOUNDS

J. K. Oh and I. Suzuki*

TABLE OF CONTENTS

* This work was supported by a grant from the National Research Council of Canada.

I. INTRODUCTION

The thiobacilli, well-known chemoautotrophic or chemolithotropic bacteria, have a unique mode of life among living organisms in that they have the capacity to derive their energy and reducing power for growth from the oxidation of inorganic sulfur compounds and to synthesize all the cellular carbon by assimilation of atmospheric CO_2.

These sulfur bacteria can be generally classified as obligate, facultative, or mixotrophic chemoautotrophs, depending on the nutritional response to organic compounds in the growth medium. Some thiobacilli, such as *Thiobacillus thiooxidans*, *Thiobacillus concretivorus*, and *Thiobacillus ferrooxidans*, grow under an extremely acid environment (pH 1.0 to pH 4.5), while others grow near the neutral pH of mineral-salt media (pH 6.5 to pH 7.5). In nature, they live freely in the soil or in water. In spite of the apparent simplicity of the mode of life, thiobacilli are known to be very complex and sophisticated microorganisms physiologically or biochemically.

The chemolithotropic sulfur bacteria oxidize reduced or partially reduced inorganic sulfur compounds, namely elemental sulfur (S°), sulfide (S^{2-}), sulfite (SO_3^{2-}), thiosulfate (SSO_3^{2-}), and possibly polythionate ($^-O_3SS_nSO_3^-$), to sulfate (SO_4^{2-}), with molecular oxygen or nitrate. ATP is produced by oxidative phosphorylation and substrate-level phosphorylation. It is generally believed that the NADH needed for CO_2 assimilation is supplied by the energy-driven reversal of electron flow in the respiratory chain. Although our understanding of this biological oxidation process has substantially advanced during the last two decades, especially in the sulfite-oxidizing system, only a basic outline for the oxidative mechanism of sulfur compounds has been elucidated. In particular, relatively little is known about the biochemical details of electron transport, energy coupling, and reduction of pyridine nucleotides associated with the oxidation of inorganic sulfur compounds.

A number of general reviews and monographs relating to sulfur metabolism have been published in the recent years.[1-5] The more specialized topics of the mechanism of oxidation and energy-coupling have been reviewed by Peck,[6] Suzuki,[7] and Aleem.[8] Reviews of related topics, such as autotrophy, respiratory chain, evolution, and microbial biogeochemistry, have also appeared in the literature.[9-19]

The purpose of this review is to discuss the current state of knowledge on the oxidative mechanism of inorganic sulfur compounds in thiobacilli.

II. SOME ASPECTS OF RESPIRATORY CHAIN AND ENERGY COUPLING

The oxidation of inorganic sulfur compounds is ultimately linked by the electron transport system to oxygen as a terminal electron acceptor or to nitrate in the case of *Thiobacillus denitrificans*.[5-8] The respiratory chain in thiobacilli seems to resemble those in the inner mitochondrial membrane of higher organisms and in some bacteria, such as *Mycobacterium phlei*, *Bacillus subtilis*, and *Paracoccus denitrificans*, possessing mitochondrion-like respiratory chains.[20-23]

Much of the evidence for determining the sequence and role of respiratory chain components, however, has been obtained from spectrophotometric and inhibition studies with intact cells and cell-free extracts of thiobacilli. In such studies, we have to realize that the mechanism of action of many inhibitors is not yet fully understood, and a selective blocking of electron transport at certain sites by certain inhibitors may not be universal among all electron transport systems. Also in most such studies on thiobacilli, there is still much uncertainty regarding the stoichiometry and nature of sulfur compounds involved in the oxidation.

More direct evidence, the isolation and characterization of individual electron transport components of thiobacilli, is rather limited except for some c-type cytochromes and adenosine phosphosulfate (APS) reductase at the flavin level. The nature of specific electron transport components studied concerning the oxidation of individual inorganic sulfur compounds will be discussed in their respective sections later. The only well-characterized system, the components involved in sulfite oxidation by *Thiobacillus novellus*, will be described in this section in addition to some aspects of general importance.

A large number of c-type cytochromes have been isolated from various thiobacilli. Their physicochemical properties are in Table 1.

T. novellus is the most extensively studied organism, with two types of cytochrome c, cytochromes c_{550}, and c_{551}.[24,34-36] Highly purified cytochrome c_{550} shows physicochemical and enzymatic properties similar to mammalian cytochrome c.[24] It reacts fairly rapidly with cow cytochrome oxidase (ferrocytochrome c: oxygen oxidoreductase, EC 1.9.3.1), very rapidly with either yeast cytochrome c peroxidase (ferrocytochrome c: hydrogen-peroxide oxidoreductase, EC 1.11.1.5) or sulfite:cytochrome c oxidoreductase (sulfiterricytochrome c oxidoreductase, EC 1.8.2.1) of *T. novellus* and poorly with *Pseudomonas aeruginosa* cytochrome oxidase (nitrite reductase, ferrocytochrome c_2:oxygen oxidoreductase, EC 1.9.3.2). It differs from mammalian cytochrome c in having a lower isoelectric point (IEP = 7.5). Among many c-type cytochromes tested by Yamanaka,[24,34,35] this cytochrome c_{550} is the only bacterial cytochrome c that reacts rapidly with the mammalian cytochrome oxidase, but does not react with the bacterial-type terminal oxidase, nitrite reductase. In contrast, cytochrome c_{551} does not react with any of the enzymes mentioned above, except sulfite:cytochrome c oxidoreductase. It has an IEP of pH 5.2.

The sulfite:cytochrome c oxidoreductase of *T. novellus*, originally studied by Charles and Suzuki,[37] can use mammalian cytochrome c as well as native cytochrome c_{550} and c_{551} as electron acceptor, but not bacterial c-type cytochromes, namely *P. aeruginosa* c_{551} and *Pseudomonas stutzeri* c_{552}.[24] This reactivity with c-type cytochromes is identical to that of cow cytochrome oxidase, yeast cyochrome peroxidase, or native *T. novellus* cytochrome oxidase (cytochrome a).[35,36,38] Yamanaka and co-workers,[17,24,36,38] based on this informations, have suggested that *T. novellus* is an evolutionally advanced organism among bacteria.

The following electron transfer system is envisaged for the oxidation of sulfite to sulfate by *T. novellus*:[24] SO_3^{2-} → sulfite:cytochrome c oxidoreductase → cytochrome c_{551} → cytochrome c_{550} → cytochrome a → O_2.

Although some c-type cytochromes act as natural electron acceptors for the purified thiosulfate-oxidizing enzyme (*Thiobacillus neapolitanus* $c_{553.5}$),[26] sulfite-linked nitrate reductase (*T. denitrificans* c_{554}),[29] sulfide-linked nitrite reductase (*T. denitrificans* cd complex),[30] and sulfite-oxidizing system (*T. novellus* c_{550}, c_{551}, and *T. concretivorus* c_{550}),[24,33] the physiological functions of other cytochromes of c-type have not yet been determined.

A dissimilatory sulfite reductase (hydrogen sulfideceptor oxidoreductase, EC 1.8.99.1), containing siroheme as a prosthetic group, has been partially purified from *T. denitrificans*[39] and *Thiobacillus thioparus*[40] grown on thiosulfate. The siroheme is known to be involved in the six-electron reduction of both sulfite and nitrite in many microorganisms and higher plants.[41-43] The physiological role of the hemoprotein in thiobacilli is not known at this time. Since both organisms contain relatively high concentrations of the hemoprotein, this multielectron carrier may somehow be involved in the oxidation of reduced sulfur compounds in vivo.

It is now well recognized that electron transport during the aerobic or anaerobic oxidation of inorganic sulfur compounds in thiobacilli is coupled to energy generation

Table 1
PHYSICOCHEMICAL PROPERTIES OF C-TYPE CYTOCHROMES ISOLATED FROM VARIOUS THIOBACILLI

| Organism | Name of c type | Absorption bands | | | | Molecular weight | IEP | E_0' (mV) | Ref. |
| | | Reduced | | | Oxidized | | | | |
		α	β	γ	γ				
Thiobacillus novellus	c_{550}	550	520	414.5	410	13,270	7.5	+276	24
	c_{551}[a]	551	522	416	410.5	N.D.[b]	5.1	+260	24
Thiobacillus thioparus	c_{551} (s)	551[c]	526[c]	414[c]	410[c]	N.D.[b]	N.D.[b]	+145	25
Thiobacillus x (Thiobacillus neapolitanus)	c_{550}	550	521	416	409.5	N.D.[b]	>7.0	+200	26
	$c_{553.5}$	553.5	524	418	410.5	N.D.[b]	<7.0	+210	26
	c_{557}	557	525	419	409	N.D.[b]	N.D.[b]	N.D.[b]	26
Thiobacillus denitrificans	c_{552}	552	522	416	410	N.D.[b]	10.2	+273	27, 28
	c_{551}	551	521	412	408	N.D.[b]	N.D.[b]	N.D.[b]	27
	c_{554}[d]	554	524	417	N.D.[b]	13,000	N.D.[b]	+223	29
	cd[e]	549—554	523	418	405	N.D.[b]	N.D.[b]	N.D.[b]	30
Thiobacillus thiooxidans	c_{552}	552	523	418	412	12,600	9.2	+247	31
	c_{550}	550	521	415	413	N.D.[b]	10.6	+253	32
Thiobacillus concretivorus	c_{550}	550	522	414	N.D.[b]	N.D.[b]	N.D.[b]	N.D.[b]	33

Note: IEP, isoelectric point. E_0', standard redox potential at pH 7.

[a] Sulfite:cytochrome c oxidoreductase component.
[b] N.D., Not determined.
[c] Estimated from the graphs in the original literature.
[d] Sulfite-linked nitrate reductase component.
[e] Sulfide-linked nitrite reductase component.

by oxidative phosphorylation and to generation of reduced pyridine nucleotide necessary for cellular biosynthesis. The coupling mechanism has been extensively discussed,[6-8] and only a brief discussion on the more recent work will be presented here.

Oxidative phosphorylation has been demonstrated by measuring ATP formation in cell-free extracts of *T. novellus*,[44-46] *T. neapolitanus*,[47,48] *T. thioparus*,[49] and *T. concretivorus*[50] during the oxidation of thiosulfate, sulfite, sulfide, NADH, succinate, or ascorbate. The reported P/O ratios, however, appear to vary greatly. The P/O ratios with thiosulfate, sulfite, sulfide, NADH, succinate, and ascorbate are 0.4 to 0.96, 0.1 to 0.13, 0.3 to 1.4, 0.3 to 3.4, 1.9, and 0.76 to 0.9, respectively. The lower level of P/O ratios obtained compared to the mitochrondrial system is reminiscent of other bacterial oxidative phosphorylation systems.[21,51] In general, ATP synthesis coupled to sulfur compound oxidation is sensitive to uncouplers of oxidative phosphorylation,[45,47,48,52] but the question of the location and number of energy coupling sites for each sulfur compound still remains open, since the point of entry for electrons for many substrates and the nature of intermediates or primary products are not yet fully known. Here again caution is necessary in interpreting the results based solely on inhibitor studies. It seems to be generally accepted, however, that electrons from either sulfite with sulfite:cytochrome *c* oxidoreductase or thiosulfate with thiosulfate-oxidizing enzyme enter at the cytochrome *c* level and are transferred to the terminal site of the respiratory chain for oxidative phosphorylation.[6-8]

It has been reported by Cole and Aleem that a "soluble" oxidative phosphorylation system of *T. novellus* does not carry out an acid-induced ATP formation,[45,46] while *T. neapolitanus* cells as well as cells and extracts of *T. novellus* carry out such a reaction,[46,48] first reported in chloroplasts by Jagendorf and Uribe.[53] It is possible, however, that the "soluble" system may have contained small membrane fragments with a high content of lipid which could be reassociated into larger particles during incubation with substrate and cofactors.[7]

Recently respiration-driven proton translocation, according to the Mitchell's chemiosmotic hypothesis[54-56] of oxidative phosphorylation, has been demonstrated in intact cells of *T. neapolitanus* with thiosulfate, sulfide, sulfite, or ascorbate-TMPD as electron donor.[57,58] The H^+/O ratio approaches a value of 2.0, i.e., P/O ratio of 1.0, assuming an H^+/P of 2.0, with all these compounds. The results suggest the operation of one energy coupling site. An outward proton translocation coupled to thiosulfate oxidation has also been shown in *T. denitrificans* cells using a quinacrine fluorescence technique.[59] There is a close coupling between the electron transport and proton translocation, although uncouplers of oxidative phosphorylation inhibit only the latter. Thus, these preliminary studies favor the proton-translocation mechanism of energy coupling in the oxidation of sulfur compounds by thiobacilli, basically similar to that of the mitochrondrial system. Various questions regarding the quantitative energy relationship, stoichiometry, membrane energization, and actual coupling device, however, have to be answered in future work.

T. thiooxidans contains two membrane-bound ATPase (AP phosphohydrolase, EC 3.6.1.3), namely Mg-ATPase and sulfite-dependent ATPase.[60] The sulfite-dependent ATPase can be separated by treatment with trypsin from the Mg-ATPase.[61] A soluble ATPase has also been isolated and characterized from *T. ferrooxidans*.[62] It should be interesting to study their possible roles in oxidative phosphorylation and active ion transport. A soluble inorganic pyrophosphatase (pyrophosphate phosphohydrolase, EC 3.6.1.1) and a membrane-bound, sulfate-dependent acid phosphatase (ortho-phosphoric-monoester phosphohydrolase, EC 3.1.3.2) have also been purified and characterized in *T. thiooxidans*, as well as *T. ferrooxidans*, although their roles in thiobacilli are not certain.[63-65]

Substrate-level phosphorylation coupled to sulfite oxidation (APS reductase pathway) which is insensitive to uncouplers will be discussed in the sulfite oxidation section.

The energy-dependent reduction of pyridine nucleotides coupled to the oxidation of thiosulfate and sulfide was first demonstrated in *T. novellus* and later in *T. neapolitanus.*[66-69] The process generates NADH or NADPH, only under anaerobic conditions or in the presence of azide, by the ATP-driven reversed electron flow. The rate of reduction, however, is low compared to that of mitochondria[70] and NAD^+ is reduced faster than $NADP^+$. NADH formed is reoxidized during the reduction of 3-phosphoglyceric acid to 3-phosphoglyceradehyde. The system is inhibited by flavoprotein inhibitors, antimycin A, and uncouplers of oxidative phosphorylation. One of the unexplained features of these reports is the amount of energy required for NAD reduction by thiosulfate; 2 to 3 mol of ATP for each mole of NAD reduced in *T. neapolitanus*, but only 1 mol of ATP per mole of NAD in *T. novellus*. In addition "soluble" fractions can catalyze these reactions in *T. novellus* and *T. neapolitanus.*[66,68] Sadler and Johnson[71] have reported that in *T. neapolitanus* NADH oxidase, containing the electron transport chain, resides in the particulate fraction. Since the organism contains a low-potential *c*-type cytochrome (c_{554}), a possibility for direct reduction of pyridine nucleotides coupled to the oxidation of inorganic sulfur compounds has been suggested. Low potential c_3-type cytochromes ($E_o' = -250$ mV) are present in some photosynthetic and sulfate-reducing bacteria.[72] The proposal, however, is believed unlikely from a thermodynamic consideration.[5] In addition the standard redox potential (E_o') of the cytochrome c_{554} has not been determined.[5]

Since these organisms are strict aerobes and NADH can also be oxidized by NADH oxidase located in the same respiratory chains,[10,66,71,73] most thiobacilli may not have reversed electron flow under physiological conditions unless they have a well-controlled mechanism in the cell. For example, such a regulatory mechanism operates in *T. denitrificans*, where NAD and 5'-AMP inhibit a particulate NADH-linked nitrate reductase, but not the sulfite-linked nitrate reductase.[74] Obviously much more work is necessary to clarify the mechanism of reduction of pyridine nucleotides.

III. SULFUR OXIDATION

Although many thiobacilli are able to obtain energy for growth through the oxidation of elemental sulfur to sulfuric acid, the most intensive investigations on the mechanism of sulfur oxidation have been conducted with *T. thiooxidans* which was first isolated by Waksman and Joffe in 1922.[75] The early work with the intact cells of *T. thiooxidans* established that elemental sulfur is oxidized aerobically to sulfuric acid as follows:[76-79]

$$S^o + 1\tfrac{1}{2}O_2 + H_2O \longrightarrow SO_4^{2-} + 2H^+$$

It is generally agreed that sulfite is the key intermediate in this oxidation:[6,7]

$$S^o + O_2 + H_2O \longrightarrow SO_3^{2-} + 2H^+$$
$$SO_3^{2-} + \tfrac{1}{2}O_2 \longrightarrow SO_4^{2-}$$

Three types of sulfur-oxidizing systems have been prepared from cell-free extracts of the organism:[7]

1. Sulfur-oxidizing enzyme (sulfur:oxygen oxidoreductase, EC 1.13.11.18) which catalyzes the oxidation of elemental sulfur to sulfite in the presence of a catalytic amount of reduced glutathione (GSH)[80] (It is found in a soluble fraction of cell-free extracts.)

$$S^0 + O_2 + H_2O \xrightarrow{GSH} SO_3^{2-} + 2H^+$$

2. A large cell-wall membrane complex which catalyzes the oxidation of elemental sulfur (presumably to sulfate) as in intact cells[81,82]
3. The sulfur-oxidizing system which catalyzes the oxidation of sulfur to sulfate and requires both soluble and membrane fractions[83,84]

A sulfur-oxidizing enzyme has been partially purified, and the following mechanism has been proposed for the reaction catalyzed by the enzyme:[80,85]

$$S_n + GSH \longrightarrow GSS_nH$$

$$GSS_nH + O_2 + H_2O \xrightarrow{enzyme} GSS_{n-1}H + SO_3^{2-} + 2H^+$$

where S_n represents a polymerized molecular state (S_8 ring structure) of elemental sulfur. S_n is first converted to GSS_nH, glutathione polysulfide, by a nucleophilic attack with GSH. GSS_nH is then oxidized to sulfite by the sulfur-oxidizing enzyme. Sulfite has been identified as the initial product of sulfur oxidation in *T. thiooxidans* and *T. thioporus* by formaldehyde trapping.[85] Finally, thiosulfate is formed through a nonenzymatic condensation of sulfur and sulfite under the assay conditions. In the cells, sulfite would be further oxidized to sulfate via sulfite-oxidizing systems, as discussed later. In this proposed reaction mechanism of sulfur oxidation, sulfur atoms can be converted successively to sulfite, while GSH will be regenerated for a further attack on S_n. *T. thiooxidans* has glutathione reductase which would regenerate GSH from any oxidized glutathione (GSSG) which might be formed during sulfur oxidation:[86]

$$GSSG + NADPH + H^+ \longrightarrow 2\,GSH + NADP^+$$

The sulfur-oxidizing enzyme contains iron as a cofactor, apparently nonheme iron in nature, and removal of the metal with 2,2'-dipyridyl results in relatively inactive enzyme, suggesting that the iron is essential for the enzyme action.[85] An oxygenase nature of the sulfur-oxidizing enzyme has been tentatively identified with the use of $^{18}O_2$, although the amount of ^{18}O incorporated into thiosulfate during sulfur oxidation is relatively low.[87] Whether or not the sulfur-oxidizing enzyme functions in vivo as a dioxygenase (as in vitro experiments indicate) remains to be established in future studies. It is conceivable that this enzyme may couple in the cells to the respiratory chain by some unknown mechanism (before the final formation of sulfite). Unfortunately, little is known about the interactions between this soluble system and the respiratory chain of intact cells. Further investigation with the enzyme of high purity in sufficient quantities appears to be necessary to identify the type of reaction and to understand better the mechanism of the sulfur oxidation process at the molecular level.

The sulfur-oxidizing enzyme has also been isolated and characterized in *T. thiopa-*

rus, *T. novellus*, and *T. ferrooxidans*, and the enzymes properties appear to be similar to that of *T. thiooxidans.*[80,85,88,89] The enzyme has been resolved as one of the components for the membrane-bound thiosulfate-oxidizing complex of *T. novellus.*[90]

These findings seem to elevate elemental sulfur or colloidal sulfur to an important position in the oxidation of the outer sulfur moiety, after the cleavage of thiosulfate to sulfur and sulfite by all thiobacilli, as discussed later.[85]

In contrast to the soluble system, sulfur oxidation by the cell wall-membrane system from *T. thiooxidans* and *T. neapolitanus* prepared by a French Pressure Cell is similar to the oxidation by intact cells in not requiring addition of a sulfhydryl compound such as GSH.[81,82] The inhibition by thiol-binding agents, however, suggests that they have protein-bound active sulfhydryl groups which may substitute for the GSH requirement of the soluble sulfur-oxidizing enzyme. Cyanide and azide may be affecting overall sulfur oxidation by stopping the oxidation of sulfite (the product of the sulfur-oxidizing enzyme) to sulfate which involves the cytochrome chain, as described in the later section.[81] Ubiquinone has been proposed as a component of the complex in *T. thiooxidans.*[91]

A sulfur-oxidizing system which requires both soluble and membrane fractions has been prepared by sonication of *T. thiooxidans* cells under a nitrogen atmosphere.[83,84]

The soluble fraction has been fractionated into collodion membrane-permeable and membrane-impermeable components, the former being replaceable by NAD$^+$ or NADP$^+$, but not by GSH or cysteine. Only the membrane fraction has an ability to oxidize sulfite, NADH, and NADPH. In this system therefore it appears that the sulfur-oxidizing enzyme capable of oxidizing sulfur to sulfite is probably present in the soluble fraction and upon reconstitution, is coupled to the sulfite-oxidizing system in the membrane fraction to achieve the oxidation of sulfur to sulfuric acid.

Recently the soluble membrane-impermeable fraction has been purified into components A and B.[92] The larger component A is a nonheme iron protein with a molecular weight of 120,000 and an absorption maximum at 410 nm in the oxidized form which shifts to 420 nm upon reduction. The smaller component B has an absorption maximum at 410 nm, with a shoulder at 485 nm in the oxidized state. The shoulder disappears upon reduction. Component B has been identified as a flavoprotein containing nonheme iron with a molecular weight of 23,000.

In reconstitution experiments, removal of iron by KCN or diethyldithiocarbamate decreases the enzyme activity, indicating that non-heme iron is necessary for sulfur oxidation. This is similar to the purified sulfur-oxidizing enzyme of *T. thioparus* which contains tightly bound nonheme iron, as well as labile sulfur,[85] although the soluble sulfur-oxidizing enzyme activity and the content of labile sulfur have not been determined in components A and B. Inhibition and spectrophotometric studies have clearly indicated the involvement of the electron transport system containing *a*, *b*, and *c*-type cytochromes in the oxidation of sulfite to sulfate by the membrane fraction, while the nature of the electron transport in the oxidation of elemental sulfur to sulfite has not been elucidated. One of the most interesting features of the system is that the electron transport mechanism operating in sulfur oxidation distinctly differs from the classical cytochrome oxidase in that carbon monoxide inhibition is not reversed by light whereas the binding of carbon monoxide to the terminal oxidase in sulfite oxidation is photo-reversible.

More recently the sulfur-oxidizing enzyme has been reinvestigted in our laboratory by measuring the oxygen uptake polarographically with a Clark oxygen electrode.[94] Compared to the Warburg manometric method originally used for the investigation,[80,85] this method is more sensitive and rapid. Catalase and 2,2'-dipyridyl, used to protect GSH from nonenzymatic oxidation, are no longer required. The sulfur-oxidizing activity is found both in the soluble (66%) and membrane fractions (33%), whereas the sulfite-oxidizing activity is associated solely with the $150,000 \times g$ membrane fraction. The property of the soluble fraction is basically the same as that of the sulfur-oxidizing enzyme previously described.[80,85] It has been shown that GSH can be replaced by dithiothreitol or dithioerythritol. The membrane fraction, on the other hand, appears to resemble other membrane-bound sulfur-oxidizing systems, with the exception of the GSH requirement for sulfur oxidation.[81,82] Presumably, this membrane system catalyzes a complete oxidation of sulfur to sulfate, since no thiosulfate is produced and both sulfur and sulfite are oxidized by the same fraction.

In summarizing our present knowledge of the oxidation of elemental sulfur to sulfuric acid with sulfite as intermediate, the central role of the sulfur-oxidizing enzyme seems to remain unaffected by the isolation of these membrane-associated sulfur-oxidizing systems. Differences may simply be due to the presence or absence of an essential sulfhydryl group-generating system. In its absence, external addition of GSH or a similar compound will be required. In addition the membrane systems are capable of oxidizing sulfite, the product of sulfur oxidation, leading to complete oxidation of sulfur to sulfate. It will be interesting to study the connection between sulfite oxidation and generation or maintenance of essential sulfhydryl groups.

Despite extensive studies for a long period of time, very little information is available with regard to the mechanism of attack on elemental sulfur particles by bacterial cells in cultures of *T. thiooxidans* and the mobilization of this insoluble substrate to the essential enzyme systems. Umbreit and co-workers[95-98] reported that direct contact between terminal fat globules of bacterial cells and sulfur particles was necessary in order to dissolve the elemental sulfur for oxidation. In electron microscopic studies, however, they failed to confirm the presence of such fat globules, and a later work by Knaysi[99] indicated that the fat globules actually consisted of volutin and sulfur. Another concept is that both the attachment of the organisms to sulfur and solubilization of the sulfur particles by phospholipids and other extracellular compounds released by the cells are obligatory for oxidation.[100-102] This theory has been supported by the fact that cells have been photographed clustered around eroded sulfur particles in the microscopic studies, and a brief stationary phase after incubation of a culture appears to be necessary for the production of phospholipids or other materials required for adhesion.[103] The mechanism based on this theory, however, appears to be a complex one, and the basic problem has not yet been thoroughly explored. A further complication comes from the fact that whether the sulfur is oxidized directly at the outer membrane surface or after its translocation into the cell is still unknown.

IV. SULFIDE OXIDATION

The oxidation of sulfide to sulfate is a process requring an eight-electron transfer:

$$S^{2-} + 2 O_2 \longrightarrow SO_4^{2-}$$

As in the case of the sulfur-oxidizing system, it is apparent that polysulfide and sulfite are involved as intermediates in the process:

$$S^{2-} + \tfrac{1}{2}O_2 + 2H^+ \longrightarrow [S] + H_2O$$

$$[S] + O_2 + H_2O \longrightarrow SO_3^{2-} + 2H^+$$

$$SO_3^{2-} + \tfrac{1}{2}O_2 \longrightarrow SO_4^{2-}$$

where [S] represents a polysulfide-sulfur. Thus, the mechanism of sulfide oxidation in thiobacilli is analogous to that for the oxidation of colloidal sulfur (polysulfide) or elemental sulfur once sulfide is converted to the oxidation level of sulfur.

Early studies on growth and the mode of sulfide oxidation with both intact cells and crude extracts led to the hypothesis that elemental sulfur, thiosulfate, and polythionates were the intermediate products during the oxidation of sulfide to sulfate in *T. thiooxidans*.[79,104-106] While the cell-free extracts of *T. thiooxidans* were shown to oxidize sulfide with a concomitant reduction of cytochromes,[107] the oxidation of sulfide was considered by some workers to be a nonenzymatic process not involving cytochromes.[81] The chemical reactivity and instability of sulfide was an inherent problem in the enzymatic study of its oxidation.

Nevertheless, a recent study by Moriarty and Nicholas[33,108] has demonstrated that intact cells as well as cell-free extracts of *T. concretivorus*, *T. thiooxidans*, and *T. thioparus* grown on elemental sulfur as energy source catalyze an enzymatic oxidation of sulfide. The sulfide oxidase from *T. concretivorus* extracts prepared in a French Pressure Cell is associated with the membrane fraction (the pellet of 144,000 × g for 1 hr). Sulfide is rapidly converted in two stages to a membrane-bound polysulfide, with the electrons being transferred through the cytochrome system to oxygen. The first stage (disappearance of soluble sulfide) is inhibited by a copper-chelating agent, sodium diethyldithiocarbamate, and Tris-HCl, but not by carbon monoxide and azide. The second stage is photoreversibly inhibited by carbon monoxide. A native cytochrome *c* purified from the extracts has been identified as the CO-binding site for the inhibition of polysulfide formation. The membrane-bound polysulfide is slowly oxidized to sulfate upon the addition of dialyzed extracts. In a subsequent study on the membrane system, two electron-transfer pathways have been identified;[50] cytochromes of the *b*, *c*, and *d* types involved in sulfide oxidation and cytochromes of *b*, *c*, and *a* types in sulfite oxidation. Oxidative phosphorylation occurs concomitantly with sulfide or sulfite oxidation. The involvement of both flavin and ubiquinone as possible components of the respiratory chain in the sulfide oxidation is suggested, and a copper protein is implicated as a binding site for sulfide. From these results, a tentative electron transfer scheme has been presented for sulfide oxidation to sulfur (polysulfide):[50]

$$S^{2-} \rightarrow \text{Cu protein} \rightarrow \text{(flavin?)} \rightarrow \text{cytochromes } b \rightarrow c \rightarrow d \rightarrow O_2$$

A similar sulfide-oxidizing system has been reported in *T. neopolitanus*,[48] where sulfide oxidation proceeds also in two steps; a rapid step catalyzed by a membrane fraction and a second slower step requiring a soluble fraction. As has been suggested previously,[7] the second step of sulfide oxidation is probably similar to the reaction catalyzed by the soluble sulfur-oxidizing enzyme system in *T. thiooxidans*, *T. thioparus*, and *T. ferrooxidans*.[80,85,89]

All the investigations described above seem to indicate that the membrane-bound sulfur polymer would be oxidized to sulfate by a sulfur-oxidizing enzyme and sulfite-oxidizing system through a mechanism similar to those described by Suzuki and associates.[80,85,88] A further study on the resolution and identification of individual enzymes involved will be required before a definite conclusion can be made.

In *T. denitrificans*, sulfide is oxidized by intact cells with either molecular oxygen or nitrate as the terminal electron acceptor, although the anaerobic oxidation rate of sulfide is only half that of thiosulfate, and the aerobic rate is 40% of thiosulfate oxidation rate.[109]

A recent work by Aminuddin and Nicholas[110] with intact cells and extracts of the organism shows the sulfide oxidation to be linked to the reduction of nitrite and nitrate. Cell suspensions reduce nitrite under anaerobic conditions to nitrogenous gases, such as NO, N_2O, and N_2, with sulfide as electron donor. The work with extracts shows that sulfide is first oxidized to the membrane-bound polyslfide, as in *T. concretivorus*, then to sulfite and sulfate. Sulfide is oxidized to polysulfide only in air or under anaerobic conditions in the presence of nitrite by a nitrite reductase associated with the membrane fraction. A nitrate reductase, which catalyzes the reduction of nitrate to nitrite with a concomitant oxidation of sulfite to sulfate, is also located in the same membrane. The scheme for the mechanism of sulfide oxidation in this system is shown below:

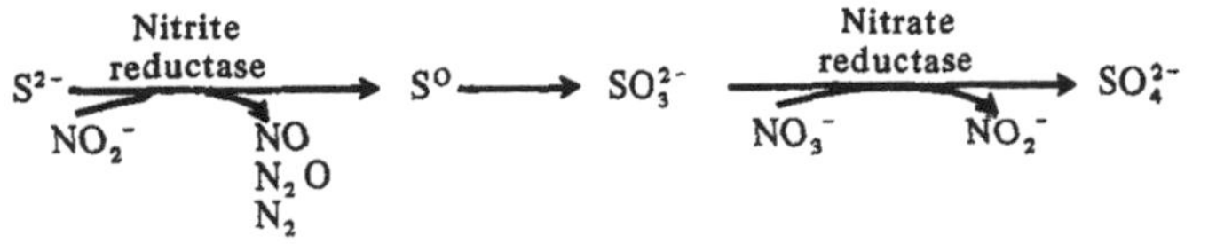

$$S^{2-} \xrightarrow[\substack{NO_2^- \quad NO \\ N_2O \\ N_2}]{\substack{\text{Nitrite} \\ \text{reductase}}} S^0 \longrightarrow SO_3^{2-} \xrightarrow[\substack{NO_3^- \quad NO_2^-}]{\substack{\text{Nitrate} \\ \text{reductase}}} SO_4^{2-}$$

Sulfite may also be oxidized by the APS reductase present in this organism.[110,111] Although the soluble and membrane fractions have not been further fractionated and the specific enzymes involved have not been identified completely, the sulfur-oxidizing enzyme is probably present in the soluble fraction as described earlier. In a later study, an electron transfer mechanism has been proposed for sulfide oxidation that involves flavin, copper, and cytochrome *cd* complex as a terminal oxidase.[112] Cytochromes of the *c* and *d* types in the membrane fraction are reduced by sulfide and reoxidized by either air or nitrite, and inhibitor studies support the view that sulfide oxidation is linked to oxygen or nitrite via a respiratory chain with the above components.

In a very recent study, Sawhney and Nicholas[30] have succeeded in solubilization and purification of the sulfide-linked nitrite reductase from *T. denitrificans*. The purified enzyme reduces nitrite to NO and N_2O with sulfide as the electron donor. It has a molecular weight of 120,000 and contains cytochromes *c* and *d* in the ratio of 1:1 which can be resolved into each component with sodium dodecyl sulfate. A kinetic study indicates that cytochrome *c* is reduced by sulfide before cytochrome *d*, while cytochrome *d* is oxidized by nitrite before cytochrome *c*. The enzyme also functions as a cytochrome oxidase as do the nitrite reductases from *Micrococcus denitrificans*[113] and *P. aeruginosa*.[114,115] Thus, a native cytochrome, c_{551}, stimulates the rate of nitrite reduction or oxygen uptake by the enzyme with sulfide as the electron donor, while another native cytochrome, c_{554}, or mammalian cytochrome *c* has no effect. A possible scheme for the electron transfer in this system therefore is:

$$S^{2-} \longrightarrow \text{cyt. } c_{551} \longrightarrow [\text{cyt. } c \longrightarrow \text{cyt. } d] \longrightarrow O_2 \text{ or } NO_2^-$$

Nitrite reductase

V. SULFITE OXIDATION

The oxidation of sulfite to sulfate in thiobacilli is the final event in the reaction sequence of the oxidation of all inorganic sulfur compounds. Two different mechanisms are basically functional in the oxidation of sulfite. The APS reductase pathway involves adenosine phosphosulfate (APS) as an energy-rich phosphosulfate bond intermediate, and the sulfite oxidase pathway is a direct oxidation of sulfite to sulfate without involving APS.[88,116] The former is coupled to substrate level phosphorylation and the latter to oxidative phosphorylation. Thus, ATP synthesis linked to sulfite oxidation involves both of the two general methods of biological energy conservation.

In studying the oxidation of reduced sulfur compounds in *T. thioparus*, Peck[116,118] first proposed a phosphorylating pathway involving APS reductase (AMP, sulfite:(acceptor)oxidoreductase, EC 1.8.99.2), ADP sulfurylase (ADP:sulfate adenylyltransferase, EC 2.7.7.5), and adenylate kinase (ATP:AMP phosphotransferase, EC 2.7.4.3) for sulfite oxidation as follows:

$$2SO_3^{2-} + 2AMP \xrightarrow{\text{APS reductase}} 2APS + 4e^-$$

$$2APS + 2P_i \xrightarrow{\text{ADP sulfurylase}} 2ADP + 2SO_4^{2-}$$

$$2ADP \xrightarrow{\text{adenylate kinase}} AMP + ATP$$

$$2SO_3^{2-} + AMP + 2P_i \longrightarrow 2SO_4^{2-} + ATP + 4e^-$$

where APS is the first product of sulfite oxidation, and an energy-rich phosphate bond is generated during conversion of APS to sulfate and ADP. Adenylate kinase regenerates AMP. The pathway is apparently the same as the reversal of the dissimilatory sulfate-reducing system in species of *Desulfovibrio* and *Desulfotomaculum*.[117,119-121]

APS reductase has been purified and characterized from *T. denitrificans*,[111] *T. thioparus*,[122] and photosynthetic sulfur bacteria, such as *Thiocapsa roseopersicina*[123] and *Chlorobium limicola*.[124] In spite of the oxidative biological role, the enzymes from these sources appear to resemble APS reductase isolated from *Desulfovibrio desulfuricans*[125] in that all have a similar molecular weight of between 170,000 and 218,500 and contain 1 FAD, 4 to 13 nonheme irons, and 4 to 12 acid-labile sulfur per mole. A difference exists, however, with respect to the specificity of the enzymes toward different types of cytochrome *c* as electron acceptor. The *T. thioparus* enzyme can be coupled with either ferricyanide or *Candida krusei* cytochrome *c*, whereas horse heart cytochrome *c* is only 30% as active as the yeast cytochrome *c*. The situation is similar with the enzyme from *T. roseopersicina*, except that horse heart cytochrome *c* is completely inactive. The enzyme from *T. roseopersicina* functions well with native cytochrome c_{552}. The *C. limicola* APS reductase does not react with either horse heart cytochrome *c* or *C. krusei* cytochrome *c*. It has been suggested that the apparent difference in specificities of APS reductase from different organisms to cytochrome *c* may well be due to the different responses of the enzymes toward the salt concentration in the assay system.[126] It should be noted that horse heart cytochrome *c* is an excellent electron acceptor for sulfite:cytochrome *c* oxidoreductase.[37] Meanwhile, the *Desulfovibrio vulgaris* enzyme can also use both horse heart cytochrome *c* and yeast cytochrome *c* as electron acceptor, but superoxide radicals are responsible for the cytochrome reduction as in xanthine oxidase system, since the enzyme activity is inhibited 100% by anaerobiosis and 84% with 1 μM superoxide dismutase.[127] In the *T. thioparus* enzyme, the affinity for sulfite as substrate is over 100-fold higher with the yeast cytochrome *c* than with ferricyanide as electron acceptor, although the enzyme activity with the cytochrome *c* is much lower than that with ferricyanide. The pH optimum shifts also from 7.4 (ferricyanide) to 9.5 (yeast cytochrome *c*).[122] An ordered Quad Ter mechanism for APS reductase using cytochrome *c* as electron acceptor ($2SO_3^{2-}$ + 2 yeast ferricytochrome *c* + AMP → 2 yeast ferrocytochrome *c* + APS) has been proposed from the kinetic studies.[128]

A further study on the mechanism of APS reductase has been carried out by means of difference spectrum and stopped-flow techniques.[129] The results of the study suggest the reduction of FAD to $FADH_2$ by sulfite followed by the partial reoxidation of $FADH_2$ to a red semiquinone (FADH) upon the adition of AMP with concomitant reduction of nonheme iron:

$$E-FAD-Fe^{3+} + SO_3{}^{2-} + H_2O \longrightarrow E-FADH_2-Fe^{3+} (SO_4{}^{2-})$$

$$E-FADH_2-Fe^{3+} (SO_4{}^{2-}) + AMP \longrightarrow E-FADH^{\bullet}-Fe^{2+} (APS) + OH^-$$

This mechanism differs from a three step mechanism postulated for the *D. vulgaris* system,[130,131] where the formation of a flavin-sulfite adduct is an important step of the mechanism as reported for many flavoproteins.[132-134]

Furthermore, in contrast to the *D. vulgaris* system, a direct binding or reduction of cytochrome *c* in the *T. thioparus* enzyme system has been proposed, since the enzyme in the presence of superoxide dismutase or under anaerobic conditions retained 65 to 75% of the aerobic rate of cytochrome *c* reduction without superoxide dismutase.[129]

APS reductase with a cytochrome *c* as physiological electron acceptor should be able to conserve energy through oxidative phosphorylation in addition to substrate level phosphorylation.[7] The high affinity for substrates in the reaction would ensure the efficiency of the system. It appears essential, therefore, to search for the actual physiological acceptor in order to elucidate the detailed mechanism of this reaction. Preliminary studies carried out in our laboratory with *T. thioparus* indicate that a native soluble cytochrome c_{552} can function as actual electron acceptor for the purified APS reductase.[40] The purified cytochrome c_{552} is free of all known intermediate enzymes involved in the oxidation of sulfur compounds and has properties similar to those of horse-heart cytochrome *c*. The cytochrome c_{552} isolated from *T. roseopersicina* has been found to act as the electron acceptor for APS reductase.[123] Under aerobic dark conditions, the purple sulfur bacterium has the same ability for chemolithotrophic growth on thiosulfate as do the thiobacilli.[135]

The observation that APS reductase constitutes up to 4 to 5% of the total cellular protein in the cells of *T. thioparus* and *T. denitrificans* is consistent with the suggestion that the APS pathway for sulfite oxidation plays a significant role in the energy metabolism of thiobacilli with respect to the generation of ATP through substrate level phosphorylation, as well as possible oxidative phosphorylation linked with the respiratory chain.[7,111,122]

The finding of a sulfite:cytochrome *c* oxidoreductase in *T. novellus* distinct from the APS reductase has led to the proposal that an AMP-independent system,[37,44,88] which neither needs AMP nor produces APS, functions as an additional mechanism of sulfite oxidation. The purified enzyme oxidizes sulfite to sulfate with a concomitant reduction of cytochrome *c*:

$$SO_3^{2-} + 2 \text{ ferricytochrome } c + H_2O \rightarrow SO_4^{2-} + 2 \text{ ferrocytochrome } c + 2 H^+$$

The reduced cytochrome *c* is then oxidized with molecular oxygen by cytochrome oxidase (cytochrome *c*:O_2 oxidoreductase, EC 1.9.3.1):

$$2 \text{ ferrocytochrome } c + \tfrac{1}{2}O_2 + 2H^+ \rightarrow 2 \text{ ferricytochrome } c + H_2O$$

The enzyme has been isolated and characterized from *T. thioparus*,[136] *Thiobacillus intermedius*,[137] and *T. ferrooxidans*.[138]Recently it has been identified as one of the components in the membrane-bound thiosulfate-oxidizing complex system of *T. novellus* and the sulfite-oxidizing activity (SO_3^{2-} + ½ O_2 → SO_4^{2-}) has been successfully reconstituted.[90]

The properties of the enzyme from thiobacilli are similar to those of liver sulfite oxidase[139,140] in that thiol-binding agents and anions such as phosphate and chloride are strongly inhibitory. The enzyme can couple with either horse heart cytochrome *c* or native cytochrome *c*, as well as with ferricyanide as electron acceptor, but not with molecular oxygen.

The *T. thioparus* enzyme with a molecular weight of 54,000 contains nonheme iron and acid-labile sulfur as essential components. The pH has a marked effect on the K_m value for sulfite; 40 μM at pH 8.0 and 2 μM at pH 6.0, suggesting that bisulfite (HSO_3^-) may be the actual substrate for the enzyme instead of sulfite ion (SO_3^{2-}). From a kinetic study, a Bi Bi Uni Uni Ping Pong mechanism for the enzyme reaction has been proposed. It differs from a Ping Pong mechanism postulated for bovine liver sulfite oxidase.[140]

Cytochrome oxidase of *T. novellus* has been purified recently.[141] The bacterial oxidase resembles mammalian cytochrome oxidase in that it has a similar absorption spectrum and heme *a* as the prosthetic group. The enzyme reacts slowly with horse and cow cytochrome *c*, but reacts rapidly with native, yeast, and tuna cytochrome *c*. It does not react with bacterial cytochromes *c*, such as *P. aeruginosa* c_{551} and *Nitrosomonas europea* c_{552}.

It is interesting from the bioenergetic point of view on evolution that so far APS reductase has not been detected in any of the facultative chemolithotrophic thiobacilli, but strict chemolithotrophs contain both APS reductase and sulfite:cytochrome *c* oxidoreductase.[15,16,123]

Intact cells of *T. denitrificans* oxidize sulfite either with oxygen or nitrate as the terminal electron acceptor.[109] A sulfite-dependent nitrate reductase has been partially purified from the organism and shown to carry out the reduction of nitrate to nitrite by a Ping Pong mechanism.[112,142,143] The enzyme is AMP-independent, associated with the membrane fraction, and uses nitrate, oxygen, or ferricyanide as an electron acceptor. From inhibitor studies, the involvement of flavin and cytochromes of the *a*, *b*, and *c* types in the respiratory chain is indicated. NADH in addition to sulfite is also an effective electron donor for the respiratory chain. A later work[29] shows that the property of sulfite-linked nitrate reductase differs from that of NADH-linked enzyme in its stability to storage at 0°C and, after sodium deoxycholate treatment, in its pH optimum and sensitivities to various inhibitors. Both FAD and cytochrome c_{554} separated from the solubilized enzyme have been identified as essential components of the enzyme system. Either native or horse heart cytochrome *c* mediates electron flow between sulfite or NADH and nitrate. Finally, the sulfite-linked nitrate reductase is markedly inhibited by 0.05 *M* phosphate buffer, a property similar to that of sulfite:cytochrome *c* oxidoreductase from *T. novellus* and *T. thioparus*.[37,136]

Another interesting sulfite-oxidizing membrane system has been obtained from *T. thiooxidans*.[83,144] The enzyme catalyzes the oxidation of sulfite with either O_2 or bacterial cytochrome c_{552} prepared from *P. stutzeri* as electron acceptor. The presence of cytochromes of the *a*, *b*, and *c* types in the membrane fraction, and the study with respiratory inhibitors indicate the participation of a complete cytochrome system.

A more recent study of this membrane system suggests a possible involvement of flavin in addition to the cytochromes.[93] The situation is complicated, however, by the presence of cyanide- and azide-insensitive sulfite oxidation in this system. Two CO-binding pigments, cytochrome *a* and *o* types, are found spectrophotometrically. The sulfite-oxidizing and ascorbate-reduced TMPD oxidase systems are inhibited photoreversibly by CO.[83,144]

Two similar membrane-bound sulfite-oxidizing systems from *T. concretivorus* and *T. thiooxidans* have been reported.[50,94] The *T. concretivorus* system is associated with flavoprotein, ubiquinone, and cytochromes of the *b*, *c*, and *a* types, and oxidative phosphorylation occurs during electron transfer via the cytochrome chain. In the *T. thiooxidans* system, the sulfite:cytochrome *c* oxidoreductase and cytochrome oxidase are also detected in the partially purified membrane fraction.

These studies on the membrane-bound sulfite-oxidizing systems suggest that the mechanism of sulfite oxidation is essentially the same and experimental results are consistent with the basic idea that all sulfite-oxidizing systems consist of sulfite oxidases (sulfite:cytochrome *c* oxidoreductase or possibly APS reductase) and terminal oxidases (cytochrome oxidase or nitrate reductase).

Caution has to be exercised in interpreting some experimental results on aerobic sulfite oxidation where a possibility exists of sulfite oxidation by superoxide anion generated by some electron transport components such as flavin or nonheme iron.[139,145,146]

VI. THIOSULFATE OXIDATION

Thiosulfate has been widely used as a preferred growth substrate in the routine cultivation of many thiobacilli that grow at neutral pH conditions.[1,4] Both sulfur atoms of thiosulfate are normally oxidized to sulfate as follows:

$$S_2O_3^{2-} + 2\,O_2 + H_2O \longrightarrow 2\,SO_4^{2-} + 2\,H^+$$

where sulfur and sulfite are well-documented intermediates.

A great deal of research has been centered on the mechanism of thiosulfate oxidation in thiobacilli, and various theories have been proposed.[1,3,6,7,118,147,148]

The first theory is that the oxidation of thiosulfate to sulfate involves tetrathionate and other polythionates as intermediates:[1,3,104,106,149,150]

$$S_2O_3^{2-} \longrightarrow S_4O_6^{2-} \longrightarrow S_3O_6^{2-} \longrightarrow SO_3^{2-} \longrightarrow SO_4^{2-}$$

The second theory is that thiosulfate is oxidized to sulfur and sulfate without formation of polythionates; a reductive cleavage of thiosulfate with GSH to sulfide and sulfite, followed by the oxidations of sulfide to elemental sulfur and sulfite to sulfate via the APS pathway:[116,118,151]

$$S\!-\!SO_3^{2-} \xrightarrow{\text{GSH}} \begin{array}{l} S^{2-} \longrightarrow S^0 \\ SO_3^{2-} \longrightarrow APS \longrightarrow SO_4^{2-} \end{array}$$

The third theory is that thiosulfate oxidation is initiated by its cleavage to sulfur (colloidal sulfur), [S], and sulfite, followed by their oxidation through sulfur-oxidizing and sulfite-oxidizing systems:[7,88,152]

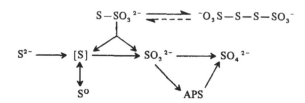

When the oxidation rate of sulfur is slower than the rate of thiosulfate cleavage, elemental sulfur (S°) may be excreted outside bacterial cells (e.g., *T. thioparus*). When thiosulfate cleavage is disturbed, some thiobacilli obtain energy by its oxidation to tetrathionate which accumulates in the medium. Once sulfite is produced, however, tetrathionate formation is inhibited.[152] Tetrathionate is oxidized after a reductive cleavage (electrons supplied by sulfite oxidation) to thiosulfate, sulfur, and sulfite.[7]

In early studies, the production of polythionates consisting mainly of tetrathionate along with elemental sulfur and sulfate was observed during thiosulfate oxidation in growing cultures or intact cells of many thiobacilli. The metabolism of polythionates has been well documented by Trudinger[1] and Roy and Trudinger.[4] The significance of the polythionate pathway has been widely questioned for the following reasons:

1. Studies on the pathway have led to a variety of conflicting results, the oxidation often stopping at tetrathionate or trithionate. Possible causes for such variations are strains of organism and experimental conditions used.[1,3] Some discrepancies in the experimental results, however, remain to be clarified.
2. Intact cells or cell-free extracts of some thiobacilli oxidize thiosulfate completely to sulfate without polythionate accumulation and are unable to metabolize polythionates.[3,6,88,118]
3. A catabolic repression study suggests that tetrathionate may not be a necessary intermediate in the thiosulfate oxidation.[155]
4. Most importantly there is no knowledge of specific enzymes responsible for thiosulfate oxidation through this pathway except a thiosulfate-oxidizing enzyme which catalyzes the following reaction:

$$2S_2O_3^{2-} \longrightarrow {}^-O_3SSSSO_3^- + 2e^-$$

Trudinger[26,156,157] first isolated and characterized the thiosulfate-oxidizing enzyme from *T. neapolitanus*. The native cytochrome $c_{553.5}$ is the natural electron acceptor. Other native *c*-type cytochromes, c_{550} and c_{557}, do not react directly with the enzyme, and mammalian cytochrome *c* is only slowly reduced. Ferricyanide, on the other hand, is a very good electron acceptor.

The enzyme has also been purified and characterized from *T. ferroxidans* grown on sulfur[158] and *T. thioparus*.[152] The *T. ferrooxidans* enzyme, unlike the thiosulfate-oxidizing enzyme of other thiobacilli, couples only with ferricyanie as the electron acceptor. The 160-fold purified enzyme from *T. thioparus* contains nonheme iron and is active over a wide pH range (4.5 to 10.0). Although ferricyanide is a better electron acceptor than horse heart cytochrome *c*, the K_m for thiosulfate is lowered from 0.1 mM to 6 μM, and the pH optimum shifts from 5.0 to 7.0 with the latter electron acceptor. The enzyme is sensitive to sulfhydryl inhibitors. Interestingly sulfite is a very potent inhibitor, causing a loss of 50% of the activity in 5 min at 5 μM. The inhibition is time dependent and irreversible. From a kinetic study, a Theorell-Chance mechanism for the enzyme reaction is proposed.[128]

A soluble thiosulfate:Ocytochrome c reductase has been reported in *T. novellus*,[159] but the oxidation product of thiosulfate has not been identified. The reported effect caused by cyanide suggests an enzyme system consisting of rhodanese (thiosulfate:cyanide sulfurtransferase, EC 2.8.1.1) and sulfite:cytochrome c oxidoreductase that is responsible for the observed activity.[88]

The first cell-free system from *T. thioparus* capable of oxidizing thiosulfate to sulfate without tetrathionate formation was isolated by Peck, and the mechanism of thiosulfate oxidation without formation of polythionates (the second theory) was proposed.[116,118] In this proposal, the initial reductive cleavage catalyzed by thiosulfate reductase requires substrate amounts of reduced glutathione ($S_2O_3^{2-}$ + 2GSH → GSSG + H_2S + SO_3^{2-}) and for elemental sulfur to be produced from sulfide (H_2S + ½O_2 → $S°$ + H_2O). Sulfate is formed rom sulfite via the APS pathway. Since the sulfur-oxidizing enzyme and sulfite:cytochrome c oxidoreductaseare present in the same organism in addition to the enzymes involved in the APS pathway,[85,136] elemental sulfur could be further oxidized to sulfite, and the oxidation of sulfite could involve both pathways of sulfite oxidation as suggested by Lyric and Suzuki.[152] Thus, the original theory may be modified to account for further oxidation of elemental sulfur to sulfate. Also thiosulfate reductase has yet to be purified, and the specific activity of the enzyme is apparently very low.[116] It is possible to formulate the oxidation of outer sulfane sulfur without its initial reduction to sulfide:$S_2O_3^{2-}$ + GSH → SO_3^{2-} + GSSH, GSSH + O_2 + H_2O → GSH + SO_3^{2-} + $2H^+$.[85] Thiosulfate reductase may catalyze the first reaction.

In contrast to *T. neapolitanus* and *T. thioparus*, the cell-free thiosulfate-oxidizing system of *T. novellus* does not contain the thiosulfate-oxidizing enzyme and APS reductase, but contains thiosulfate-cleaving enzyme (rhodanese), sulfur-oxidizing enzyme, sulfite:cytochrome c oxidoreductase, and cytochrome oxidase.[88] This finding led to an alternative mechanism for thiosulfate oxidation as follows:

$$S_2O_3^{2-} \xrightarrow[\text{(Rhodanese)}]{\text{Thiosulfate-cleaving enzyme}} [S] + SO_3^{2-}$$

$$[S] + O_2 + H_2O \xrightarrow{\text{Sulfur-oxidizing enzyme}} SO_3^{2-} + 2H^+$$

$$2SO_3^{2-} + O_2 \xrightarrow{\text{Sulfite-oxidizing system}} 2SO_4^{2-}$$

$$S_2O_3^{2-} + 2O_2 + H_2O \longrightarrow 2SO_4^{2-} + 2H^+$$

where thiosulfate is first cleaved to sulfur and sulfite by thiosulfate-cleaving enzyme. Sulfur derived from the sulfane group (outer position of thiosulfate) is oxidized to sulfite by sulfur-oxidizing enzyme, and sulfite is finally oxidized to sulfate by sulfite:cytochrome c oxidoreductase and cytochrome oxidase. The same thiosulfate-oxidizing system has been reported in *Thiobacillus intermedius* and *T. thioparus*.[137,152]

Recently, a membrane-associated thiosulfate-oxidizing system capable of oxidizing thiosulfate to sulfate has been isolated and characterized from *T. novellus*.[90,160] The system seems to require an initial reducing power which can be supplied by either sulfite, NADH, GSH, or endogenous electron flow. The multienzyme complex system is sensitive to oxygen, high temperature, freezing, metal-binding agents, thiol-binding agents, and electron transport and oxidative phosphorylation inhibitors. The oxygen-sensitive components are possibly involved in the initial thiosulfate-cleaving step and sulfur-oxidizing enzyme. A spectrophotometric study indicates the presence of flavin and cytochromes *b*, *c*, *a*, and *d*. Although cytochromes of the *c* and *a* types are definitely involved in the sulfite-oxidizing system, the functions of the flavin and cytochromes *b* and *d* types remain unclear at the present time. The isolated thiosulfate-oxidizing complex is phospholipoprotein in nature and contains thiosulfate-cleaving e n z y m e ,   s u l f u r - o x i d i z i n g   e n z y m e ,   s u l f i t e : c y t o - chrome *c* oxidoreductase, and cytochrome oxidase. In this study, a modified scheme for the thiosulfate-cleaving enzyme and sulfur-oxidizing enzyme has been presented as follows:

$$RSH + SSO_3^{2-} \longrightarrow RSSH + SO_3^{2-}$$

$$RSSH + O_2 + H_2O \longrightarrow RSH + SO_3^{2-} + 2H^+$$

where RSH is the thiosulfate-cleaving enzyme, and RSSH is an actual substrate for sulfur-oxidizing enzyme. When RSH is in its oxidized inactive form, RSSR, activation by either GSH (RSSR + GSH → RSH + GSSR) or by electrons generated from sulfite or NADH (RSSR + 2e⁻ + 2H⁺ → 2RSH) is necessary for thiosulfate oxidation. Thus, the initial step of thiosulfate cleavage may play a very importat role for the overall reaction.

The scheme for the oxidation of inorganic sulfur compounds in all thiobacilli, as shown in the third theory, is consistent with most observations and based on the enzymes that have been isolated and characterized from various species of *Thiobacillus*, with an exception of tetrathionate reductase.[7]

In many thiobacilli and phototrophic sulfur bacteria, rhodanese ($S_2O_3^{2-}$ + CN^- → SCN^- + SO_3^{2-}) most likely catalyzes the first reaction in the oxidation of thiosulfate, i.e., the cleavage of thiosulfate,[88,137,161] although cynide is unlikely to be the physiological sulfur acceptor. The enzyme activity has been detected in many thiosulfate oxidizing bacteria, and the enzyme has been purified and characterized from *T. denitrificans*, *T. ferrooxidans*, *Thiobacillus A2* and *Chromatium*.[4,161-165] The enzyme properties seem to be analogous to that of mammalian rhodanese which is widely distributed in the mitochondria of liver and kidney tissues, where it probably serves a variety of functions related to its capacity to act as a general sulfane sulfur transferase by catalytic double-replacement mechanism.[3,4,166] The rhodanese activity is found in sulfur-oxidizing enzyme preparations of *T. thiooxidans* and *T. thioparus* and in the isolated membrane-associated thiosulfate-oxidizing system of *T. novellus*.[88,90,94] It has been suggested that the membrane-bound rhodanese is the thiosulfate-cleaving enzyme, closely associated with the sulfur-oxidizing enzyme in the membrane fraction. In the presence of a sulfite-oxidizing system, thiosulfate can be oxidized to sulfate by a series of reactions initiated by rhodanese.[90]

The *Thiobacillus A2* rhodanese is able to use dihydrolipoate as an acceptor for the sulfane moiety of thiosulfate producing *a*-lipoate, with the intermediate formation of persulfide.[165,167] Exactly the same reaction was originally reported for beef rhodanese by Volini and Westley.[168] Whether such a compound is a natural acceptor for the rhodanese of thiobacilli, however, has yet to be proven. There is no clear agreement as to whether the cyanide, GSH, and dihydrolipoate-dependent rhodanese activities in all cases are due to a single enzyme. A separate existence for different rhodaneses in both mammalian and bacterial systems has been claimed.[169,170]

Bovine rhodanese has a role in forming the labile sulfur of a nonheme iron protein, ferredoxin, from thiosulfate and apoferredoxin.[171,172] Crystallized rhodanese from beef kidney is able to accept cleaved sulfur in the form of a persulfide group (RSSH) in the active site of the enzyme with a maximum absorbance at 335 nm.[173] Recently a similar sulfur-transferring role of rhodanese, in its sulfane sulfur-containing form, has been reported in the iron-sulfur flavoprotein, succinate dehydrogenase.[174,175] In summary rhodanese, as thiosulfate sulfur, transferase seems to have properties required of its proposed role in the oxidation of thiosulfate.[85,88,137] The exact function and mechanism of the sulfur transferase in thiobacilli, however, obviously need further elucidation.

VII. CONCLUDING REMARKS

Our knowledge of the mechanism of biological oxidation of inorganic sulfur compounds by thiobacilli has markedly increased in the past 20 years. This review has outlined such advances and highlighted some of the outstanding problems. A detailed knowledge of the molecular mechanism for this complex system, however, is still missing and a large number of fundamental questions remain unanswered.

More information is particularly needed in the following areas:

1. The pathway for the complete oxidation of inorganic sulfur compounds requires further work on the mechanism of cleavage of thiosulfate and polythionates and the roles played by rhodanese and the sulfur-oxidizing enzyme.
2. More coordinated studies of whole cells, cell-free preparations, and purified enzymes involved in sulfur metabolism are required to bridge the gaps in our knowledge of these different systems.
3. An extensive investigation is necessary on the isolation, identification, and characterization of respiratory components and on their role in the oxidation of sulfur compounds and in the generation of energy and reducing power.
4. The energy-coupling mechanism has to be studied further to determine the number of potential energy-coupling sites, stoichiometry, and possible roles of the ATPase and membrane structure. Although the preliminary evidence supports the chemiosmotic coupling mechanism as discussed earlier, at present it is not possible to state that this is the only mechanism of oxidative phosphorylation linked to the oxidation of sulfur compounds in thiobacilli.
5. Finally further research is needed to study interactions between the energy- and reducing power-generating systems and the cellular biosynthetic processes, especially the control mechanisms that regulate these energy-providing and energy-utilizing systems.

It is hoped that future work on the biochemical mechanism in these areas will eventually provide new avenues to an understanding of the challenging problems of these physiologically unique and sophisticated microorganisms.

REFERENCES

1. **Trudinger, P. A.**, The metabolism of inorganic sulfur compounds by thiobacilli, *Rev. Pure Appl. Chem.*, 17, 1, 1967.
2. **Kelly, D. P.**, Biochemistry of oxidation of inorganic sulfur compounds by micro-organisms, *Aust. J. Sci.*, 31, 165, 1969.

3. **Trudinger, P. A.**, Assimilatory and dissimilatory metabolism of inorganic sulfur compounds by microorganisms, in *Advances in Microbial Physiology*, Vol. 3, Rose, A. H. and Wilkinson, J. F., Eds., Academic Press, New York, 1969, 111.
4. **Roy, A. B. and Trudinger, P. A.**, *The Biochemistry of Inorganic Compounds of Sulfur*, Cambridge University Press, London, 1970.
5. **Aleem, M. I. H.**, Biochemical reaction mechanisms in sulfur oxidation by chemosynthetic bacteria, *Plant Soil*, 43, 587, 1975.
6. **Peck, H. D., Jr.**, Energy-coupling mechanism in chemolithotrophic bacteria, *Ann. Rev. Microbiol.*, 22, 489, 1968.
7. **Suzuki, I.**, Mechanisms of inorganic oxidation and energy coupling, *Ann. Rev. Microbiol.*, 28, 85, 1974.
8. **Aleem, M. I. H.**, Coupling of energy with electron transfer reactions in chemolithotrophic bacteria, in *Microbial Energetics*, Haddock, B. A. and Hamilton, W. A., Eds., Cambridge University Press, London, 1977, 351.
9. **Rittenberg, S. C.**, The roles of exogenous organic matter in the physiology of chemolithotrophic bacteria, in *Advances in Microbial Physiology*, Vol. 3, Rose, A. H. and Wilkinson, J. F., Eds., Academic Press, New York, 1969, 159.
10. **Kelly, D. P.**, Autotrophy: concepts of lithotrophic bacteria and their organic metabolism, *Ann. Rev. Microbiol.*, 25, 177, 1971.
11. **Rittenberg, S. C.**, The obligate autotroph — the demise of a concept, *Antonie van Leeuwenhoek J. Microbiol. Serol.*, 38, 457, 1972.
12. **Whittenbury, R. and Kelly, D. P.**, Autotrophy: a conceptual phoenix, in *Microbial Energetics*, Haddock, B. A. and Hamilton, W. A., Eds., Cambridge University Press, London, 1977, 121.
13. **Lemberg, R. and Barrett, J.**, *Cytochromes*, Academic Press, New York, 1973.
14. **Yamanaka, T. and Okunuki, K.**, Cytochromes, in *Microbial Iron Metabolism*, Neilands, J. B., Eds., Academic Press, New York, 1974, 349.
15. **Broda, E.**, the evolution of bioenergetic process, *Progr. Biophys. Mol. Biol.*, 21, 143, 1970.
16. **Broda, E.**, *The Evolution of the Bioenergetic Processes*, Pergamon Press, Toronto, 1975.
17. **Yamanaka, T. and Fukumori, Y.**, An evolutionary aspect of the reactivity of cytochrome *c* with cytochrome oxidase, in *Evolution of Protein Molecules*, Matsubara, H. and Yamanaka, T., Eds., Japan Scientific Societies Press, Tokyo, 1978, 387.
18. **Zajic, J. E.**, *Microbial Biogeochemistry*, Academic Press, New York, 1969.
19. **Murr, L. E., Torma, A. E., and Brierley, J. A.**, *Metallurgical Applications of Bacterial Leaching and Related Microbiological Phenomena*, Academic Press Inc., New York, 1978.
20. **Asano, A., Cohen, N. S., Baker, R. F., and Brodie, A. F.**, Orientation of the membrane in ghosts and electron transport particles of *Mycobacterium phlei*, *J. Biol. Chem.*, 248, 3386, 1973.
21. **Gel'man, N. S., Lukoyanova, M. A., and Ostrovskii, D. N.**, *Bacterial Membranes and the Respiratory Chain*, Plenum Press, New York, 1975.
22. **Haddock, B. A. and Jones, C. W.**, Bacterial respiration, *Bacteriol. Rev.*, 41, 47, 1977.
23. **John, P. and Whatley, F. R.**, The bioenergetics of *Paracoccus denitrificans*, *Biochem. Biophys. Acta*, 403, 129, 1977.
24. **Yamanaka, T., Takenami, S., Akiyama, N., and Okunuki, K.**, Purification and properties of cytochrome c-550 and cytochrome c-551 derived from the facultative chemoautotroph, *Thiobacillus novellus, J. Biochem. (Tokyo)*, 70, 349, 1971.
25. **Skarżyński, B., Klimek, R., and Szczepkowski, T. W.**, Cytochrome in *Thiobacillus thioparus, Bull. Acad. Polon. Sci. Cl. 2*, 4, 299, 1956.
26. **Trudinger, P. A.**, Thiosulfphate oxidation and cytochromes in *Thiobacillus X*. I. Fractionation of bacterial extracts and properties of cytochromes, *Biochem. J.*, 78, 673, 1961.
27. **Aubert, J. P., Milhaud, G., Moncel, C., and Millet, J.**, Existence, isolement et proprietes physicochimiques d'un cytochrome c de *Thiobacillus denitrificans*, *Compt. Rend. Acad. Sci.*, 246, 1616, 1958.
28. **Aubert, J. P., Millet, J., and Milhaud, G.**, Isolement et proprietes du cytochrome c de *Thiobacillus denitrificans*, *Ann. Inst. Pasteur.*, 96, 640, 1959.
29. **Sawhney, V. and Nicholas, D. J. D.**, Sulphite- and NADH-dependent nitrate reductase from *Thiobacillus denitrificans*, *J. Gen. Microbiol.*, 100, 49, 1977.
30. **Sawhney, V. and Nicholas, D. J. D.**, Sulphide-linked nitrite reductase from *Thiobacillus denitrificans* with cytochrome oxidase activity: purification and properties, *J. Gen. Microbiol.*, 106, 119, 1978.
31. **Takakuwa, S.**, Purification and some properties of cytochrome c-552 from a sulfur-oxidizing bacterium, *Thiobacillus thiooxidans, J. Biochem. (Tokyo)*, 78, 181, 1975.
32. **Tano, T., Kagawa, H., and Imai, K.**, Physiological studies on thiobacilli. Part IV. Purification and certain properties of cytochrome c from *Thiobacillus thiooxidans*, *Agric. Biol. Chem.*, 32, 279, 1968.

33. **Moriarty, D. J. W. and Nicholas, D. J. D.**, Enzymatic sulphide oxidation by *Thiobacillus concretivorus, Biochim. Biophys. Acta,* 184, 114, 1969.
34. **Yamanaka, T.**, Evolution of cytochrome c molecule, *Adv. Biophys.,* 3, 227, 1972.
35. **Yamanaka, T. and Kimura, K.**, Eucaryotic cytochrome c-like properties of cytochrome c-550 (*Thiobacillus novellus*), *FEBS Lett.,* 48, 253, 1974.
36. **Yamanaka, T.**, A comparative study on the redox reactions of cytochrome c with certain enzymes, *J. Biochem. (Tokyo),* 77, 493, 1975.
37. **Charles, A. M. and Suzuki, I.**, Purification and properties of sulfite: cytochrome c oxidoreductase from *Thiobacillus novellus, Biochim. Biophys. Acta,* 128, 522, 1966.
38. **Yamanaka, T., Shinra, M., and Kimura, K.**, A comparison between *Nitrosomonas europea* and *Thiobacillus novellus* on the basis of their oxidation systems of inorganic compounds, *BioSystems,* 9, 155, 1977.
39. **Schedel, M., LeGall, J., and Baldensperger, J.**, Sulfur metabolism in *Thiobacillus denitrificans.* Evidence for the present of a sulfite reductase activity, *Arch. Microbiol.,* 105, 339, 1975.
40. **Oh, J. K. and Suzuki, I.**, unpublished data, 1978.
41. **Murphy, M. L. and Seigel, L. M.**, Siroheme and sirohydrochlorin. The basis for a new type of porphyrin-related prosthetic group common to both assimilatory and dissimilatory sulfite reductases, *J. Biol. Chem.,* 248, 6911, 1973.
42. **Murphy, M. L., Siegel, L. M., Tove, S. R., and Kamin, H.**, Siroheme: a new prosthetic group participating in six-electron reduction reactions catalyzed by both sulfite and nitrite reductases, *Proc. Natl. Acad. Sci. U.S.A.,* 71, 612, 1974.
43. **Vega, J. M., Garrett, R. H., and Siegel, L. M.**, Siroheme: a prosthetic group of the *Neurospora crassa* assimilatory nitrite reductase, *J. Biol. Chem.,* 250, 7980, 1975.
44. **Charles, A. M. and Suzuki, I.**, Sulfite oxidase of a facultative autotroph, *Thiobacillus novellus, Biochem. Biophys. Res. Commun.,* 19, 686, 1965.
45. **Cole, J. S. and Aleem, M. I. H.**, Oxidative phosphorylation in *Thiobacillus novellus, Biochem. Biophys. Res. Commun.,* 38, 736, 1970.
46. **Cole, J. S. and Aleem, M. I. H.**, Electron transport-linked compared with proton-induced ATP generation in *Thiobacillus novellus, Proc. Natl. Acad. Sci. U.S.A.,* 70, 3571, 1973.
47. **Ross, A. J., Schoenhoff, R. L., and Aleem, M. I. H.**, Electron transport and coupled phosphorylation in the chemoautotroph *Thiobacillus neapolitanus, Biochem. Biophys. Res. Commun.,* 32, 301, 1968.
48. **Saxena, J. and Aleem, M. I. H.**, Oxidation of sulfur compounds and coupled phosphorylation in the chemoautotroph *Thiobacillus neapolitanus, Can. J. Biochem.,* 51, 560, 1973.
49. **Davis, E. A. and Johnson, E. J.**, Phosphorylation coupled to the oxidation of sulfite and 2-mercaptoethanol in extracts of *Thiobacillus thioparus, Can. J. Microbiol.,* 13, 873, 1967.
50. **Moriarty, J. W. and Nicholas, D. J. D.**, Electron transfer during sulphide and sulphite oxidation by *Thiobacillus concretivorus, Biochim. Biophys. Acta,* 216, 130, 1970.
51. **Gel'man, N. S., Lukoyana, M. A., and Ostrovskii, D. N.**, *Respiration and Phosphorylation of Bacteria,* Plenum Press, New York, 1967.
52. **Kelly, D. P. and Syrett, P. J.**, Inhibition of formation of adenosine triphosphate in *Thiobacillus thioparus* by 2:4-dinitrophenol, *Nature (London),* 202, 597, 1964.
53. **Jagendorf, A. and Uribe, E.**, ATP formation caused by acid-base transition of spinach chloroplasts, *Proc. Natl. Acad. Sci. U.S.A.,* 55, 170, 1966.
54. **Mitchell, P.**, Chemiosmotic coupling in oxidative and photosynthetic phosphorylation, *Biol. Rev. Proc. Cambridge Philos. Soc.,* 41, 445, 1966.
55. **Mitchell, P.**, Chemiosmotic coupling in energy transduction: a logical development of biochemical knowledge, *Bioenergetics,* 3, 5, 1972.
56. **Mitchell, P.**, Vectorial chemistry and the molecular mechanics of chemiosmotic coupling: power transmission by proticity, *Biochem. Soc. Trans.,* 4, 399, 1976.
57. **Drozd, J. W.**, Respiration-driven proton translocation in *Thiobacillus neapolitanus c, FEBS Lett.,* 49, 103, 1974.
58. **Drozd, J. W.**, Energy conservation in *Thiobacillus neapolitanus c:* sulphide and sulphite oxidation, *J. Gen. Microbiol.,* 98, 309, 1977.
59. **Tuovinen, O. H., Nicholas, D. J. D., and Aleem, M. I. H.**, Proton translocation in intact cells of *Thiobacillus denitrificans, Arch. Microbiol.,* 113, 11, 1977.
60. **Marunouchi, T. and Mori, T.**, Studies on the sulfite-dependent ATPase of a sulfur-oxidizing bacterium, *Thiobacillus thiooxidans, J. Biochem. (Tokyo),* 62, 401, 1967.
61. **Marunouchi, T.**, Separation of sulfite-dependent ATPase from other ATP hydrolyzing enzymes, *J. Biochem. (Tokyo),* 66, 113, 1969.
62. **Adapoe, C. and Silver, M.**, The soluble adenosine triphosphatase of *Thiobacillus ferrooxidans, Can. J. Microbiol.,* 21, 1, 1975.

63. **Tominaga, N. and Mori, T.,** Purification and characterization of inorganic pyrophosphatase from *Thiobacillus thiooxidans, J. Biochem. (Tokyo),* 81, 477, 1977.

64. **Tominaga, N. and Mori, T.,** A sulfate-dependent acid phosphatase of *Thiobacillus thiooxidans.* Its partial purification and some properties, *J. Biochem. (Tokyo),* 76, 397, 1974.

65. **Howard, A. and Lundgren, D. G.,** Inorganic pyrophosphatase from *Ferrobacillus ferrooxidans (Thiobacillus ferrooxidans), Can. J. Biochem.,* 48, 1302, 1970.

66. **Aleem, M. I. H.,** Generation of reducing power in chemosynthesis. III. Energy-linked reduction of pyridine nucleotides in *Thiobacillus novellus, J. Bacteriol.,* 91, 729, 1966.

67. **Aleem, M. I. H.,** Generation of reducing power in chemosynthesis. VI. Energy-linked reactions in the chemoautotroph, *Thiobacillus neapolitanus, Antonie van Leeuwenhoek J. Microbiol. Serol.,* 35, 379, 1969.

68. **Saxena, J. and Aleem, M. I. H.,** Generation of reducing power in chemosynthesis. VII. Mechanism of pyridine nucleotide reduction by thiosulfate in the chemoautotroph *Thiobacillus neapolitanus, Arch. Microbiol.,* 84, 317, 1972.

69. **Roth, C. W., Hempfling, W. P., Conners, J. N., and Vishniac, W. V.,** Thiosulfate- and sulfide-dependent pyridine nucleotide reduction and gluconeogenesis in intact *Thiobacillus neapolitanus, J. Bacteriol.,* 114, 592, 1973.

70. **Low, H. and Vallin, I.,** Reduction of added DPN from the cytochrome *c* level in submitochondrial particles, *Biochem. Biophys. Res. Commun.,* 9, 307, 1962.

71. **Sadler, M. H. and Johnson, E. J.,** A comparison of the NADH oxidase electron transport systems of two obligately chemolithotrophic bacteria, *Biochim. Biophys. Acta,* 283, 167, 1972.

72. **Meyer, T. E., Bartch, R. G., and Kamen, M. D.,** Cytochrome c_3. A class of electron transfer heme proteins found in both photosynthetic and sulfate-reducing bacteria, *Biochem. Biophys. Acta,* 245, 453, 1971.

73. **Trudinger, P. A. and Kelly, D. P.,** Reduced nicotinamide adenine dinucleotide oxidation by *Thiobacillus neapolitanus* and *Thiobacillus* strain C, *J. Bacteriol.,* 95, 1962, 1968.

74. **Sawhney, V. and Nicholas, D. J. D.,** Regulation of NADH-linked nitrate reductase by NAD^+ and 5′-AMP in *Thiobacillus denitrificans, FEMS Microbiol. Lett.,* 3, 211, 1978.

75. **Waksman, S. A. and Joffe, J. S.,** Microorganisms concerned in the oxidation of sulfur in the soil. II. *Thiobacillus thiooxidans,* a new sulfur-oxidizing organism isolated from the soil, *J. Bacteriol.,* 7, 239, 1922.

76. **Waksman, S. A.,** Microorganisms concerned in the oxidation of sulfur in the soil. III. Media used for the isolation of sulfur bacteria from the soil, *Soil Sci.,* 13, 329, 1922.

77. **Waksman, S. A. and Starkey, R. L.,** On the growth and respiration of autotrophic bacteria, *J. Gen. Physiol.,* 5, 285, 1923.

78. **Starkey, R. L.,** Concerning the physiology of *Thiobacillus thiooxidans,* an autotrophic bacterium oxidizing sulfur under acid conditions, *J. Bacteriol.,* 10, 135, 1925.

79. **Parker, C. D. and Prisk, J.,** The oxidation of inorganic compounds of sulfur by various sulfur bacteria, *J. Gen. Microbiol.,* 8, 344, 1953.

80. **Suzuki, I.,** Oxidation of elemental sulfur by an enzyme system of *Thiobacillus thiooxidans, Biochim. Biophys. Acta,* 104, 359, 1965.

81. **Adair, F. W.,** Membrane-associated sulfur oxidation by the autotroph *Thiobacillus thiooxidans, J. Bacteriol.,* 92, 899, 1966.

82. **Taylor, B. F.,** Oxidation of elemental sulfur by an enzyme system from *Thiobacillus neapolitanus, Biochim. Biophys. Acta,* 170, 112, 1968.

83. **Kodama, A. and Mori, T.,** Studies on the metabolism of a sulfur-oxidizing bacterium. V. Comparative studies on sulfur and sulfite oxidizing systems of *Thiobacillus thiooxidans, Plant Cell Physiol.,* 9, 725, 1968.

84. **Kodama, A.,** Studies on the metabolism of a sulfur-oxidizing bacterium. VI. Fractionation and reconstitution of the elementary sulfur-oxidizing system of *Thiobacillus thiooxidans, Plant Cell Physiol.,* 10, 645, 1969.

85. **Suzuki, I. and Silver, M.,** The initial product and properties of the sulfur-oxidizing enzyme of thiobacilli, *Biochim. Biophys. Acta,* 122, 22, 1966.

86. **Suzuki, I. and Werkman, C. H.,** Glutathione reductase of *Thiobacillus thiooxidans, Biochem. J.,* 74, 359, 1960.

87. **Suzuki, I.,** Incorporation of atmospheric oxygen-18 into thiosulfate by the sulfur-oxidizing enzyme of *Thiobacillus thiooxidans, Biochim. Biophys. Acta,* 110, 97, 1965.

88. **Charles, A. M. and Suzuki, I.,** Mechanism of thiosulfate oxidation by *Thiobacillus novellus, Biochim. Biophys. Acta,* 128, 510, 1966.

89. **Silver, M. and Lundgren, D. G.,** Sulfur-oxidizing enzyme of *Ferrobacillus ferrooxidans (Thiobacillus ferrooxidans), Can. J. Biochem.,* 46, 457, 1968.

90. Oh, J. K. and Suzuki, I., Resolution of a membrane-associated thiosulfate-oxidizing complex of *Thiobacillus novellus*, *J. Gen. Microbiol.*, 99, 413, 1977.
91. Adair, F. W., Inhibition of oxygen utilization and destruction of ubiquinone by ultraviolet irradiation of *Thiobacillus thiooxidans*, *J. Bacteriol.*, 95, 147, 1968.
92. Takakuwa, S., Studies on the metabolism of a sulfur-oxidizing bacterium. VII. Purification and characterization of soluble components indispensable for sulfur oxidation by *Thiobacillus thiooxidans*, *Plant Cell Physiol.*, 16, 1027, 1975.
93. Takakuwa, S., Studies on the metabolism of a sulfur-oxidizing bacterium. IX. Electron transfer and the terminal oxidase system in sulfite oxidation of *Thiobacillus thiooxidans*, *Plant Cell Physiol.*, 17, 103, 1976.
94. Lukow, O. M., Studies on the Sulfur Metabolism of *Thiobacillus thiooxidans*, M. Sci. thesis, University of Manitoba, Winnipeg, 1977.
95. Vogler, K. G. and Umbreit, W. W., The necessity for direct contact in sulfur oxidation by *Thiobacillus thiooxidans*, *Soil Sci.*, 51, 331, 1941.
96. Umbreit, W. W., Vogel, H. R., and Vogler, K. G., The significance of fat in sulfur oxidation by *Thiobacillus thiooxidans*, *J. Bacteriol.*, 43, 141, 1942.
97. Umbreit, W. W. and Anderson, T. F., A study of *Thiobacillus thiooxidans* with the electron microscope, *J. Bacteriol.*, 44, 317, 1942.
98. Schaeffer, W. I., Holbert, P. E., and Umbreit, W. W., Attachment of *Thiobacillus thiooxidans* to sulfur crystals, *J. Bacteriol.*, 85, 137, 1963.
99. Knaysi, G., A cytological and microchemical study of *Thiobacillus thiooxidans*, *J. Bacteriol.*, 46, 451, 1943.
100. Schaeffer, W. I. and Umbreit, W. W., Phosphatidylinositol as a wetting agent in sulfur oxidation by *Thiobacillus thiooxidans*, *J. Bacteriol.*, 85, 492, 1963.
101. Jones, G. E. and Benson, A. A., Phosphatidyl glycerol in *Thiobacillus thiooxidans*, *J. Bacteriol.*, 89, 260, 1965.
102. Shively, J. M. and Benson, A. A., Phospholipids of *Thiobacillus thiooxidans*, *J. Bacteriol.*, 94, 1679, 1967.
103. Cook, T. M., Growth of *Thiobacillus thiooxidans* in shaken culture, *J. Bacteriol.*, 88, 620, 1964.
104. Vishniac, W. and Santer, M., The thiobacilli, *Bacteriol. Rev.*, 21, 195, 1957.
105. Suzuki, I. and Werkman, C. H., Glutathione and sulfur oxidation by *Thiobacillus thiooxidans*, *Proc. Natl. Acad. Sci. U.S.A.*, 45, 239, 1959.
106. London, J. and Rittenberg, S. C., Path of sulfur in sulphide and thiosulphate oxidation by thiobacilli, *Proc. Natl. Acad. Sci. U.S.A.*, 52, 1183, 1964.
107. London, J., Cytochrome in *Thiobacillus thiooxidans*, *Science*, 140, 409, 1963.
108. Moriarty, D. J. W. and Nicholas, D. J. D., Products of sulphide oxidation in extracts of *Thiobacillus concretivorus*, *Biochim. Biophys. Acta*, 197, 143, 1970.
109. Peeters, T. and Aleem, M. I. H., Oxidation of sulfur compounds and electron transport in *Thiobacillus denitrificans*, *Arch. Microbiol.*, 71, 319, 1970.
110. Aminuddin, M. and Nicholas, D. J. D., Sulphide oxidation linked to the reduction of nitrate and nitrite in *Thiobacillus denitrificans*, *Biochim. Biophys. Acta*, 325, 81, 1973.
111. Bowen, T. J., Happold, F. C., and Taylor, B. F., Studies on the adenosine-5'-phosphosulphate from *Thiobacillus denitrificans*, *Biochim. Biophys. Acta*, 118, 566, 1966.
112. Aminuddin, M. and Nicholas, D. J. D., Electron transfer during sulphide and sulphite oxidation in *Thiobacillus denitrifcans*, *J. Gen. Microbiol.*, 82, 115, 1974.
113. Lam, Y. and Nicholas, D. J. D., A nitrite reductase with cytochrome oxidase activity from *Micrococcus denitrificans*, *Biochim. Biophys. Acta*, 180, 459, 1969.
114. Yamanaka, T. and Okunuki, K., Crystalline *Pseudomonas* cytochrome oxidase. I. Enzymatic properties with special reference to the biological specificity, *Biochim. Biophys. Acta*, 67, 379, 1963.
115. Yamanaka, T., Identity of *Pseudomonas* cytochrome oxidase with *Pseudomonas* nitrite reductase, *Nature (London)*, 204, 253, 1964.
116. Peck, H. D., Jr., Adenosine-5-phosphosulfate as an intermediate in the oxidation of thiosulfate by *Thiobacillus thioparus*, *Proc. Natl. Acad. Sci. U.S.A.*, 46, 1053, 1960.
117. Peck, H. D., Jr., Evidence for the reversibility of the reaction catalyzed by adenosine-5'-phosphosulfate reductase, *Biochim. Biophys. Acta*, 49, 621, 1961.
118. Peck, H. D., Jr., Symposium of metabolism of inorganic compounds. V. Comparative metabolism of inorganic sulfur compounds in microorganisms, *Bacteriol. Rev.*, 26, 67, 1962.
119. Ishimoto, M. and Fujimoto, D., Biochemical studies on sulfate-reducing bacteria. X. Adenosine-5'-phosphosulfate reductase, *J. Biochem. (Tokyo)*, 50, 299, 1961.
120. Peck, H. D., Jr., The role of adenosine-5'-phosphosulfate in the reduction of sulfate to sulfite by *Desulfovibrio desulfuricans*, *J. Biol. Chem.*, 237, 198, 1962.

121. **Peck, H. D., Jr.**, Enzymatic basis for assimilatory and dissimilatory sulfate reduction, *J. Bacteriol.*, 82, 933, 1961.

122. **Lyric, R. M. and Suzuki, I.**, Enzymes involved in metabolism of thiosulfate in *Thiobacillus thioparus.* II. Properties of adenosine-5'-phosphosulfate reductase, *Can. J. Biochem.*, 48, 344, 1970.

123. **Trüper, H. G. and Rogers, L. A.**, Purification and properties of adenylyl sulfate reductase from the phototrophic sulfur bacterium, *Thiocapsa roseopersicina*, *J. Bacteriol.*, 108, 1112, 1971.

124. **Kirchhoff, J. and Trüper, H. G.**, Adenylylsulfate reductase of *Chlorobium limicola*, *Arch. Microbiol.*, 100, 115, 1974.

125. **Peck, H. D., Jr., Deacon, T. E., and Davidson, J. T.**, Studies on adenosine-5'-phosphosulfate reductase from *Desulfovibrio desulfuricans* and *Thiobacillus thioparus.* I. The assay and purification, *Biochim. Biophys. Acta*, 96, 429, 1965.

126. **Labeyrie, F.**, Discussion on the mechanism of adenylyl sulfate reductase, in *Flavins and Flavoproteins*, Kamin, H., Ed., University Park Press, Baltimore, 1971, 579.

127. **Bramlett, R. N. and Peck, H. D., Jr.**, Some physical and kinetic properties of adenylyl sulfate reductase from *Desulfovibrio vulgaris*, *J. Biol. Chem.*, 250, 2979, 1975.

128. **Lyric, R. M. and Suzuki, I.**, Kinetic studies of sulfite:cytochrome *c* oxidoreductase, thiosulfate-oxidizing enzyme, and adenosine-5'-phosphosulfate reductase from *Thiobacillus thioparus*, *Can. J. Biochem.*, 48, 594, 1970.

129. **Adachi, K. and Suzuki, I.**, A study on the reaction mechanism of adenosine 5'-phosphosulfate reductase from *Thiobacillus thioparus*, an iron-sulfur flavoprotein, *Can. J. Biochem.*, 55, 91, 1977.

130. **Michaelis, G. B., Davidson, J. T., and Peck, H. D., Jr.**, A flavin-sulfite adduct as an intermediate in the reaction catalyzed by adenylyl sulfate reductase from *Desulfovibrio vulgaris*, *Biochem. Biophys. Res. Commun.*, 39, 321, 1970.

131. **Michaelis, G. B., Davidson, J. T., and Peck, H. D., Jr.**, Studies on the mechanism of adenylyl sulfate reductase from the sulfate-reducing bacterium, *Desulfovibrio vulgaris*, in *Flavins and Flavoproteins*, Kamin, H., Ed., University Park Press, Baltimore, 1971, 555.

132. **Swoboda, B. E. P. and Massey, V.**, On the regulation of the glucose oxidase from *Aspergillus niger* with bisulfite, *J. Biol Chem.*, 241, 3409, 1966.

133. **Massey, V., Muller, F., Feldberg, R., Schuman, M., Sullivan, P. A., Howell, L. G., Mayhew, S. G., Matthews, R. G., and Foust, G. P.**, The reactivity of flavoprotein with sulfite, *J. Biol. Chem.*, 244, 3999, 1969.

134. **Muller, F. and Massey, V.**, Flavin-sulfite complexes and their structures, *J. Biol. Chem.*, 244, 4007, 1969.

135. **Kondratieva, E. N., Zhukov, V. G., Ivanovsky, R. N., Petushkova, Y. P., and Monosov, E. Z.**, The capacity of phototrophic sulfur bacterium *Thiocapsa roseopersicina* for chemosynthesis, *Arch. Microbiol.*, 108, 287, 1976.

136. **Lyric, R. M. and Suzuki, I.**, Enzymes involved in the metabolism of thiosulfate by *Thiobacillus thioparus*, I. Survey of enzymes and properties of sulfite:cytochrome *c* oxidoreductase, *Can. J. Biochem.*, 48, 334, 1970.

137. **Charles, A. M.**, Mechanism of thiosulfate oxidation by *Thiobacillus intermedius*, *Arch. Biochem. Biophys.*, 129, 124, 1969.

138. **Vestal, J. R. and Lundgren, D. G.**, The sulfite oxidase of *Thiobacillus ferrooxidans (Ferrobacillus ferrooxidans)*, *Can. J. Biochem.*, 49, 1125, 1971.

139. **MacLeod, R. M., Farkas, W., Fridovich, I., and Handler, P.**, Purification and properties of hepatic sulfite oxidase, *J. Biol. Chem.*, 236, 1841, 1961.

140. **Howell, L. G. and Fridovich, I.**, Sulfite:cytochrome *c* oxidoreductase, *J. Biol. Chem.*, 243, 2941, 1968.

141. **Yamanaka, T. and Fukumori, Y.**, *Thiobacillus novellus* cytochrome oxidase can separate some eucaryotic cytochrome *c*, *FEBS Lett.*, 77, 155, 1977.

142. **Adams, C. A., Warnes, G. M., and Nicholas, D. J. D.**, A sulphite-dependent nitrate reductase from *Thiobacillus denitrificans*, *Biochim. Biophys. Acta*, 235, 398, 1971.

143. **Aminuddin, M. and Nicholas, D. J. D.**, An AMP-independent sulphite oxidase from *Thiobacillus denitrificans*: purification and properties, *J. Gen. Microbiol.*, 82, 103, 1974.

144. **Kodama, A., Kodama, T., and Mori, T.**, Studies on the metabolism of a sulfur-oxidizing bacterium. VII. Oxidation of sulfite by a cell-free extract of *Thiobacillus thiooxidans*, *Plant Cell Physiol.*, 11, 701, 1970.

145. **Fridovich, I. and Handler, P.**, Detection of free radicals generated during enzymic oxidations by the initiation of sulfite oxidation, *J. Biol. Chem.*, 236, 1836, 1961.

146. **Nakamura, S.**, Initiation of sulfite oxidation by spinach ferredoxin-NADP reductase and ferredoxin system: a model experiment on the superoxide anion radical production by metalloflavoproteins, *Biochem. Biophys. Res. Commun.*, 41, 177, 1970.

147. Lees, H., Energy metabolism in chemolithotrophic bacteria, *Ann. Rev. Microbiol.*, 14, 83, 1960.
148. Vishniac, W. and Trudinger, P. A., Carbon dioxide fixation and substrate oxidation in the chemo-synthetic sulfur and hydrogen bacteria, *Bacteriol. Rev.*, 26, 168, 1962.
149. Trudinger, P. A., The effect of thiosulfate and oxygen concentration on tetrathionate oxidation by *Thiobacillus X* and *Thiobacillus thioparus*, *Biochem. J.*, 90, 640, 1964.
150. Trudinger, P. A., Evidence for a four-sulfur intermediate in thiosulfate oxidation by *Thiobacillus X*, *Aust. J. Biol. Sci.*, 17, 577, 1964.
151. Peck, H. D., Jr., and Fisher, E., Jr., The oxidation of thiosulfate and phosphorylation in extracts of *Thiobacillus thioparus*, *J. Biol. Chem.*, 237, 190, 1962.
152. Lyric, R. M. and Suzuki, I., Enzymes involved in the metabolism of thiosulfate by *Thiobacillus thioparus*. III. Properties of thiosulfate-oxidizing enzyme and proposed pathway of thiosulfate oxidation, *Can. J. Biochem.*, 48, 355, 1970.
153. Vishniac, W., The metabolism of *Thiobacillus thioparus*. I. The oxidation of thiosulfate, *J. Bacteriol.*, 64, 363, 1952.
154. Pankhurst, E. S., Polarographic evidence of the production of polythionates during the bacterial oxidation of thiosulfate, *J. Gen. Microbiol.*, 34, 427, 1964.
155. LéJohn, H. B., Van Caeseele, L. A., and Lees, H., Catabolic repression in the facultative chemoau-totroph *Thiobacillus novellus*, *J. Bacteriol.*, 94, 1484, 1967.
156. Trudinger, P. A., Thiosulfate oxidation and cytochromes in *Thiobacillus X*. II. Thiosulfate-oxidizing enzyme, *Biochem. J.*, 78, 680, 1961.
157. Trudinger, P. A., Effect of thiol-binding reagents on the metabolism of thiosulfate and tetrathionate by *Thiobacillus neapolitanus*, *J. Bacteriol.*, 89, 617, 1965.
158. Silver, M. and Lundgren, D. G., The thiosulfate-oxidizing enzyme of *Ferrobacillus ferrooxidans* (*Thiobacillus ferrooxidans*), *Can. J. Biochem.*, 46, 1215, 1968.
159. Aleem, M. I. H., Thiosulfate oxidation and electron transport in *Thiobacillus novellus*, *J. Bacteriol.*, 90, 95, 1965.
160. Oh, J. K. and Suzuki, I., Isolation and characterization of a membrane-associated thiosulphate-oxi-dizing system of *Thiobacillus novellus*, *J. Gen. Microbiol.*, 99, 397, 1977.
161. Smith, A. J. and Lascelles, J., Thiosulphate metabolism and rhodanese in *Chromatium* sp. strain D, *J. Gen. Microbiol.*, 42, 357, 1966.
162. Yoch, D. C. and Lindstrom, E. S., Survey of the photosynthetic bacteria for rhodanese (thiosulfate: cyanide sulfur transferase) activity, *J. Bacteriol.*, 106, 700, 1971.
163. Bowen, T. J., Butler, P. J., and Happold, F. C., Some properties of the rhodanese system of *Thiobacillus denitrificans*, *Biochem. J.*, 97, 651, 1965.
164. Tabita, R., Silver, M., and Lundgren, D. G., The rhodanese enzyme of *Ferrobacillus ferrooxidans* (*Thiobacillus ferrooxidans*), *Can. J. Biochem.*, 47, 1141, 1969.
165. Silver, M. and Kelly, D. P., Rhodanese from *Thiobacillus A2*: catalysis of reactions of thiosulphate with dihydrolipoate and dihydrolipoamide, *J. Gen. Microbiol.*, 97, 277, 1976.
166. Westley, J., Sulfane-transfer catalysis by enzymes, in *Bioorganic Chemistry*, Vol. 1, Van Tamelen, E. E., Eds., Academic Press, New York, 1977, 371.
167. Silver, M., Howarth, O. W., and Kelly, D. P., Rhodanese from *Thiobacillus A2*: determination of activity by proton nuclear magnetic resonance spectroscopy, *J. Gen. Microbiol.*, 97, 285, 1976.
168. Volini, M. and Westley, J., The mechanism of the rhodanese-catalyzed thiosulfate-lipoate reactions, *J. Biol. Chem.*, 241, 5168, 1966.
169. Koj, A., Enzymatic reduction of thiosulphate in preparations from beef liver, *Acta Biochim. Pol.*, 15, 161, 1968.
170. Burtner, C. P. and Akagi, J. M., Observations on the rhodanese activity of *Desulfotomaculum ni-grificans*, *J. Bacteriol.*, 107, 375, 1971.
171. Finazzi Argò, A., Cannella, C., Graziani, M. T., and Cavallini, D., A possible role for rhodanese: the formation of "labile" sulfur from thiosulfate, *FEBS Lett.*, 16, 172, 1971.
172. Bonomi, F., Pagani, S., and Cerletti, P., Insertion of sulfide into ferredoxins catalyzed by *FEBS Lett.*, 84, 149, 1977.
173. Cannella, C., Pecci, L., Pensa, B., Costa, M., and Cavallini, D., The titration of rhodanese with substrate, *FEBS Lett.*, 49, 22, 1974.
174. Pagani, S., Cannella, C., Cerletti, P., and Pecci, L., Restoration of reconstitutive capacity of succi-nate dehydrogenase by rhodanese, *FEBS Lett.*, 51, 112, 1975.
175. Bonomi, F., Pagani, S., Cerletti, P., and Cannella, C., Rhodanese-mediated sulfur transfer to suc-cinate dehydrogenase, *Eur. J. Biochem.*, 72, 17, 1977.

Chapter 7

HEME-REQUIRING BACTERIAL RESPIRATORY SYSTEMS

C. J. Knowles

TABLE OF CONTENTS

I. INTRODUCTION

Cytochromes, hemoglobin, myoglobin, catalase, peroxidase, and several other proteins contain iron-heme prosthetic groups. Formation of these proteins requires a supply of both the apoprotein and the prosthetic group.

A wide range of bacteria form cytochromes, catalase, peroxidase, and various oxygenases. Most of these organisms are able to biosynthesize both the apoprotein and the heme prosthetic groups. However, a few species, especially those found in habitats that contain hemoproteins, heme, or related compounds, are unable to form hemoproteins unless supplied with exogenous heme or a precursor.

It is the aim of this review to discuss the respiratory systems of bacteria that require heme or a related compound for cytochrome synthesis. Heme-requiring mutants of many bacteria have been isolated;[1] in general their properties will not be discussed here.

II. HEMES AND HEMOPROTEINS

Hemes are planar quadridentate ferrous-iron chelates of porphyrins (substituted porphins).[2] The structures of some hemes are in Figure 1.

Protoheme IX is usually referred to as protoheme or simply heme. When the iron is in the ferric state, it is called hemin or hematin, depending on whether it is liganded to a chloride or hydroxyl group. Protoheme is the prosthetic group of cytochromes b and o, catalase, and some peroxidases and oxygenases, as well as hemoglobin and myoglobin in higher organisms.[3-8] Heme c is found in c-type cytochromes and some peroxidases.[3-8] It is the only heme group to have covalent linkages to the protein (between the vinyl side chains of the heme and cysteine sulfhydryl groups of the protein). Heme a is found in cytochromes of the a-, a_1-, and a_3-types. Heme d (formerly heme a_2), which has a partially saturated pyrrole ring and is therefore really an iron dihydroporphyrin or chlorin, is found in d-type cytochromes (previously cytochrome a_2).[3-5]

The fifth and sixth coordination positions of the iron in hemes are perpendicular to the plane of the porphyrin ring. They are usually occupied by imidazole groups of histidine residues of the associated protein. In the case of oxidases, the sixth position is occupied by oxygen or the inhibitors carbon monoxide, cyanide, or azide.[3-5] Cytochromes d, o, and a_3 are oxidases.[3-5,8,9] Cytochrome a_1 and some c-type cytochromes are also able to bind CO and are potential oxidases.[9-11]

III. THE PATHWAY OF HEME BIOSYNTHESIS

The pathway of heme biosynthesis and its regulation has been reviewed several times,[12-16] and only a brief discussion will be included in this article. The pathway is outlined in Figure 2, and the structures of the intermediates are in Figure 3.

Synthesis of protoporphyrin IX from coproporphyrinogen is oxidative (Steps 5 and 6, Figure 2). In mitochondria two enzymes, coproporphyrinogenase and protoporphyrinogenase, are involved, and molecular oxygen is required for the conversion.[17-22] Extracts of some aerobically and anaerobically grown bacteria can carry out this process via a soluble, oxygen-requiring enzyme system.[23-25]

Many strictly anaerobic bacteria and facultatively anaerobic bacteria growing anaerobically are able to synthesize cytochromes. Clearly formation of protoporphyrin from coproporphyrinogen under these conditions cannot involve oxygen. Extracts derived from *Escherichia coli* grown anaerobically in nitrate- or fumarate-containing media can form protoporphyrin from protoporphyrinogen in the presence of these elec-

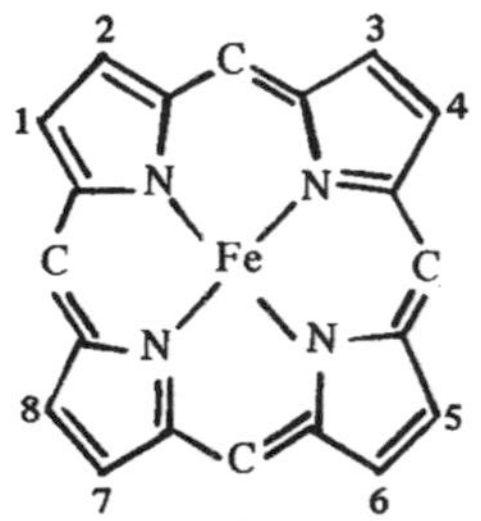

	Side chains at positions					
Compound	1,3,5	6,7	2	4	8	Comments
Protoheme IX	$-CH_3$	$-CH_2 CH_2 COOH$	$-CH=CH_2$	$-CH=CH_2$	$-CH_3$	Protoheme, heme
Mesoheme	$-CH_3$	$-CH_2 CH_2 COOH$	$-CH_2 CH_3$	$-CH_2 CH_3$	$-CH_3$	
Hematoheme	$-CH_3$	$-CH_2 CH_2 COOH$	$-CH(OH)CH_3$	$-CH(OH)CH_3$	$-CH_3$	
Deuteroheme	$-CH_3$	$-CH_2 CH_2 COOH$	$-H$	$-H$	$-CH_3$	
Heme a	$-CH_3$	$-CH_2 CH_2 COOH$	$-CH(OH)CH_2 R$	$-CH=CH_2$	$-CHO$	$R = -\langle CH_2 CH=C(CH_3)CH_2 \rangle_2 CH_2 CH=C(CH_3)_2$
Heme c	$-CH_3$	$-CH_2 CH_2 COOH$	$-CH(CH_3)SR$	$-CH(CH_3)SR$	$-CH_3$	$-SR$ = cysteine residue (i.e., covalent linkage to the cytochrome protein)
Heme d	$-CH_3$	$-CH_2 CH_2 COOH$	$-CH(OH)CH_3$	$-CH=CH_2$	$-CH_3$	Saturated carbons at Positions 7 and 8, partially reducing one of the pyrrole rings (i.e., an iron dihydroporphyrin, or chlorin, derivative)

FIGURE 1. The structures of iron tetrapyrrole compounds.

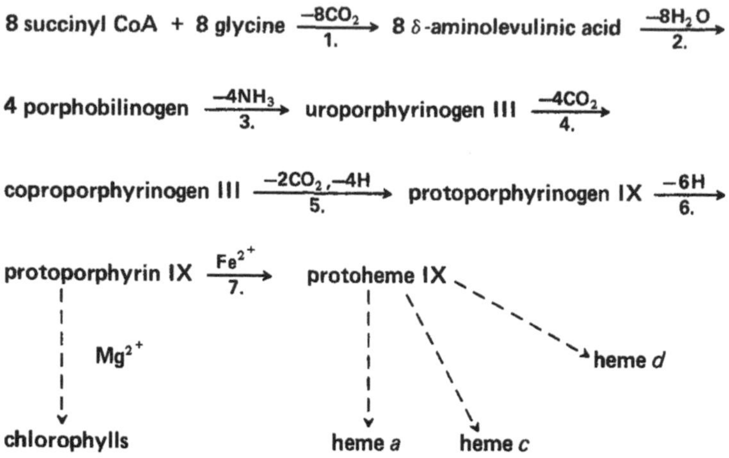

$$\text{8 succinyl CoA} + \text{8 glycine} \xrightarrow[1.]{-8CO_2} \text{8 } \delta\text{-aminolevulinic acid} \xrightarrow[2.]{-8H_2O}$$

$$\text{4 porphobilinogen} \xrightarrow[3.]{-4NH_3} \text{uroporphyrinogen III} \xrightarrow[4.]{-4CO_2}$$

$$\text{coproporphyrinogen III} \xrightarrow[5.]{-2CO_2, -4H} \text{protoporphyrinogen IX} \xrightarrow[6.]{-6H}$$

$$\text{protoporphyrin IX} \xrightarrow[7.]{Fe^{2+}} \text{protoheme IX}$$

protoporphyrin IX —(Mg²⁺)→ chlorophylls

protoheme IX → heme a, heme c, heme d

FIGURE 2.   The pathway of heme biosynthesis.

tron acceptors.[26,27] With fumarate as the electron acceptor, the conversion is linked to respiration, since the activity is strongly inhibited by (1) 2-heptyl-4-hydroxyquinoline-N-oxide (HOQNO), an inhibitor of electron transport,[28] and (2) UV radiation, which destroys quinones involved in respiration; the latter inhibition can be relieved by addition of menadione.[29] Although the nature of the anaerobic coproporphyrinogenase activity of *E. coli* is not known, formation of protoheme from δ-aminolevulinic acid has been demonstrated, with fumarate as an electron acceptor.[26] In *Rhodopseudomonas spheroides, Paracoccus denitrificans*, and yeast, anaerobic coproporphyrinogenase activity occurs in the presence of ATP, Mg²⁺, NAD(P)⁺, and S-adenosyl-methionine or methionine.[30-32] Interestingly, extracts of *Staphylococcus epidermidis* are unable to oxidize protoporphyrinogen under anaerobic conditions.[27] This organism is unable to form cytochromes under anaerobic conditions. Under aerobic conditions, nonenzymic oxidation of protoporphyrinogen may enable heme synthesis to occur.

Ferrous iron is incorporated into protoporphyrin IX to give protoheme by ferrochelatase[33-41] which is inhibited by protoheme,[39] preventing excess heme synthesis. In bacteria this enzyme is located in the cytoplasmic membrane.[39,40] Ferric iron is not a substrate for ferrochelatase and must first be reduced to the ferrous form.[39,33] In mitochondria and *Spirillum itersonii*, NADH or succinate may act as respiration-linked electron donors for the reduction of iron.[42,43] In *S. itersonii*, the reduction of iron is inhibited by oxygen.[43] This may be a control mechanism, since the cytochrome concentrations in many aerobic or facultative anaerobic bacteria are optimal at low oxygen tensions during growth and partially repressed at higher oxygen levels.[44]

Little is known about the synthesis of the heme groups of cytochromes a, c, and d (see Jacobs).[14] Protoheme-requiring bacteria (see following sections) and mutants of other bacteria[1] form these cytochromes, suggesting that protoheme is a precursor. However, there is only limited evidence for the steps involves.[45-49] The ability of many bacteria growing anaerobically to synthesize cytochromes a, c, and d indicates that molecular oxygen is not required for their synthesis.

IV. BACTERIA REQUIRING PROTOHEME FOR RESPIRATION

A. *Bacteriodes*

Bacteriodes spp. are strictly anaerobic Gram negative, nonsporing rods.[50] They are isolated from the natural orifices of man and other animals and from infections. Some

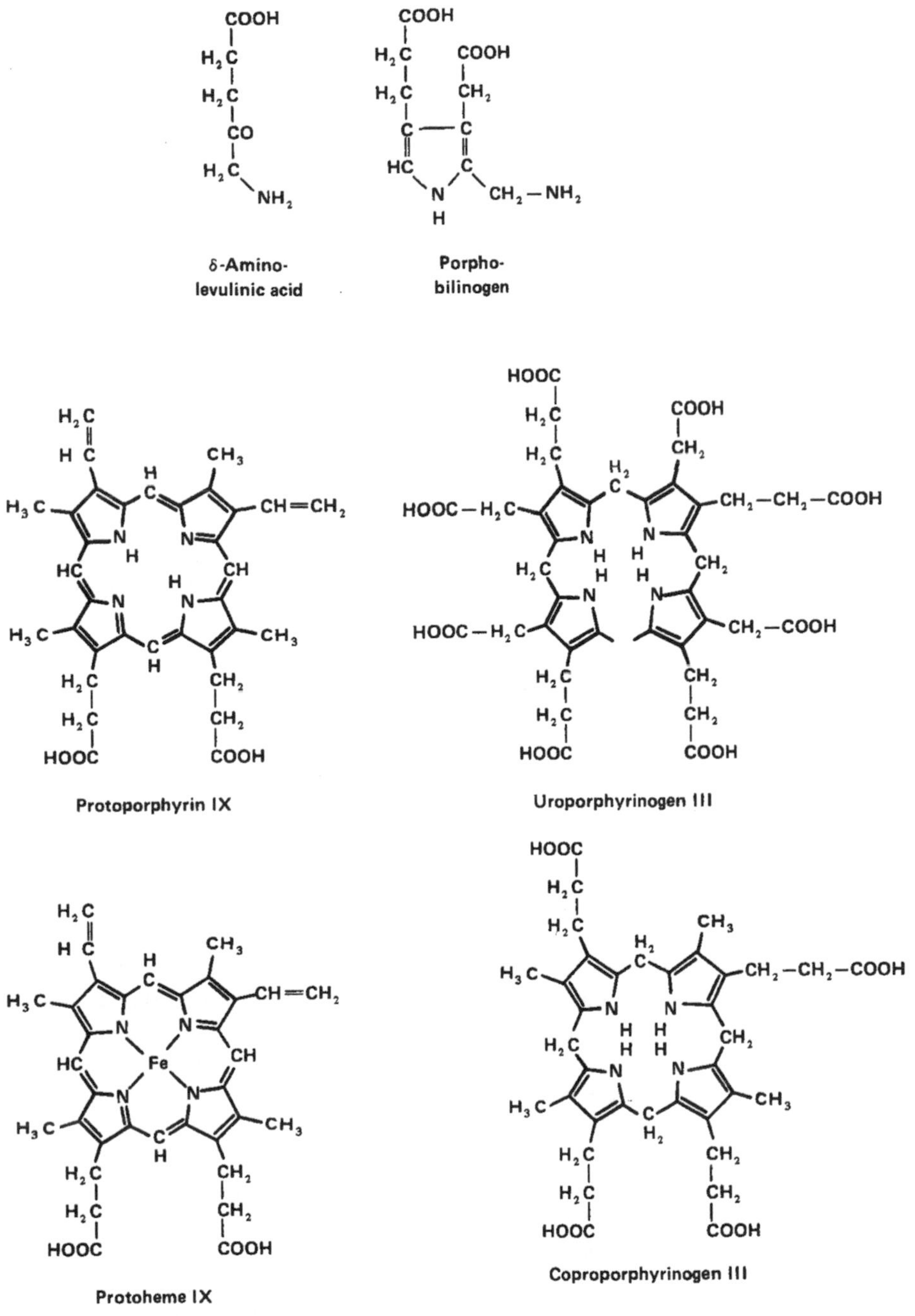

FIGURE 3. Structure of intermediates of the heme biosynthetic pathway.

species are pathogenic. *Bacteriodes ruminicola* is numerically predominant in the rumen.[51]

Bacteriodes melaninogenicus requires 0.4 to 1 μg/mℓ hematin for optimal growth.[52,53] Some strains also require menadione or other analogues of vitamin K for growth.[52,54] Dicoumarol, which is an antagonist of vitmin K, inhibits growth; the inhibition can be overcome by menadione.[52] On blood agar plates *B. melaninogenicus* forms characteristic shiny black colonies. The black coloration is due to protoheme[55] which constitutes up to 43% of the dry weight.[53]

TABLE 1

HEME-REPLACEMENT FACTORS FOR GROWTH OF
BACTERIODES RUMINICOLA SUBSP. *RUMINICOLA*
(STRAIN 23)

Compounds able to replace hemin

Protoporphyrin, mesoprophyrin, deuteroporphyrin, hematoporphyrin, coproporphyrinogen, uroporphyrinogen, deuteroheme, mesoheme, Mn-protoheme, Zn-protoheme,
hemoglobin, catalase, peroxidase, heme *c*

Compounds unable to replace hemin

Tetrapyrroles: coproporphyrin, uroporphyrin, Cu-protoheme
Related compounds: chlorophyll, pheophytin, bilirubin, phycoerythrin
Biosynthetic precursors: α-oxoglutarate, serine, porphobilinogen, δ-aminolevulinic acid
Hemoproteins: cytochrome *c*

Bryant and Robinson[56] isolated 89 strains of *Bacteriodes* from the rumen of cows
of which 31% of the bacteria of one morphological group required hemin or heme-
replacing factors in the ruminal fluid for growth. Two strains (23 and GA20) were
shown to be *B. ruminicola* subsp. *ruminicola*. The optimal concentration of hemin for
both strains was reported to be 1 μg/ml, but a later paper suggests that only about
0.05 μg/ml hemin is required.[57]

An extensive investigation of the nutritional requirements of ruminal and nonruminal *Bacteriodes* sp. confirmed that *B. ruminicola* subsp. *ruminicola* strain 23 required hemin for growth and that only *Bacteriodes hypermagas* of the nonruminal
species tested did not require heme.[58] The heme-requiring strains were *Bacteriodes fragilis* subsp. *fragilis*, *B. fragilis* subsp. *distanosis*, *B. fragilis* subsp. *thetaiotomicron*,
and *Bacteriodes oralis*. The growth stimulation caused by hemin was greatest for *B.
fragilis*.

There has been some discussion whether *B. fragilis* has a requirement for heme for
growth. In a glucose-containing, heme-free medium, it had a generation time of 8 hr,
and the fermentation end-products were mainly lactate, fumarate, and some acetate,
but in the presence of hemin, the cell yield increased dramatically, the generation time
was 2 hr, and the end-products were succinate and acetate plus some propionate and
formate.[59] However, it has recently been shown that *B. fragilis* has an absolute requirement for heme.[60] Growth in "heme-free" media was probably due to residual heme
in the glassware, carry-over of heme with the inoculum or traces of it in the constitu-
ents of complex media. About 2 to 10 ng/ml and 50 ng/ml hemin is required for half-
maximal and maximal cell yields, respectively.[60] Growth rates and end-product formation are also dependent on the hemin concentration.

For growth of several *Bacteriodes* spp.,[58] protoheme could be replaced by protoporphyrin and hematoporphyrin or with reduced growth rates and yields, by protoheme
containing proteins (peroxidase, catalase, hemoglobin, or cytochrome *c*). It was shown
that tetrapyrrole precursors of protoheme (protoporphyrin, coproporphyrinogen, and
uroporphyrinogen) are able to replace heme as growth factors for *B. ruminicola* subsp.
ruminicola (Table 1), but earlier precursors (α-oxoglutarate, δ-aminolevulinic acid, and
porphobilinogen) are inactive.[57] This suggests that the bacterium synthesises only enzymes for conversion of the tetrapyrrole precursors (Figure 2) but it is also possible
that the tetrapyrroles are able to enter the cell, whereas δ-aminolevulinic acid, etc. do
not have access to the appropriate enzymes. However, δ-aminolevulinic acid dehydrase
(Enzyme 2, Figure 2) is not formed, and the organism is unable to convert δ-aminole-

vulinic acid to porphobilinogen or coproporphyrinogen, whereas a heme-independent strain of *B. ruminicola* is able to do so.[57]

Porphobilinogen and δ-aminolevulinic acid failed to act as replacements for hemin for growth of *B. fragilis*, but protoporphyrin was a growth factor; unfortunately other precursors of protoheme have not been tried as growth factors.[60] Nonetheless, it seems reasonable to suppose that, like *B. ruminicola*, it is able to synthesize protoheme from tetrapyrrole precursors. It is not known whether any precursors of protoheme act as growth factors for *B. melaninogenicus*.

B. melaninogenicus synthesizes a cytochrome with an α-peak at 555 nm in room temperature spectra of intact cells and at 550 nm in $-190°C$ spectra of a particulate fraction.[53] It was reduced by endogenous substrates in intact cells and oxidized by fumarate, and despite *B. melaninogenicus* being a strict anaerobe, oxygen. NADH reduced the cytochrome in the particulate fraction and fumarate oxidized it; the reduction was inhibited by 2-*n*-nonyl-4-hydroxyquinoline *N*-oxide. Pyridine hemochrome spectra of the residue of acid-acetone extracts have confirmed that it is a *c*-type cytochrome. A CO-binding pigment with peaks at 578, 540, and 416 nm, and minima at 553 and 425 nm is also formed; this could be either a cytochrome *o* or a CO-binding cytochrome *c*. Although protoheme was found in the cell, it does not appear to be in an enzymatically reducible state (i.e., not a *b*-type cytochrome), so the CO-binding pigment is more likely to be a cytochrome c_{co}.

In contrast, a different strain of *B. melaninogenicus* which requires both vitamin K and heme for growth has been briefly reported to form a *b*-type cytochrome.[61] Succinate could replace the growth requirement for hemin in the presence of vitamin K when no cytochrome was formed.

Fumarate reductase activity has been detected in extracts of *B. fragilis*,[59] as has reduction of fumarate coupled to oxidation of NADH or molecular hydrogen.[62] The latter process required the presence of a catalytic quantity of clostridial ferredoxin, FMN^+, FAD^+, or benzyl viologen. Added ferredoxin also mediated reduction of NAD^+, but not $NADP^+$ by H_2, suggesting that NAD^+ is the physiological mediator of hydrogenase in *B. fragilis*. The overall oxidation of H_2 by fumarate was partially inhibited by rotenone, acriflavin, HOQNO, antimycin A, and UV light.[62] *B. fragilis* also forms at least two cytochromes *b*, cytochrome *o*, and possibly a cytochrome *c*.[59]

The overall respiratory pathway is probably as follows:

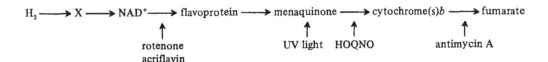

where X is a low-potential electron carrier, such as flavodoxin or ferredoxin. Although *b*-type cytochromes have been shown to be involved in respiration to fumarate in other bacterial species (see Kroger, this volume, and Reference 62), cytochromes *c* and *o* (the latter is usually thought to be an oxidase)[3-5,8,9] have not been implicated in this process, and their roles in *B. fragilis* remain unclear.

Sugar catabolism by *B. fragilis* is by glycolysis to phosphoenol pyruvate (PEP).[59,64] This is converted to either pyruvate, lactate, and acetate, or to oxaloacetate, malate, fumarate, and succinate and then via several stages to propionate. Energy is conserved as ATP by substrate level phosphorylation at the usual stages of catabolism. Maximal energy-conserving efficiency is attained by the use of a pyrophosphate-dependent 6-phosphofructokinase, rather than the more usual ATP-linked enzyme, and by conversion of PEP to oxaloacetate by PEP carboxykinase with the generation of an extra ATP.

It is probable that ATP is also formed by respiration-coupled oxidation of the NADH generated during catabolism. The difference in redox potentials of the NADH + H$^+$/NAD$^+$ and fumarate/succinate couples are sufficient to enable one, or less likely, two molecules of ATP to be generated by oxidative phosphorylation per molecule of fumarate reduced.[65] ATP may also be generated by the coupling of H$_2$ oxidation to reduction of fumarate. To date no direct evidence is available for oxidative phosphorylation (for example by measurements of proton extrusion).

Data from studies on growth yields should be accepted with caution unless proper account is taken of changes in growth rates, maintenance energy, etc. (see Stouthamer[66] and Neijssel and Tempest, Chapter 1, Volume I). Nonetheless, measurements of growth yields of *B. fragilis* appear to provide indirect support for the conclusion that ATP is generated by oxidative as well as substrate-level phosphorylation.[59,60]

When grown on "heme-free" media (probably containing limited concentrations of heme; see the preceding discussion), the growth yields of *B. fragilis* were much lower, no cytochromes or fumarate reductase were formed, and fumarate rather than succinate was the end product of catabolism.[59] Therefore, little or no energy conservation by oxidative phosphorylation occurred under these growth conditions. Synthesis of catalase but not superoxide dismutase (which is not a heme-containing enzyme) by *Bacteriodes* spp. was dependent on the hemin content of the medium.[67,68]

B. ruminicola ferments glucose to PEP by glycolysis, followed by formation of succinate, acetate, and formate by CO$_2$-requiring pathways similar to those of *B. fragilis*.[69-71] Growth yield studies suggest that ATP synthesis occurs in *B. ruminicola* strain B$_1$4 by oxidative and substrate phosphorylation.[69]

A heme-requiring strain (23) of *B. ruminicola* has been reported to form flavoprotein, cytochromes *b*, and *o* and to contain low levels of catalase.[72] Pyridine hemochrome spectra indicated that no heme *c* was formed. In intact cells, flavoprotein and cytochrome *b* were reduced by N$_2$ and oxidized by air, but no aerobic oxidase activity was detected. Fumarate or its immediate metabolic precursors, malate and oxaloacetate, caused oxidation of reduced cytochrome *b*. Oxidation of fumarate was not inhibited by HOQNO.

In contrast, another heme-requiring strain (B$_1$4) of *B. ruminicola* does not synthesize cytochrome *o*, but forms membrane-bound cytochrome *b* and vitamin K$_2$ (menaquinone).[73] HOQNO-sensitive NADH:fumarate oxidoreductase activity was detected in particulate preparations. Activity was lost when the quinone was extracted with *n*-pentane and restored on readdition of vitamin K$_1$ or the *n*-pentane extract. The cytochrome *b* underwent oxidation-reduction when fumarate and NADH were added to the particulate fraction, but the rate of reduction in preparations partially inactivated by HOQNO was insufficient to account for overall respiratory activity. A scheme for the respiratory pathway similar to the following was proposed:[73]

$$\text{NADH} \longrightarrow \text{flavoprotein} \longrightarrow \text{menaquinone} \longrightarrow \text{fumarate}$$

$$\text{cytochrome } b$$

$$\text{HOQNO}$$

where the solid arrows indicate the major pathways.

The nonphysiological tetrapyrroles, mesoporphyrin, deuteroporphyrin, and hematoporphyrin, as well as Mn- and Zn-protoheme satisfied the growth requirements of *B. ruminicola* strain 23 (Table 1). Deuteroporphyrin, mesoporphyrin, and Mg-, Mn-, and Zn-protoheme also substituted for protoheme in a heme-requiring strain (GA 20) of *B. ruminicola*.[74] When deutero- or mesoporphyrin were the growth factors, cytochrome *b* was replaced by a cytochrome containing deutero- or mesoheme, respectively, as the prosthetic group. Substitution of iron for the nonferrous metal occurred when Mg-, Mn-, or Zn-protoheme were used, resulting in formation of normal *b*-type cytochrome. The unusual cytochromes were apparently functional, since the organism grew, and they were readily reduced by glucose.[74]

B. *Haemophilus*

Bacteria of the *Haemophilus* sp. require V-factor (NAD⁺) and/or X-factor (protoheme) for growth.[50,75,76]

Granick and Gilder[77,78] observed that protoporphyrin could replace hemin for growth of several heme-requiring strains of *Haemophilus influenzae*. Other porphyrins (meso-, hemato-, deutero-, and coproporphyrin and their methylated derivatives, and porphin and dimethyl protoporphyrin) were unable to substitute for protoheme, whereas the iron-containing derivatives, mesoheme, hematoheme, and deuteroheme, acted as growth factors for some of the bacteria. In a rough strain (Turner), mesoporphyrin acted as a growth factor. Cytochrome *c* and catalase were unable to act as sources of heme.

The reasons for the ability to utilize the various iron-containing porphyrin derivatives (hemes), but not the parental porphyrins, has not been investigated. A possible explanation is that the ferrochelatases of these bacteria are unable to insert iron into porphyrins other than protoporphyrin IX (except for mesoporphyrin in the case of the Turner strains). It has yet to be determined whether the various hemes are converted to the appropriate heme prosthetic group before insertion into the cytochromes (see McCall and Caldwell).[74]

In proteose-peptone, the growth yields and generation times of *Haemophilus aegyptius*, *H. influenzae*, and *Haemophilus canis* were proportional to hemin content of the growth medium up to a maximum of 0.1 μg/mℓ, whereas growth of *Haemophilus parainfluenzae* and other protoheme-independent species was unaffected by the presence of hemin.[79,80]

As mentioned above, protoporphyrin can substitute for hemin for growth of *H. influenzae*. It was shown to be similarly active with *H. canis*, but was not a growth factor for *H. aegyptius*; the latter organism must lack ferrochelatase (Enzyme 7, Figure 2). Aminolevulinic acid, porphobilinogen, coproporphyrin, and coproporphyrinogen were inactive as growth factors for all three species. This was shown,[79] using cell free extracts, to be due to an inability of the heme-requiring strains to convert aminolevulinic acid to porphobilinogen, uroporphyrinogen, and coproporphyrinogen, i.e., they are deficient in Enzymes 1 to 4 and probably Enzyme 5 (Figure 2).

During the 1960s, White and colleagues intensively investigated the respiratory system of protoheme-independent *H. parainfluenzae;* this research has been reviewed by White and Sinclair.[81] *H. parainfluenzae* is able to respire with oxygen, nitrate, or fumarate as terminal electron acceptors. The respiratory system consists of flavoproteins, desmethyl menaquinone, and cytochromes of the a_1, *b*, *c*, *d*, and *o* types. Cytochrome a_1 appears to be associated with the nitrate reductase activity. It is a pity that more recently there has been less interest in the respiratory system of this bacterium. Little is known about the spatial organization of the respiratory system across the cytoplasmic membrane, the structure and function of the ATPase, or efficiency of oxidative phosphorylation. *H. parainfluenzae* has the unusual property of being able to oxidize readily exogenous NADH, probably due to permeability to it rather than the dehydrogenase being located on the outer side of the membrane. This property should be extremely useful when studying the efficiency of proton extrusion during oxidation of NAD-linked substrates, respiratory carrier location, etc.

Less is known about the respiratory systems of heme-requiring *Haemophilus* spp.[80] *H. influenzae*, *H. canis*, and *H. aegyptius* were shown to form respiratory systems apparently similar to that of *H. parainfluenzae* when grown in the presence of high concentrations of hemin.[80] Catalase was also formed.[80,82] At a limiting hemin concentration (2 ng/ml) growth was slow for all three species, no cytochromes were formed, and there was no respiratory activity, but low levels of catalase were present.[80] In the presence of glucose, an HOQNO-insensitive NADH oxidase flavoprotein was formed.

The composition of the respiratory systems of *Haemophilus* spp. have recently been used as the basis of a taxonomic survey.[83]

C. *Staphylococcus*

When *S. epidermidis* AT2 was grown with limited aeration in a glucose- and nitrate-containing medium, nitrate was converted quantitatively to nitrite.[84] With vigorous aeration or under anaerobic growth conditions, nitrate was not reduced. The lack of reduction of nitrate during anaerobic growth was not due to decreased formation of nitrate reductase. In resting cell suspensions of anaerobically grown bacteria, nitrate was not reduced by physiological substrates such as glucose. No cytochromes could be detected, there was no catalase formed, and coproporphyrin accumulated in the medium for bacteria grown anaerobically in the presence of nitrate.[84,85] Addition of hemin plus nitrate to the anaerobic growth medium enabled conversion of nitrate to nitrite to occur, catalase was formed, coproporphyrin accumulation was abolished, and there was an increase in growth yield. Several other *Staphylococcus* spp. were also unable to utilize nitrate under anaerobic growth conditions, unless hemin plus nitrate was added to the growth medium.[86]

It appeared, therefore, that during anaerobic growth, there was a deficiency of an intermediary component(s) involved in respiration to nitrate, probably due to an inability to synthesise the cytochrome heme-prosthetic group(s). The build-up of coproporphyrin was due to an inefficient feedback regulation of an early step in protoheme biosynthesis, and added hemin was presumably able to exert greater control of this enzyme. The oxygen-requiring step for protoheme biosynthesis is at a stage subsequent to coproporphyrinogen synthesis, i.e., the steps involving the oxidative enzymes, coproporphyrinogenase or protoporphyrinogenase (Figure 1, Enzymes 5 and 6).

Aerobic growth of *S. epidermidis* resulted in formation of cytochromes *b*, *o*, and *a* and a high rate of glucose oxidation by intact cells and NADH oxidase activity by cell-free extracts.[87-89] With anaerobic growth, only traces of catalase and cytochromes *b* and *o* were found, and the oxidases were about 10% of their aerobic activities. When grown anaerobically in the presence of hemin, catalase and the protoheme- containing cytochromes *b* and *o* were formed, but not cytochrome *a*. The rate of oxidation of glucose by intact cells and NADH by cell free extracts were 64 and 38%, respectively, of the values found in aerobically grown cells. The cytochromes were reducible by endogenous substrates, and were associated with the particulate fraction.

The apoenzymes of the cytochromes and catalase were present in anaerobically grown bacteria, as addition of hemin or aeration of washed suspensions caused a rapid increase in catalase and respiratory activity.[89,90] Protein synthesis was not involved, since chloramphenicol had no effect on the development of catalase and respiratory activity. Addition of δ-aminolevulinic acid enhanced heme synthesis that occurred on aeration of anaerobically grown *S. epidermidis*. Increases in coproporphyrin and uroporphyrin content were also observed; they were probably present in the bacteria in the prophyrinogen form and were oxidized during extraction. When δ-aminolevulinic acid was added to anaerobic suspensions, there was a much greater accumulation of coproporphyrin, but no formation of protoheme. Nitrate could not substitute for oxygen for stimulation of protoheme synthesis.

These results confirmed that there was a requirement for oxygen for synthesis of protoporphyrin from coproporphyrinogen by *S. epidermidis*. Aerobic assay of the conversion of protoporphyrinogen to protoporphyrin showed that it was deficient in the enzymes required to carry out this process. Jacobs and Jacobs[27] have suggested that *S. epidermidis* possibly makes heme aerobically by the nonenzymatic auto-oxidation of protoporphyrinogen.

An aerobic respiratory system containing menaquinone and cytochromes *b*, *o*, and *a* is formed by *Staphylococus aureus*.[91,92] With anaerobic growth, little protoheme is formed, no cytochromes can be detected, and the menaquinone content is lower.[92] Stationary cultures grown in flasks filled to the neck with a medium containing nitrate, but without rigorous exclusion of air, are able to form a respiratory system containing cytochrome *b*, nitrate reductase, and probably menaquinone.[93-96] It is not certain whether protoheme was formed by a pathway in which nitrate acted as an electron acceptor for the oxidative steps (mediated by coproporphyrinogenase and protoporphyrinogenase) or whether sufficient oxygen was present to permit it to act as an electron acceptor. Several *S. aureus* spp. require hemin in the anaerobic growth medium for nitrate reduction to occur.[86] Mutants have been isolated that require protoheme for aerobic growth; when grown in the presence of δ-aminolevulinic acid, they accumulate coproporphyrinogen, indicating that they are unable to convert it to protoporphyrin.[97] The existence of these mutants shows that coproporphyrinogenase or protoporphyrinogenase (or both enzymes) must be formed by *S. aureus*.

Heme-deficient mutants of *S. aureus* have been used to study the nitrate reductase system.[94-96] These mutants apparently form an apo-cytochrome when grown without heme, since their ability to reduce nitrate with physiological respiratory substrates can be restored by addition of hemin to either resting cell suspensions in the presence of chloramphenicol or to cell-free extracts. A mutant of *S. aureus* unable to form protoheme converted added protheme to heme *a* under aerobic growth conditions.[45]

D. Lactic Acid Bacteria

Streptococci grow fermentatively, catabolizing carbohydrates by glycolysis.[50,98] Anaerobically lactate is produced almost quantitatively, with a yield of 2 mol of ATP formed by substrate level phosphorylation per mole of glucose consumed. Under aerobic growth conditions, acetate is the primary product, with an extra ATP generated by substrate level phosphorylation per mole of acetate produced. This occurs because NADH is oxidized without the need for reduction of pyruvate to lactate by lactate dehydrogenase, thereby enabling pyruvate to be further metabolised to acetate.

Anaerobically grown *Streptococcus faecalis* 10C1 forms a potent NADH oxidizing system consisting of two soluble enzymes:[99-101]

$$NADH + H^+ + O_2 \longrightarrow NAD^+ + H_2O_2$$

$$NADH + H^+ + H_2O_2 \longrightarrow NAD^+ + 2H_2O$$

Both enzymes are FAD-containing flavoproteins. The NADH oxidase has a pH optimum of 6.8 to 7.5 and is stimulated by Mn^{2+}. The NADH peroxidase has a pH optimum of 5.4 and is specific for NADH as a reductant, but ferricyanide will replace H_2O_2 as an oxidant. It is not inhibited by cyanide or azide. It has been highly purified and its mechanism studied in detail.[102] A different NADH oxidase has been partially purified from aerobically grown *S. faecalis* 10C1.[103] This enzyme catalyzes direct reduction of oxygen to water without formation of free H_2O_2 or participation of NADH peroxidase. It is FAD-requiring and insensitive to cyanide.

Lactic acid bacteria are generally considered to be catalase negative.[104,105] Under some growth conditions, many species produce H_2O_2. In addition to the action of NADH peroxidase, H_2O_2 may be degraded by some lactic acid bacteria by a nonheme "pseudocatalase, and others, when grown in hematin or lysed red blood-cell containing media, form a more usual heme-containing catalase.[104-107]

Pseudocatalase is inactivated by acid growth conditions, but it is less sensitive to peroxide inactivation than heme containing catalases and is insensitive to azide and cyanide.[105,106] It has none of the absorbance peaks characteristic of hemoproteins or flavoproteins. Pseudocatalase has been found in *Leuconostoc mesenteroides* (slime-forming species) and several other *Leuconostoc* spp., *Pediococcus cerevisiae* and other *Pediococcus* spp., and *Lactobacillus plantarum*.[104-107] Catalase has been found in *L. mesenteroides* (nonslime-forming species), and some *Pediococcus*, *Streptococcus*, and *Lactobacillus* spp. It should be emphasized that many lactic acid bacteria do not form either enzyme, and both H_2O_2-splitting enzymes are never found in the same organism.

The formation of catalase when grown in the presence of hematin indicates the ability of some lactic acid bacteria to form hemoproteins. Whittenbury[105] reported that addition of hematin to washed suspensions of several catalase-negative strains resulted in H_2O_2-splitting activity within 2 hr, and in some instances, the activity was detectable within 5 min. It is not clear whether this infers that the apoenzyme was already present or had to be induced. Hematin could not be replaced by protoporphyrin for formation of catalase, suggesting that ferrochelatase is not formed by these bacteria, although it is possible that the bacteria are impermeable to protoporphyrin.

Bryan-Jones and Whittenbury[108] showed that aerobic growth of *S. faecalis* NCDO 581 on glucose or lactate in the presence of hematin resulted in higher glucose and lactate oxidase activities by intact bacteria than for cells grown aerobically or anaerobically in the absence of hematin. The NADH oxidase activity of cell-free extracts derived from hematin-grown cells was higher than for those derived from cells grown in its absence. Moreover, only the former activity was sensitive to inhibition by cyanide and was present mainly in the particulate fraction. These data suggested that it was linked to a respiratory system, whereas the latter activity was due to soluble NADH oxidase and NADH peroxidase flavoproteins. This conclusion is supported by the studies of Ritchey and Seeley on *S. faecalis* var. *zymogenes*.[109] A particulate fraction derived from hematin-grown cells had a higher NADH oxidase activity than the particulate fraction of cells grown in its absence. The former activity was much more sensitive to inhibition by cyanide, HOQNO, and antimycin A than the latter, but was less sensitive to azide. The supernatant fraction of hematin-grown cells contained less NADH oxidase activity. Pritchard and Wimpenny have also noted a partial repression of soluble NADH oxidase and massive induction of particulate NADH oxidase in *S. faecalis* var. *zymogenes* on addition of hematin to a medium containing lactate as the energy source.[110] Oxidation of lactate by intact cells was also greatly stimulated by inclusion of hematin in the medium.

Whittenbury[105] showed that *S. faecalis* H69D5 forms cytochromes *b*, a_1 and *d*, when grown in heated blood-containing media. Bryan-Jones and Whittenbury later found cytochrome *b* in the membrane fraction of strain 581, but no *a*- or *d*-type cytochromes.[108] *S. faecalis* var. *zymogenes* was reported to form an NADH-reducible cytochrome *b*, and in CO-spectra, an *a*-type cytochrome was observed, but no cytochrome *o*.[109] Further studies on *S. faecalis* var. *zymogenes*, grown on lactate under oxygen limitation to cause maximum induction of the respiratory system, showed that it is also able to form cytochromes *d* and *o*.[110] In 77°K difference spectra, the cytochrome *b* was resolved into two components (α-peaks at 558 and 562 nm).

In each of these studies, it was shown that growth in the absence of hematin resulted in cells devoid of cytochromes.[105,108-110]

S. faecalis 10C1 forms membrane-bound 2-solanesyl-1,4-naphthoquinone when grown in a heme-free medium.[111] The effect of inclusion of hematin in the medium on the level of formation of this quinone has not been studied, nor has its function, but it seems possible that it has a role in respiration.

Addition of hematin to the aerobic growth medium or to resting cell suspensions of *Streptococcus lactis* or *L. mesenteroides* caused changes in the pattern of glucose catabolism.[112] Oxygen uptake was stimulated, less lactate and more acetate accumulated, and cytochromes *b* and *d* were formed. Respiration became more sensitive to cyanide. *Bifidobacterium longum* also formed cytochromes *b* and *d*, when grown in the presence of lysed blood, and cytochrome a_1 was induced when nitrate was also present in the growth medium.[113] Whittenbury[105] investigated the presence of cytochromes in lactic acid bacteria grown in the presence of hematin and showed that *L. plantarum* and *L. mesenteroides* formed cytochromes *b* and *d*, whereas only cytochrome *b* was detected in *Lactobacillus brevis*. A survey of the streptococci confirmed that *S. faecalis*, *S. faecalis* var. *liquefaciens*, *S. faecalis* var. *zymogenes*, and *S. lactis* formed heme-dependent "cytochrome-like respiratory systems", but a wide range of other streptococci were unable to form such respiratory systems.[114]

There have been no reports on cytochrome-linked respiration by lactic acid bacteria to acceptors other than oxygen. *S. faecalis* 10C1 produced acetate and succinate, when grown anaerobically on glucose plus fumarate, whereas only lactate was formed when grown on glucose alone.[115] During growth in the presence of fumarate, a NADH-fumarate reductase was formed.[116] Glycerol was converted anaerobically to dihydroxyacetone, with reduction of fumarate to succinate.[117] Although energy-transducing respiratory systems with fumarate as acceptor usually involve cytochrome *b* (see Kroger,[63] and Kroger, Chapter 1, Volume II), it is possible that in *S. faecalis* ATP can be synthesized without involvement of a cytochrome (see below); alternatively, in the presence of hematin, an energy-transducing, cytochrome *b*-linked respiratory system might be formed. There have also been a few reports of nitrate reduction by lactic acid bacteria,[118] suggesting that it might act as an electron acceptor. Cytochrome a_1 is induced when *B. longum* is grown in the presence of hematin and nitrate.[113]

Gallin and Vardemark grew *S. faecalis* 10C1 aerobically in the absence of hematin and obtained a P/O of about 0.24 for oxidation of NADH by cell free extracts, whereas extracts of anaerobically grown cells exhibited no oxidative phosphorylation.[119] It was claimed that coupling of respiration to phosphorylation was occurring in the flavin region (see the Site 1 region of mitochondrial respiratory systems) of the aerobically grown cells. This respiratory activity could not be due to the NADH oxidase/peroxidase system, the aerobic NADH oxidase or, since hemin had not been added to the medium, a cytochrome-containing respiratory system.

A P/O of 0.19 was obtained for oxidation of NADH, with fumarate as electron acceptor in a particulate fraction derived from *S. faecalis* 10C1 grown anaerobically on glucose in the presence of fumarate, supporting the proposal that Site 1 coupling occurs in this organism.[120]

In contrast, no oxidative phosphorylation appears to occur in extracts of *S. faecalis* 581 or *S. faecalis* var. *zymogenes* grown aerobically in the absence of hematin.[105,109] When hematin was added to the medium under aerobic growth conditions, a P/O of about 0.3 was found for oxidation of NADH by extracts of both organisms.

There have been several attempts to determine the efficiency of oxidative phosphorylation in *S. faecalis* from measurements of growth yields.[108,109,121] For aerobic growth of *S. faecalis* 10C1, Smalley et al.[121] have estimated that there is a P/O of 0.6 for oxidation of NADH, in agreement with the claim from this laboratory that oxidative phosphorylation occurs in cell free extracts of cells grown in the absence of hematin. Ritchey and Seeley[109] were unable to show oxidative phosphorylation in growth yield measurements of *S. faecalis* var. *zymogenes* grown in the absence of hematin, in agreement with the findings of Bryan-Jones and Whittenbury[108] with *S. faecalis* 581. In the presence of hematin, *S. faecalis* var. *zymogenes* was estimated to have a P/O of 0.6 to 0.7.[109] There were increases in growth yield when hematin, but not protoporphyrin, was added to the growth medium with glucose, fructose, or lactate as the carbon source for growth of *S. faecalis* 581.[108] Unfortunately, changes in growth rate and maintenance energy have not been taken into account in all these experiments, and the assumption that the results indicate that oxidative phosphorylation is occurring may not be valid (see Stouthamer[66] and Neijssel and Tempest, Chapter 1, Volume I).

That energy transduction occurs in *S. faecalis* var. *zymogenes* grown in the presence of hematin is shown by measurements of proton translocation. An H^+/O value of 1.37 was obtained for oxidation of lactate by bacteria grown in continuous culture with lactate as the carbon source.[110] The proton pulses were eliminated by the uncoupling agent carbonyl cyanide *m*-chlorophenylhydrazone (CCCP). Bacteria grown without added hematin did not show proton translocation.

Cyanide and azide sensitive NADH oxidase activity has been found in cell free extracts of *Streptococcus agalactiae* grown aerobically on a heme-free medium.[122,123] An unconvincing spectrum was presented to indicate that no cytochromes were formed. However, the NADH oxidase activity was unlikely to be associated with a proton translocating respiratory system, since it was found mainly in the high-speed supernatant fraction of extracts of the bacterium. Dialysis resulted in loss of activity which could be restored by addition of the filtrate, menadione or ferricyanide. Phosphorylation was purported to be associated with the NADH oxidase activity (P/O of about 0.2), but nonoxidative background phosphorylation was extremely high. Growth yield studies were claimed to show that a P/O of 1 occurred in vivo.[124]

E. Other Bacteria

Hematin is required for growth or stimulates growth of *Pasteurella pestis*, *Bordetella pertussis*, and catalase negative strains of *Mycobacterium tuberculosis*.[125] Several compounds can replace hematin, including catalase, pyruvate, charcoal, sulfite, and thioglycollate. All these compounds are able to decompose H_2O_2, and their growth stimulatory effect is due to removal of H_2O_2 from the medium.

Hematin stimulated growth of the catalase negative obligate aerobe *Rickettsia quintana*.[126] Catalase, pyruvate, charcoal, sulfite, and thioglycollate could not replace hematin. It was concluded that hematin was required as a precursor for synthesis of hemoproteins.

The obligate anaerobe *Eubacterium lentum* formed cytochromes *a*, *b*, *c*, and *o* when grown in a heme-free medium.[127] Addition of hemin to the medium stimulated formation of cytochrome *c*.

Aerobic growth of *Corynebacterium pyogenes*, an animal pathogen, was stimulated 270% by addition of hemin to the growth medium.[128] In the absence of hemin, no cytochromes were formed, but cytochrome *b* was present when hemin was added to the growth medium.

V. DISCUSSION

A wide variety of bacteria have been shown to require protoheme (supplied as hemin or hematin) for formation of their respiratory systems. There is an obligatory requirement for heme for growth of several *Bacteriodes* spp. which are strict anaerobes that respire with fumarate as electron acceptor and *Haemophilus* spp. which are facultative anaerobes that respire to fumarate, nitrate, or oxygen. *S. faecalis* and various other lactic acid bacteria are able to grow both aerobically and anaerobically in the absence of heme, obtaining ATP by substrate level phosphorylation; under aerobic conditions in the presence of heme, they form a respiratory system. In contrast, *S. epidermidis* forms a respiratory system when grown aerobically in the absence of heme, but when grown under anaerobic conditions, it requires the presence of heme for synthesis of a respiratory system.

Bacteriodes spp. are able to form protoheme for incorporation into catalase and cytochromes from tetrapyrrole precursors, whereas *Haemophilus* spp. are only able to convert protoporphyrin to protoheme, and in *S. faecalis*, only protoheme itself appears to serve as a precursor for hemoprotein synthesis. *S. epidermidis* is unable to oxidise coproporphyrinogen to protoporphyrin in the absence of oxygen.

In addition to synthesis of protoheme-containing catalase and cytochromes *b* and *o*, various heme-requiring bacteria form cytochromes *a*, *c*, or *d*. Therefore, enzymes involved in biosynthesis of heme *a*, heme *c*, and chlorin must be formed. These bacteria should be of use in elucidating the pathways because of the specificity of product formation from the protoheme precursor.

Little attention has been paid to oxidative phosphorylation by heme-requiring bacteria. The development of the proton pulse technique and its adaptation for use with intact bacteria[129] should lead to an understanding of the efficiency of oxidative phosphorylation in heme-requiring bacteria. A shift from a fermentative to a respiratory mode of metabolism on addition of hematin to the medium, involves a change of role for the membrane ATPase from a catabolic function to that of, primarily, an ATP synthetase. Also, a change in the proton motive force may occur on development of a respiratory system. *S. faecalis*, grown in the absence of hematin, has already been used for studies on the proton motive force[130] and ATPase structure and function.[129] The availability of these data should aid future investigations of the effect of growth in the presence of hematin on the function of the ATPase and changes in the proton motive force.

REFERENCES

1. **Haddock, B. A.**, The isolation of phenotypic and genotypic variants for the functional characterisation of bacterial oxidative phosphorylation, in *Microbial Energetics*, Haddock, B. A. and Hamilton, W. A., Eds., Symp. 27 Soc. General Microbiology, Cambridge University Press, London, 1977, 95.
2. **Falk, J. E.**, *Porphyrins and Metalloporphyrins*, Elsevier New York, 1964.
3. **Lamberg, R. and Barrett, J.**, *Cytochromes*, Academic Press, London, 1973.
4. **Kamen, M. D. and Horio, T.**, Bacterial cytochromes. I. Structural aspects, *Annu. Rev. Biochem.*, 39, 673, 1970.
5. **Kamen, M. D. and Horio, T.**, Bacterial cytochromes. II. Functional aspects, *Annu. Rev. Microbiol.*, 24, 399, 1970.
6. **Hayaishi, O.**, *Molecular Mechanisms of Oxygen Activation*, Academic Press, New York, 1974.
7. **Meyer, D. J. and Jones, C. W.**, Distribution of cytochromes in bacteria-relationship to general physiology, *Int. J. Syst. Bacteriol.*, 23, 459, 1973.
8. **Yamanaka, T. and Okunuki, K.**, Cytochromes, in *Microbial Iron Metabolism*, Neilands, J. B., Ed., Academic Press, New York, 1974, chap. 14.
9. **Lloyd, D. and Edwards, S. W.**, Terminal oxidases: a summary, in *Federation of European Biochemical Societies*, Vol. 49, Colloquium B6, Degn, H., Lloyd, D., and Hill, G. C., Eds., Pergamon Press, Oxford, 1978, 187.

10. **Ingledew, W. J.**, Cytochrome a_I, as an oxidase (?) in *Federation of European Biochemical Societies,* Vol. 49, Colloquium B6, Degn, H., Lloyd, D., and Hill, G. C., Eds., Pergamon Press, Oxford, 1978, 79.

11. **Weston, J. A. and Knowles, C. J.**, A soluble CO-binding *c*-type cytochrome from the marine bacterium *Beneckea natriegens, Biochim. Biophys. Acta,* 305, 11, 1973.

12. **Lascelles, J.**, *Tetrapyrrole Biosynthesis and its Regulation,* W. A. Benjamin, New York, 1964.

13. **Burnham, B. F.**, Metabolism of Porphyrins and Corrinoids, in *Metabolic Pathways,* Vol. 3, 3rd ed., Greenberg, D. M., Ed., Academic Press, New York, 1969, chap. 18.

14. **Jacobs, N. J.**, Biosynthesis of heme, in *Microbial Iron Metabolism,* Neilands, J. B., Ed., Academic Press, New York, 1974, chap. 6.

15. **Maines, M. D. and Kappas, A.**, Metals as regulators of heme metabolism, *Science,* 198, 1215, 1977.

16. **Lascelles, J.**, The regulation of heme and chlorophyll synthesis in bacteria, *Ann. N.Y. Acad. Sci.,* 242, 334, 1975.

17. **Porra, R. J. and Falk, J. E.**, The enzymic conversion of coproporphyrinogen III into Protoporphyrin IX, *Biochem. J.,* 90, 69, 1964.

18. **Battle, A. M. del C., Benson, A., and Rimington, C.**, Purification and properties of coproporphyrinogenase, *Biochem. J.,* 97, 731, 1965.

19. **Hsu, W. P. and Miller, G. W.**, Coproporphyrinogenase in tobacco (*Nicotiana tabacum L.*), *Biochem. J.,* 117, 215, 1970.

20. **Poulson, R. and Polglase, W. J.**, Site of glucose repression of heme biosynthesis, *FEBS Lett.,* 40, 258, 1974.

21. **Poulson, R. and Polglase, W. J.**, The enzymic conversion of protoporphyrinogen IX to protoporphyrin IX, *J. Biol. Chem.,* 250, 1269, 1975.

22. **Poulson, R.**, The enzymic conversion of protoporphyrinogen IX to protoporphyrin IX in mammalian mitochondria, *J. Biol. Chem.,* 251, 3730, 1978.

23. **Jacobs, N. J., Jacobs, J. M., and Brent, P.**, Formation of protoporphyrin from coproporphyrinogen in extracts of various bacteria, *J. Bacteriol.,* 102, 398, 1970.

24. **Jacobs, N. J., Jacobs, J. M., and Brent, P.**, Characterization of late steps of microbial heme synthesis; conversion of coproporphyrinogen to protoporphyrin, *J. Bacteriol.,* 107, 203, 1971.

25. **Jacobs, N. J., Jacobs, J. M., and Morgan, H. E., Jr.**, Comparative effect of oxygen and nitrate on protoporphyrin and heme synthesis from δ-aminolevulinic acid in bacterial cultures, *J. Bacteriol.,* 112, 1444, 1972.

26. **Jacobs, N. J. and Jacobs, J. M.**, Fumarate as alternate electron acceptor for the late steps of anaerobic heme synthesis in *Escherichia coli, Biochem. Biophys. Res. Commun.,* 65, 435, 1975.

27. **Jacobs, N. J. and Jacobs, J. M.**, Nitrate, fumarate and oxygen as electron acceptors for a late step in microbial heme synthesis, *Biochim. Biophys. Acta,* 449, 1, 1976.

28. **Jacobs, N. J. and Jacobs, J. M.**, The late steps in anaerobic heme biosynthesis in *E. coli:* role for quinones in protoporphyrinogen oxidation, *Biochem. Biophys. Res. Commun.,* 78, 429, 1977.

29. **Jacobs, N. J. and Jacobs, J. M.**, Evidence for the involvement of the electron transport system at a late step of anaerobic microbial heme synthesis, *Biochim. Biophys. Acta,* 459, 141, 1977.

30. **Tait, G. H.**, Coproporphyrinogenase activities in extracts of *Rhodopseudomonas spheroides* and *Chromatium* strain D, *Biochem. J.,* 128, 1159, 1972.

31. **Poulson, R. and Polglase, W. J.**, Aerobic and anaerobic coproporphyrinogenase activities in extracts from *Saccharomyces cerevisiae, J. Biol. Chem.,* 249, 6367, 1974.

32. **Tait, G. H.**, Control of aminolaevulinic synthetase in *Micrococcus denitrificans, Enzyme,* 16, 21, 1973.

33. **Porra, R. J. and Jones, O. T. G.**, Studies on ferrochelatase. I. Assay and properties of ferrochelatase from a pig-liver mitochondrial extract, *Biochem. J.,* 87, 181, 1963.

34. **Porra, R. J. and Ross, B. D.**, Haem synthesis and cobalt porphyrin synthase in various microorganisms, *Biochem. J.,* 94, 557, 1965.

35. **Porra, R. J. and Lascelles, J.**, Haemoproteins and haem synthesis in facultative photosynthetic and denitrifying bacteria, *Biochem. J.,* 94, 120, 1965.

36. **Jones, O. T. G.**, Ferrochelatase of spinach chloroplasts, *Biochem. J.,* 107, 113, 1968.

37. **Jones, M. S. and Jones, O. T. G.**, Properties of ferrochelatase in microorganisms adapting to changes in growth conditions, *Biochem. J.,* 116, 19p, 1970.

38. **Jones, M. S. and Jones, O. T. G.**, The structural organisation of haem synthesis in rat liver mitochondria, *Biochem. J.,* 113, 507, 1969.

39. **Jones, M. S. and Jones, O. T. G.**, Ferrochelatase from *Rhodopseudomonas spheroides, Biochem. J.,* 119, 453, 1970.

40. **Dailey, H. A., Jr., and Lascelles, J.**, Ferrochelatase activity in wild-type and mutant strains of *Spirillum iteronii.* Solubilization with chaotropic reagents, *Arch. Biochem. Biophys.,* 160, 523, 1974.

41. Dailey, H. A., Jr., Purification and characterization of the membrane-bound ferrochelatase from *Spirillum itersonii*, *J. Bacteriol.*, 132, 302, 1977.
42. Barnes, R., Connelly, J. L., and Jones, O. T. G., The utilization of iron and its complexes by mammalian mitochondria, *Biochem. J.*, 128, 1043, 1972.
43. Dailey, H. A., Jr., and Lascelles, J., Reduction of iron and synthesis of protoheme by *Spirillum intersonii* and other organisms, *J. Bacteriol.*, 129, 815, 1977.
44. Harrison, D. E. F., The regulation of respiration rate in growing bacteria, *Adv. Microb. Physiol.*, 14, 243, 1976.
45. Sinclair, P., White, D. C., and Barnett, J., The conversion of protoheme to heme a in *Staphylococcus*, *Biochem. Biophys. Acta*, 143, 427, 1967.
46. Lascelles, J., Rittenberg, B., and Clark-Walker, G. D., Growth and cytochrome synthesis in a heme-requiring mutant of *Spirillum itersonii*, *J. Bacteriol.*, 97, 455, 1969.
47. Miyake, S. and Sugimura, T., Coproporphyrinogenase in a respiration-deficient mutant of yeast lacking all cytochromes and accumulating coproporphyrin, *J. Bacteriol.*, 96, 1997, 1968.
48. Garrard, W. T., Synthesis, assembly and localisation of periplasmic cytochrome c, *J. Biol. Chem.*, 247, 5935, 1972.
49. Charalampous, F. C. and Chen, W. L., Anaerobic synthesis of apocytochrome oxidase and assembly of the holoenzyme in yeast protoplasts, *J. Biol. Chem.*, 249, 1007, 1974.
50. Buchanan, R. E. and Gibbons, N. E., Eds., *Bergey's Manual of Determinative Bacteriology*, 8th ed., Williams and Wilkins, Baltimore, 1974.
51. Bryant, M. P., Small, P. N., Bouma, C., and Chu, H., *Bacteriodes ruminicola* sp. nov. and *Succinomonas amylolytica*, gen. nov. — species of succinic acid-producing anaerobic bacteria of the bovine rumen, *J. Bacteriol.*, 76, 15, 1958.
52. Gibbons, R. J. and MacDonald, J. B., Hemin and vitamin K compounds as required factors for the cultivation of certain strains of *Bacteriodes melaninogenicus*, *J. Bacteriol.*, 80, 164, 1960.
53. Rizza, V., Sinclair, P. R., White, D. C., and Cuorant, R., Electron transport system of the protoheme-requiring anaerobe *Bacteriodes melaninogenicus*, *J. Bacteriol.*, 96, 665, 1968.
54. Gibbons, R. J. and Engle, L. P., Vitamin K in bacteria that are obligate anaerobes, *Science*, 146, 1307, 1964.
55. Schwabacher, H., Lucas, D. R. and Rimington, C., *Bacterium melaninogenicum* — a misnomer, *J. Gen. Microbiol.*, 1, 109, 1947.
56. Bryant, M. P. and Robinson, I. M., Some nutritional characteristics of predominant culturable ruminal bacteria, *J. Bacteriol.*, 84, 605, 1962.
57. Caldwell, D. R., White, D. C., Bryant, M. P., and Doetsch, R. N., Specificity of the heme requirement of *Bacteriodes ruminicola*, *J. Bacteriol.*, 90, 1645, 1965.
58. Caldwell, D. R. and Arcand, C., Inorganic and metal-organic growth requirements of the genus *Bacteriodes*, *J. Bacteriol.*, 120, 322, 1974.
59. Macy, J., Probst, I., and Gottschalk, G., Evidence for cytochrome involvement in fumarate reduction and adenosine 5'-triphosphate synthesis by *Bacteriodes fragilis* grown in the presence of hemin, *J. Bacteriol.*, 123, 436, 1975.
60. Sperry, J. F., Appleman, M. D., and Wilkins, T. D., Requirement of heme for growth of *Bacteriodes fragilis*, *Appl. Env. Microbiol.*, 34, 386, 1977.
61. Lev, M., Kendell, K. C., and Milford, A. F., Succinate as a growth factor for *Bacteriodes melaninogenicus*, *J. Bacteriol.*, 108, 175, 1971.
62. Harris, M. A. and Reddy, C. A., Hydrogenase activity and the H_2-fumarate electron transport system in *Bacteriodes fragilis*, *J. Bacteriol.*, 131, 922, 1977.
63. Kroger, A., Phosphorylative electron transport with fumarate and nitrate as terminal hydrogen acceptors, in *Microbial Energetics*, Haddock, B. A. and Hamilton, W. A., Eds., Symp. 27 Soc. General Microbiology, Cambridge University Press, London, 1977, 61.
64. Macy, J. M., Ljungdahl, L. G., and Gottschalk, G., Pathway of succinate and propionate formation in *Bacteriodes fragilis*, *J. Bacteriol.*, 134, 84, 1978.
65. Thauer, R. K., Jungermann, K., and Decker, K., Energy conservation in chemotrophic anaerobic bacteria, *Bacteriol. Rev.*, 41, 100, 1977.
66. Stouthamer, A. H., Energetic aspects of the growth of microorganisms, in *Microbial Energetics*, Haddock, B. A. and Hamilton, W. A., Eds., Symp. 27 Soc. General Microbiology, Cambridge University Press, London, 1977, 285.
67. Gregory, E. M., Veltri, B. J., Wagner, D. L., and Wilkins, T. D., Carbohydrate repression of catalase synthesis in *Bacteriodes fragilis*, *J. Bacteriol.*, 129, 534, 1977.
68. Gregory, E. M., Kowalski, J. B., and Holdeman, L. V., Production and some properties of catalase and superoxidase dismutase from the anaerobe *Bacteriodes distasonis*, *J. Bacteriol.*, 129, 1298, 1977.
69. Howlett, M. R., Mountfort, D. O., Turner, K. W., and Roberton, A. M., Metabolism and growth yields in *Bacteriodes ruminicola* strain B_14, *Appl. Env. Microbiol.*, 32, 274, 1976.

70. Joyner, A. E., Jr., and Baldwin, R. L., Enzymatic studies of pure cultures of rumen microorganisms, *J. Bacteriol.*, 92, 1321, 1966.

71. Scardovi, V. and Chiappini, M. G., Studies in rumen bacteriology. V. Carboxylation of phosphoenolpyruvate in some rumen bacterial strains and in the cell-free extract of the total rumen flora, *Ann. Microbial. Enzymol.*, 16, 119, 1966.

72. White, D. C., Bryant, M. P., and Caldwell, D. R., Cytochrome-linked fermentation in *Bacteriodes ruminicola*, *J. Bacteriol.*, 84, 822, 1962.

73. Mountfort, D. O. and Roberton, A. M., The role of menaquinone and *b*-type cytochrome in anaerobic reduction of fumarate by NADH in membrane preparations from *Bacteroides ruminicola* strain B,4, *J. Gen. Microbiol.*, 100, 309, 1977.

74. McCall, D. R. and Caldwell, D. R., Tetrapyrrole utilization by *Bacteriodes ruminicola*, *J. Bacteriol.*, 131, 809, 1977.

75. Lwoff, A. and Lwoff, M., Studies on codehydrogenase. I. Nature of growth factor V, *Proc. R. Soc. London Ser. B*, 122, 352, 1937.

76. Lwoff, A. and Lwoff, M., Role physiologique de l'hemine pour *Hemophilus influenza*, *Ann. Inst. Pasteur Paris*, 59, 129, 1937.

77. Granick, S. and Gilder, H., The porphyrin requirements of *Haemophilus influenzae* and some functions of the vinyl and propionic acid side chains of heme, *J. Gen. Physiol.*, 30, 1, 1946.

78. Gilder, H. and Granick, S., Studies on the Hemophilus group of organisms. Quantitative aspects of growth on various porphin compounds, *J. Gen. Physiol.*, 31, 103, 1947.

79. White, D. C. and Granick, S., Hemin biosynthesis in *Haemophilus*, *J. Bacteriol.*, 85, 842, 1963.

80. White, D. C., Respiratory systems in the hemin-requiring *Haemophilus* species, *J. Bacteriol.*, 85, 84, 1963.

81. White, D. C. and Sinclair, P. R., Branched electron-transport systems in bacteria, *Adv. Microbial Physiol.*, 4, 173, 1970.

82. Biberstein, E. L. and Gills, M., Catalase activity of *Haemophilus* species grown with graded amounts of hemin, *J. Bacteriol.*, 81, 380, 1961.

83. Holländer, R. and Mannheim, W., Characterization of hemophilic and related bacteria by their respiratory quinones and cytochromes, *Int. J. Syst. Bacteriol.*, 25, 102, 1975.

84. Jacobs, N. J., Johantges, J., and Deibel, R. H., Effect of anaerobic growth on nitrate reduction by *Staphylococcus epidermidis*, *J. Bacteriol.*, 85, 782, 1963.

85. Heady, R. E., Jacobs, N. J., and Deibel, R. H., Effect of haemin supplementation on porphyrin accumulation and catalase synthesis during anaerobic growth of *Staphylococcus*, *Nature (London)*, 203, 1285, 1964.

86. Jacobs, N. J., Heady, R. E., Jacobs, J. M., Chan, K., and Deibel, R. H., Effect of hemin and oxygen tension on growth and nitrate reduction by bacteria, *J. Bacteriol.*, 87, 1406, 1964.

87. Jacobs, N. J. and Conti, S. F., Effect of hemin on the formation of the cytochrome system of anaerobically grown *Staphylococcus epidermidis*, *J. Bacteriol.*, 89, 675, 1965.

88. Jacobs, N. J., MacLosky, E. R., and Conti, S. F., Effects of oxygen and hemin on the development of a microbial respiratory system, *J. Bacteriol.*, 93, 278, 1967.

89. Jacobs, N. J., MacLosky, E. R., and Jacobs, J. M., Role of oxygen and heme in heme synthesis and the development of hemoprotein activity in an anaerobically-grown *Staphylococcus*, *J. Bacteriol.*, 148, 645, 1967.

90. Jacobs, N. J., Jacobs, J. M., and Sheng, G. S., Effect of oxygen on heme and prophyrin accumulation from *d*-aminolevulinic acid by suspensions of anaerobically grown *Staphylococcus epidermidis*, *J. Bacteriol.*, 99, 37, 1969.

91. Taber, H. W. and Morrison, M., Electron transport in Staphylococci. Properties of a particle preparation from exponential phase *Staphylococcus aureus*, *Arch. Biochem. Biophys.*, 105, 367, 1964.

92. Frerman, F. E. and White, D. C., Membrane lipid changes during formation of a functional electron transport system in *Staphylococcus aureus*, *J. Bacteriol.*, 94, 1868, 1967.

93. Sasarman, A., Purvis, P., and Portelance, V., Role of menaquinone in nitrate respiration in *Staphylococcus aureus*, *J. Bacteriol.*, 117, 911, 1974.

94. Chang, J. B. and Lascelles, J., Nitrate reductase in cell-free extracts of a haemin-requiring strain of *Staphylococcus aureus*, *Biochem. J.*, 89, 503, 1963.

95. Burke, K. A. and Lascelles, J., Nitrate reductase system in *Staphylococcus aureus* wild type and mutants, *J. Bacteriol.*, 123, 308, 1975.

96. Burke, K. A. and Lascelles, J., Nitrate reductase in heme-deficient mutants of *Staphylococcus aureus*, *J. Bacteriol.*, 126, 225, 1976.

97. Tien, W. and White, D. C., Linear sequential arrangement of genes for the biosynthetic pathway of protoheme in *Staphylococcus aureus*, *Proc. Natl. Acad. Sci. U.S.A.*, 61, 1392, 1968.

98. Stanier, R. Y., Adelberg, E. A., and Ingraham, J. L., *General Microbiology*, 4th ed., Macmillan, London, 1976, 678.

99. Dolin, M. I., The oxidation and peroxidation of DPNH$_2$ in extracts of *Streptococcus faecalis*, 10C1, *Arch. Biochem. Biophys.*, 46, 483, 1953.

100. Dolin, M. I., The DPNH-oxidising enzymes of *Streptococcus faecalis*. II. The enzymes utilizing oxygen, cytochrome *c*, peroxide and 2,6-dichlorophenol-indophenol or ferricyanide as oxidants, *Arch. Biochem. Biophys.*, 55, 415, 1955.

101. Dolin, M. I., The *Streptococcus faecalis* oxidases for reduced diphosphopyridine nucleotide. III. Isolation and properties of a flavin peroxidase for reduced diphosphopyridine nucleotide, *J. Biol. Chem.*, 54, 557, 1957.

102. Dolin, M. I., Reduced diphosphopyridine nucleotide peroxidase. Intermediates formed on reduction of the enzyme with dithionite or reduced diphosphopyridine nucleotide, *J. Biol. Chem.*, 250, 310, 1975.

103. Hoskins, D. D., Whiteley, H. R., and Mackler, B., The reduced diphosphopyridine nucleotide oxidase of *Streptococcus faecalis*: purification and properties, *J. Biol. Chem.*, 237, 2647, 1962.

104. Whittenbury, R., Two types of catalase-like activity in lactic acid bacteria, *Nature (London)*, 187, 433, 1960.

105. Whittenbury, R., Hydrogen peroxide formation and catalase activity in the lactic acid bacteria, *J. Gen. Microbiol.*, 35, 13, 1964.

106. Delwiche, E. A., Catalase of *Pediococcus cerevisiae*, *J. Bacteriol.*, 81, 416, 1961.

107. Johnston, M. A. and Delwiche, E. A., Catalase of the *Lactobacillaceae*, *J. Bacteriol.*, 83, 936, 1962.

108. Bryan-Jones, D. G. and Whittenbury, R., Haematin-dependent oxidative phosphorylation in *Streptococcus faecalis*, *J. Gen. Microbiol.*, 58, 247, 1969.

109. Ritchey, T. W. and Seeley, H. W., Jr., Cytochromes in *Streptococcus faecalis* var. *zymogenes* grown in a haematin-containing medium, *J. Gen. Microbiol.*, 85, 220, 1974.

110. Pritchard, G. G. and Wimpenny, J. W. T., Cytochrome formation, oxygen-induced proton extrusion and respiratory activity in *Streptococcus faecalis* var. *zymogenes* grown in the presence of haematin, *J. Gen. Microbiol.*, 104, 15, 1978.

111. Baum, R. H. and Dolin, M. I., Isolation of 2-solanesyl-1,4-naphthoquinone from *Streptococcus faecalis* 10C1, *J. Biol. Chem.*, 240, 3425, 1965.

112. Sijpesteijn, A. K., Induction of cytochrome formation of oxidative dissimilation by hemin in *Streptococcus lactis* and *Leuconostoc mesenteroides*, *Ant. van Leeuwenhoek J. Microbiol. Serol.*, 36, 335, 1970.

113. Van Der Wiel-Korstanje, J. A. A. and De Vries, W., Cytochrome synthesis by *Bifidobacterium* during growth in media supplemented with blood, *J. Gen. Microbiol.*, 75, 417, 1973.

114. Ritchey, T. W. and Seeley, H. W., Distribution of cytochrome-like respiration in Streptococci, *J. Gen. Microbiol.*, 93, 195, 1976.

115. Deibel, R. H. and Kvetkas, M. J., Fumarate reduction and its role in the diversion of glucose fermentation by *Streptococcus faecalis*, *J. Bacteriol.*, 88, 858, 1964.

116. Aue, B. J. and Deibel, R. H., Fumarate reductase activity of *Streptococcus faecalis*, *J. Bacteriol.*, 93, 1770, 1967.

117. Jacobs, N. J. and VanDemark, P. J., Comparison of the mechanism of glycerol oxidation in aerobically and anaerobically grown *Streptococcus faecalis*, *J. Bacteriol.*, 79, 532, 1960.

118. Langston, C. W. and Williams, P. P., Reduction of nitrate by Streptococci, *J. Bacteriol.*, 82, 603, 1962.

119. Gallin, J. I. and VanDemark, P. J., Evidence for oxidative phosphorylation in *Streptococcus faecalis*, *Biochem. Biophys. Res. Commun.*, 17, 630, 1964.

120. Faust, P. J. and VanDemark, P. J., Phosphorylation coupled to NADH oxidation with fumarate in *Streptococcus faecalis* 10C1, *Arch. Biochem. Biophys.*, 137, 392, 1970.

121. Smalley, A. J., Jahrling, P., and VanDemark, P. J., Molar growth yields as evidence for oxidative phosphorylation in *Streptococcus faecalis* 10C1, *J. Bacteriol.*, 96, 1595, 1968.

122. Mickelson, M. N., Aerobic metabolism of *Streptococcus agalactiae*, *J. Bacteriol.*, 94, 184, 1967.

123. Mickelson, M. N., Phosphorylation and the reduced adenine dinucleotide oxidase reaction in *Streptococcus agalactiae*, *J. Bacteriol.*, 100, 895, 1969.

124. Mickelson, M. N., Glucose degradation, molar growth yields, and evidence for oxidative phosphorylation in *Streptococcus agalactiae*, *J. Bacteriol.*, 109, 96, 1972.

125. Lascelles, J., Tetrapyrrole synthesis in microorganisms, in *The Bacteria*, Vol. 3, Gunsalus, I. C. and Stanier, R. Y., Eds., Academic Press, New York, 1962, chap. 7.

126. Myers, W. F., Osterman, J. V., and Wisseman, C. L., Jr., Nutritional studies of *Rickettsia quintana*: nature of the hematin requirement, *J. Bacteriol.*, 109, 89, 1972.

127. Sperry, J. F. and Wilkins, T. D., Cytochrome spectrum of an obligate anaerobe, *Eubacterium lentum*, *J. Bacteriol.*, 125, 905, 1976.

128. **Reddy, C. A., Cornell, C. P., and Kao, M.,** Hemin-dependent growth stimulation and cytochrome synthesis in *Corynebacterium pyogenes, J. Bacteriol.,* 130, 965, 1977.
129. **Haddock, B. A. and Jones, C. W.,** Bacterial respiration, *Bacteriol. Rev.,* 41, 47, 1977.
130. **Laris, P. C. and Pershadsingh, H. A.,** Estimations of membrane potentials in *Streptococcus faecalis* by means of a fluroescent probe, *Biochem. Biophys. Res. Commun.,* 57, 620, 1974.

Chapter 8

RESPIRATION IN HYDROGEN BACTERIA

Irmelin Probst

TABLE OF CONTENTS

I. INTRODUCTION

Molecular hydrogen is biologically produced in large amounts during the anaerobic decomposition of organic material. Depending on the hydrogen acceptor available in the habitat, hydrogen is oxidized by a variety of microorganisms, with the production of methane, hydrogen sulfide, or nitrogen under anaerobic conditions or water if hydrogen diffuses into aerobic zones. The oxidation of hydrogen by molecular oxygen, the "knallgas" reaction, is carried out by prokaryotic organisms as diverse as *Escherichia coli*,[1] *Azotobacter*,[2] and cyanobacteria.[3] The hydrogen bacteria are, however, unique in that they couple the process of hydrogen oxidation (= energy generation) to the synthesis of cell material from the c_1-compound, CO_2. Numerous hydrogen bacteria have been discovered belonging to the genera *Pseudomonas, Alcaligenes, Paracoccus, Bacillus, Aquaspirillum, Arthrobacter, Corynebacterium, Mycobacterium, Nocardia,* and *Streptomyces*. The organisms differ profoundly in taxonomic position, cell structure,[4,5] and properties of their hydrogen activating system and its regulation.[6,7]

The ability to use the energy derived from an inorganic energy source for the assimilation of an inorganic carbon source places hydrogen bacteria among the chemolithoautotrophic organisms. The adjective autotroph is used here to describe a completely inorganic medium or cells grown in such a medium. Hydrogen bacteria can, however, adapt equally well to a chemoorganotrophic way of life, using a wide variety of organic substrates. Their metabolism is respiratory, never fermentative. The frequently used term "facultative autotroph" implies a separation between the two opposite metabolic routes.[8] Physiological studies under mixotrophic growth conditions (e.g., the concomitant presence of an organic substrate and hydrogen plus CO_2) have shown the coexistence of both alternative modes of energy generation and/or carbon assimilation.[8,9] Due to the rapid and effective conversion of CO_2 into cellular material, hydrogen bacteria have become potential sources for single cell rotein[10] and for bioregenerative life support systems in space travel.[11] The basic outlines of metabolic processes have been extensively reviewed.[6-8, 12-14]

In the past few years, attention has focussed on the characterization of isolated respiratory proteins, namely the hydrogenases, ATP generation by respiration, and development of a genetic system. The most detailed studies have been done with *Alcaligenes (:eq Hydrogenomonas) eutrophus* and *Paracoccus (:eq Micrococcus) denitrificans*. The bioenergetics of *P. denitrificans* have been recently reviewed;[15, 16] therefore this review will focus only on its autotrophic metabolism.

II. RESPIRATION IN WHOLE CELLS

Until now no obligate chemolithoautotrophic hydrogen bacterium has been isolated. Specific growth rates on organic substrates are generally higher than those obtained under autotrophic growth conditions. Autotrophic generation times, however, tended to decrease during the years (from 5 to 1.5 hr for *A. eutrophus*) due to improved culture conditions.[6,17] The highest specific growth rate ever obtained for an autotrophically growing microorganism ($\mu = 0.68$ hr^{-1}) has been reported for a thermophilic hydrogen bacterium cultured at 52°C.[18] Since changes in specific growth rates may reflect changes in the ATP production rate and the maintenance energy,[19] the composition of the mineral salts medium and the partial pressures of H_2, O_2, and CO_2 are likely to have a profound influence on the efficiency with which energy generation by hydrogen oxidation is coupled to cell growth.

A. Growth Requirements

Heterotrophic growth of *A. eutrophus* requires supplementation with iron; autotrophic growth requires both iron and nickel.[20] Trace metal requirements for cobalt, chromium, and copper were also demonstrated for autotrophically growing cells.[21]

The optimal percentage composition of the H_2, O_2, and CO_2 gas mixture depends on the organism studied and on the physiological age of the cultre. Among the gaseous nutrients, hydrogen requires the least control.[6,12,22,23] In the special case of the nitrogen-fixing hydrogen bacterium *Xanthobacter autotrophicus* optimal growth occurs at a low hydrogen concentration (pH_2: 0.1 atm) due to the sensitivity of nitrogenase activity towards hydrogen. The increase in the H_2 oxidation rate upon exposure to higher hydrogen concentrations was interpreted as a protective mechanism against hydrogen inhibition.[24]

Hydrogen bacteria are sometimes grouped according to oxygen sensitivity. A large number of organisms show optimum growth at oxygen partial pressures between 0.03 and 0.1 atm.[6,25-27] One of the possible targets for the inhibitory action of oxygen is hydrogenase. Oxygen strongly inhibits the synthesis of the membrane-bound hydrogenase from *Aquaspirillum autotrophicum*, with complete inhibition in the presence of 0.3 atm oxygen.[28] Hydrogenase activity in whole cells and isolated membranes is not inhibited by oxygen in this organism. The inhibition of hydrogenase synthesis by oxygen is thought to operate at the level of transcription.[28] Little data are available on the effect of CO_2 concentration on cell growth. A decrease in the specific growth rate of a thermophilic bacterium was observed at pCO_2 values below 0.09 atm.[18] Growth of *A. eutrophus* specifically responds to bicarbonate ion, not to CO_2 per se.[29] The lag period of this organism can be completely abolished at certain O_2 and CO_2 concentrations. The optimum pO_2 was 0.05 atm, but the optimum pCO_2 varied according to pH and the growth phase of the culture. At pH 6.4, the pCO_2 required to obtain immediate growth of exponential, postexponential, and stationary phase inocula at equal rates was 0.02, 0.05, and 0.16 atm, respectively.

Exceptionally high cell yields (25 g dry weight per l) of *A. eutrophus* were obtained in a 23l batch culture after 25 hr of incubation by careful regulation of the quantitative nutrient requirement under an optimal gas atmosphere.[23]

B. Stoichiometry of Gas Uptake

Gas uptake by hydrogen bacteria is associated with two metabolic routes: the catabolic energy-yielding process of hydrogen oxidation and the energy-consuming assimilation of CO_2 into cellular material. Therefore, information about the

$$H_2 + \tfrac{1}{2}O_2 \longrightarrow H_2O \tag{1}$$

$$2H_2 + CO_2 \longrightarrow <CH_2O> + H_2O \tag{2}$$

degree of coupling between the two processes and the energy yield from the H_2 oxidation reaction may be obtained from stoichiometry of the gas uptake.

Ruhland[22] was the first to determine the ratios of the consumed gases and found that the molar ratio of H_2 to O_2 exceeded 2, the excess H_2 uptake corresponded in a 2:1 ratio with the assimilated CO_2. He also showed that the assimilation of CO_2 is not a function of the oxidation of H_2, but that the exploitation of the energy-providing process for the assimilation of CO_2 varied considerably. In the absence of CO_2, an idle oxidation of hydrogen occurs, the "Leerlauf-Oxidation". Studies with numerous strains have confirmed wide variations in the molar ratios of H_2, O_2, and CO_2 consumption from 6:2:1 to 10:5:1.[30-34] A gas consumption ratio of 4:1:1, the most effective energy conversion thus far reported, was determined for *A. eutrophus*[35] and *Al-*

caligenes (:eq Hydrogenomonas) ruhlandii.[25] The degree with which CO_2 fixation is coupled to the oxidation of hydrogen varies not only in different strains, but also within the same organism according to the physiological conditions of the experiment. The molar ratio of CO_2 assimilated and H_2 oxidized decreases with the age of the culture[25,33,35] and under conditions of limited nutrient supply.[36-38] The influence of nutrient limitation on the stoichiometry of the gas uptake has been studied by different groups using chemostat cultures of *A. eutrophus* strains. To define the correlation between catabolic and anabolic activities, Bongers[17] introduced the "energy-yield coefficient" O/C which relates the rate of oxygen consumption to the rate of CO_2 assimilation. Cultures of the strain H 20 growing under unlimited supply of each individual gas exhibit O/C values of 2.5. When hydrogen or oxygen are limiting, the relationship between O_2 consumption and CO_2 fixation remains constant (O/C = 2.5), indicating the O/C ratio is independent of the growth rate.[35] An O/C value of 2.6 under oxygen-limited growth was also reported for *A. eutrophus* strain H 16.[36,38] A significant increase in O/C values (up to 2.5-fold) has been observed when CO_2 is the growth-limiting factor. The energy yield coefficient becomes growth-rate dependent. In the absence of CO_2, 20 to 40% of the maximal oxidative activity remains as idle hydrogen oxidation.[35,37,38] A decrease of the O/C value under hydrogen limitation has been reported for strain H 16[36,38] and under both hydrogen and oxygen limitation for strain Z 1,[37] in contrast to the results obtained with strain H 20.[35]

Limitation in the nutrient supply for the energy generating system apparently does not affect the stoichiometry of the gas uptake, whereas limitation in energy consuming reactions decreases the cell yield to a value much lower than expected on the basis of the theoretical ATP yield from the oxidation of hydrogen.[35] This uncoupled growth[19] is also observed when cultures become N-, P-, S-, K-, Mg-limited,[17,37] under which conditions the cells actively accumulate poly-β-hydroxybutyrate. Stouthamer[19] attributes this phenomenon of uncoupled growth under unfavorable condtions to the inability of bacteria to regulate the rate of catabolism exactly to the needs of anabolism. Thus, the ATP generated by the oxidation of the energy source seems to be constantly wasted. There is at present no experimental evidence to indicate whether *A. eutrophus* possesses two separate pathways of hydrogen oxidation, one of which is not coupled to oxidative phosphorylation, or whether the generated ATP is immediately destroyed by an ATPase or other energy-spilling reactions.

Values of Y$_{Substrate}^{Max}$ have been calculated from chemostat cultures of *A. eutrophus*. A total of 1 mol of hydrogen is required for the production of 6.2 to 6.6 g cells (strain H 20). Under these physiological conditions, (H_2 to O_2 to CO_2 = 4:1:1, μ_{max} = 0.4 hr^{-1}) , 6 to 8 mmol of hydrogen per gram cells per hour are consumed for cell maintenance purposes.[17,35] For strain H 16 (H_2 to O_2 to CO_2 = 6:2:1, μ_{max} = 0.3 hr^{-1}), values obtained for Y$_{H2}^{Max}$, Y$_{O2}^{Max}$, and Y$_{CO2}^{Max}$ are 3.86, 16.83, and 26.49 g cells per mole, respectively.[38] The maintenance coefficients mfor H_2, O_2, and CO_2 are 34, 90, and 35 mmol gas per gram/hour.[38]

C. Growth on Formate and Carbon Monoxide

In the autotrophic culture medium for *A. eutrophus*, formate can substitute for carbon dioxide.[39] Thus, in addition to their "classic" autotrophic metabolism, hydrogen bacteria are capable of assimilating c_1-compounds other than CO_2. Both the c_1-substrates formate[40,41] and CO[42,43] may serve s sole energy and carbon sources. Both substrates are converted to CO_2 which is assimilated via the ribulose bisphosphate carboxylase pathway. In the case of formate degradation by *A. eutrophus*, the in vivo operation of the reductive pentose phosphate cycle has been confirmed through studies with mutant strains.[41] Neither hydrogen plus CO_2 nor formate support growth in these strains which are impaired in the CO_2 fixation system.

III. MEMBRANE MORPHOLOGY AND ISOLATION

Bacterial membrane systems consist of the cytoplasmic membrane and internal membrane structures, the latter of which are usually only poorly developed in Gram-negative bacteria. Although hydrogen bacteria do not contain the extremely well-developed lamellar membrane systems that nitrifying chemoautotrophic bacteria do, electronmicroscopic investigations have shown ultrastructural diversity.[4,5] Two completely different types of internal membranes are observed. Mesosomal-like membranes are present in strains belonging to the genera *Alcaligenes*, *Pseudomonas*, and *Aquaspirillum*. The second type consists of a more or less extensive membrane system sometimes in regular parallel layers in the cortical region of the cell. This multilayer system is visible in *Corynebacterium autotrophicum* and in two as yet unnamed strains. Nothing is known about the biochemical and enzymatic properties of these structural features.

Cytoplasmic membranes consist of proteins and lipids, mainly phospholipids. Bacterial membranes include a wide variety of different phospholipids, but in individual organisms, the range of species is limited. The phospholipid content of *A. eutrophus* strain Z 1 constitues 73 to 74% of the total cellular lipids. Phosphatidylethanolamine is the major fraction (65 to 66%), whereas phosphatidylglycerol, cardiolipin, and phosphatidylcholine content are 20 to 22, 9 to 10, and 3%, respectively.[44] This composition, especially the high cardiolipin content, closely resembles that reported for *E. coli*.[45]

The mildest method commonly used to break bacterial cells is hydrolysis of the peptidoglycan layer by lysozyme in isotonic solution and subsequent osmotic shock of the spheroplasts. This method has been successfully applied to *P. denitrificans*[46] and *A. eutrophus*.[47] Strains grouped in the genera *Corynebacterium*, *Arthrobacter*, and *Nocardia* are disintegrated by ultrasound or mechanical means. As might be expected, membrane-bound enzyme activities are injuried by harsh cell breakage. The advantage of a gentle procedure for the lysis of *A. eutrophus* cells over mechanical cell disruption methods is indicated by:

1. Lysozyme-prepared membranes exhibit considerably higher respiratory activities than membranes prepared from cells broken in the French pressure cell; 93% of the in vivo activity of the H_2 oxidase is recovered in the membrane fraction, in contrast to an activity loss of 70% in membranes from cells disintegrated by ultrasound.
2. Membrane-bound oxidase activities are greatly stimulated by the uncoupler CCCP.
3. The ATPase is not solubilized, but remains membrane-bound. Lysozyme treatment results in the formation of large vesicles of greatly varying size (100 to 600 nm diameter) with particles of 9 to 10 nm diameter attached to the membrane.[47]

IV. RESPIRATORY PROTEINS

A. Hydrogenases

Electrons generated during the oxidation of oxidizable substrates can join the respiratory chain, either through the intermediate involvement of NAD-dependent dehydrogenases or directly through substrate-specific dehydrogenases that are firmly membrane-bound. In hydrogen bacteria, the enzyme responsible for the initial oxidation of the inorganic substrate, hydrogen, is directly connected with the respiratory chain. This enzyme which cannot donate electrons to NAD, transfers electrons to membrane respiratory carriers, and thus differs from a second type of hydrogenase which is soluble and reduces NAD (hydrogen to NAD^+ oxidoreductase, EC1.12.1.2). The membrane-bound hydrogenase initiates the ATP-generating oxidation of hydrogen via the electron transport chain, while the soluble enzyme is thought to primarily provide NADH for CO_2 assimilation. With respect to their hydrogenase pattern, three groups of hydrogen bacteria can so far be distinguished:

1.   Both the soluble and membrane-bound enzymes are present in *Pseudomonas saccharophila*,[48] *A. eutrophus*,[49,50] and *A. ruhlandii*.[51]
2.   The majority of the strains contain only the membrane-bound hydrogenase: *Pseudomonas facilis*,[52] *P. denitrificans*,[53] *A. autotrophicum*,[28] and several other pseudomonads and coryneform bacteria.[53]
3.   *Nocardia opaca* 1 b contains only an NAD-reducing enzyme.[54,55]

The ability to reduce redox dyes is common to all hydrogenases. However, in most cases the physiological electron acceptor is still unknown.

The possession of an NAD-linked hydrogenase in advantageous for the cell, as no additional oxidation of hydrogen is necessary for the energy-consuming generation of NADH by reverse electron transport. The presence of only the membrane-bound hydrogenase in *C. autotrophicum* 12/60/x is reflected by the comparatively high molar ratio of the hydrogen to CO_2 uptake (7.4:1).[31] Both the soluble and membrane-bound hydrogenase of *A. eutrophus* H 16 have been purified recently, and their properties are under investigation. This is a promising start to elucidating respiratory mechanisms and their regulation in this organism.

1. Soluble Hydrogenase

Soluble hydrogenases have been repeatedly purifed and characterized from *A. ruhlandii* [56,57] and *A. eutrophus*.[58-61] Much information exists concerning kinetic and catalytic characteristics of partially purified preparations. Molecular, structural, and spectral properties, however, have been reported for only the hydrogenase of *A. eutrophus* strain H 16. This enzyme has been purified with a final specific activity of 54 μmol H_2 oxidized per minute per milligram protein.[61] Molecular properties are listed in Table 1. The tetrameric enzyme is composed of three different subunits[62] (68,000; 60,000; 29,000 D; molar ratio, 1:1:2) and contains both flavin mononucleotides and Fe-S centers.[62-64] The molar ratio of FMN to iron to sulfide is 2:12:12. Up to now, this enzyme represents the first hydrogenase to be assigned to the group of conjugated Fe-S proteins. Moreover, with respect to the composition of Fe-S centers, soluble hydrogenase turned out to represent a new type of Fe-S protein, containing the (2Fe-2S) as well as the low-potential (4Fe-4S) type of Fe-S center, probably two of each type.[62,64] FMN, which is needed for maximal hydrogenase activity partially dissociates during the process of purification. The enzymatic activity, is stimulated by exogenous FMN, the percentage of activation corresponding to the amount of FMN removed from the enzyme.[63] The stimulatory effect of FMN on hydrogenase activity has also been shown for the enzymes from *A. eutrophus* ATCC 17697[58] and *Nocardia opaca* 1 b.[55]

The stability and activity of hydrogenases from various other *A. eutrophus* strains have been reported to depend on a hydrogen atmosphere and reducing agents.[53,58,65] Surprisingly, like the hydrogenase from *A. ruhlandii*,[57] the enzymes from *A. eutrophus* H 16[61] and Z 1[66] are not affected by storage under oxygen. In its oxidized state, the enzyme is inactive, but highly stable. Reduction of the enzyme by NAD(P)H or reducing agents or incubation under hydrogen resulted in complete loss of activity within 5 days.[61,67] Transformation of the oxidized inactive form into the reduced state occurs either upon incubation with reducing agents or catalytic amounts of NAD(P)H in the presence of hydrogen or by complete removal of oxygen from the hydrogen-saturated solution.[57,60,61] The enzyme from *A. eutrophus* H 16 is the only hydrogenase thus far

Table 1
PROPERTIES OF THE SOLUBLE AND MEMBRANE-BOUND HYDROGENASE FROM ALCALIGENES EUTROPHUS H 16

Enzyme property	Soluble hydrogenase[61-64]	Purified membrane-bound hydrogenase[72,73]	Membrane-bound H_2 oxidase activity[a,78]
pH optimum	8.0	5.5	7.0—7.4
Temperature optimum (°C)	33	52.0	45
Activation energy (kJ/mol)	59.9	26.2	78.0 (15-30°C) 58.1 (30—42°C)
K_m (μM)			
H_2	37	32	24 (5% O_2) 60 (20% O_2)
NAD	560		
Molecular weight	205,000	98,000	
Prosthetic groups (mole/mole enzyme)	2FMN, 12Fe, 12S^{2-}	6Fe, 6S^{2-}	
Isoelectric point	4.85	6.25	
Inhibition by oxygen	−	+	+

a Enzyme activity was measured manometrically, using lysozyme-prepared membranes in 50 mM phosphate buffer, pH 7.0. The gas atmosphere consisted of 60% H_2, 5% O_2, and N_2.

reported to be extraordinarily oxygen-tolerant under reaction conditions. Oxygen concentrations up to 0.7 mM in the assay solution corresponding to a partial pressure of 0.6 atm cause no obvious inhibition of enzyme activity.[67] Oxygen insensitivity was also confirmed by coupling the soluble hydrogenase to chloroplasts for photoproduction of H_2 from H_2O, with NAD and methyl viologen as electron mediators. The enzyme is at present the only hydrogenase that evolves H_2 continuously in the presence of O_2 and at the same rate, whether oxygen scavengers are present or not.[68] Hydrogen evolution from reduced NAD and methyl viologen has also been reported in other soluble hydrogenase preparations.[56,65] Neither NAD or NADH, previously considered to be essential for mediating electrons to secondary acceptors,[60,69] nor other carrier systems are required for the reduction of substrates. Reduced pyridine nucleotides were found to be necessary only for the reductive activation of the enzyme.[61] In its active form, the enzyme reduces a large number of physiological and artificial electron acceptors directly (FMN, FAD, quinones, cytochrome c, and methylene blue). Therefore, apart from its function to supply NADH for the CO_2 fixation, the soluble hydrogenase might channel electrons directly into the respiratory chain.

N. opaca 1 b is unusual among hydrogen bacteria in that it has been reported to catalyze hydrogenase-dependent NAD-reduction in both the soluble and particle fraction.[54] The membrane-associated enzyme can be solubilized by repeated washings with phosphate buffer, pH 6.5.[70] It remains to be established whether this enzyme becomes adsorbed to or is partly released from the membranes during cell disintegration or whether it is an ambiquitous enzyme which reversibly partitions in vivo into distinct soluble and membrane-bound forms.[71] The purified enzyme shows a complete dependence on either nickel or magnesium ions. Maximum activity is achieved by simultaneous addition of Ni, Mg, and FMN. The enzyme is relatively stable under air. High stability is observed in the presence of Ni and Mg under hydrogen.[55]

2. Membrane-Bound Hydrogenase

Membrane-bound hydrogenase activities are commonly measured by hydrogen-dependent reduction of nonphysiological dyes. The highest oxidation rates are exhibited by some organisms that lack the soluble NAD-reducing hydrogenase.[53] The tightly membrane-bound hydrogenase from *A. eutrophus* H 16 has recently been purified and characterized.[72,73] A maximal yield of 22% of the hydrogenase activity from intact membranes was solubilized by Triton® X-100 and deoxycholate which corresponds to the total amount of enzyme present. The apparent loss of activity is due to a change in the pH-optimum after solubilization. At pH 7.0, the solubilized enzyme has only 24% of its maximal activity measured at pH 5.5. The enzyme has been purified by ammonium sulfate fractionation and chromatography on CM-cellulose with a yield of 36% and a final specific activity of 170 μmol H_2 oxidized per minute per milligram protein. Molecular and kinetic properties are in Table 1. The enzyme consists of two types of subunits, one of 67,000 daltons and one of 31,000 daltons at a 1:1 ratio; the sum of these molecular weights, 98,000 daltons, is in accordance with the values obtained for the total enzyme by sucrose density gradient centrifugation. A distinct absorption around 400 nm and the presence of 6 mol labile sulfur and 6 mol iron per mole enzyme characterize the enzyme as an Fe-S protein. Among a variety of artificial and physiological electron acceptors, only few react with the purified enzyme, both in the presence or absence of lecithin: methylene blue, phenazine methosulfate, menadione, pyocyanine, and ferricyanide. Ubiquinone Q_8 and vitamin K_2 isolated from *A. eutrophus* H 16 do not serve as electron acceptors. A comparison of the K_m values for hydrogen of the soluble and membrane-bound enzyme in either its solubilized or particulate form show that the two enzymes exhibit almost identical substrate affinities.

Like the soluble hydrogenase, the membrane enzyme is stable during storage under air and unstable under reducing conditions. It is activated by the removal of oxygen.[73] All other membrane-bound hydrogenases tested so far in hydrogen bacteria require reducing conditions for stability.[53,74] In the solubilized state, the membrane-bound *A. eutrophus* hydrogenase activity undergoes strong reversible inhibition by oxygen (50% inhibition at 30 μM O_2).[75] The effect of oxygen on hydrogenase activity or the oxyhydrogen reaction has long been a subject of investigation. Maximal rates of the H_2 oxidase in whole cell suspensions are obtained at 0.05 atm (*P. facilis*)[76,77] and between 0.05 to 0.09 atm oxygen (*A. ruhlandii*).[25]

Oxygen concentrations higher than 60 μM (pO_2 = 0.05 atm, 30°C) reversibly inactivate the H_2 oxidase reaction in membrane preparations from *A. eutrophus* H 16[78] which is in accordance with the optimal growth of this organism t pO_2 = 0.05 atm.[29,79]

The activities of both soluble and membrane-bound hydrogenase were found to vary in a parallel manner, depending on culture conditions.[80] Therefore, the question was raised, whether these two enzymes were two different proteins or two different active forms of one ambiquitous enzyme. Although they share a remarkable stability to oxygen, the pure proteins reveal significant differences in their molecular properties. This, together with the result that both purified enzymes do not show any cross reactions in immunodiffusion experiments,[75] is strong evidence for two different native proteins.

The membrane-bound hydrogenase from *P. denitrificans* has been partially purified recently.[81] The solubilization by Triton® X-100 resulted in a fourfold increase in specific activity. Benzyl viologen and methyl viologen accept electrons from the enzyme in both its membrane-bound and solubilized form which points to a direct interaction of hydrogenase with these redox dyes. Hydrogenase activity is not significantly increased by disintegration of the cells. Therefore, it is likely that the enzyme reacts with the viologen dyes at the outer surface of the cytoplasmic membrane. Reduction of ferricyanide by hydrogen is catalyzed by the enzyme only in its membrane-bound state. Inhibitor studies revealed that electron flow from hydrogen to ferricyanide is mediated by the electron transport chain.

B. Dehydrogenases

Succinate, formate, and CO, as well as hydrogen and NAD(P)H, donate electrons directly to the respiratory chain via membrane-bound dehydrogenases.[47,82-85] Growth on formate[41] or CO[43] specifically induces the respective dehydrogenases without significant repression of hydrogenase activity. In all strains investigated (grown on H_2 + CO_2 or on CO), the H_2 oxidase activity is the highest of all the oxidase activities present. Analogous to growth on hydrogen, formate-grown *A. utrophus* contains a soluble NAD-dependent formate dehydrogenase[40,41] and a membrane-bound enzyme activity.[47] Formate therefore is a disguised H_2/CO_2 substrate, and growth on this compound is supposedly ubiquitous among hydrogen bacteria.

The mechanism of CO oxidation has been investigated for *Pseudomonas carboxydovorans*.[85] The oxidation of CO by crude extract preparations could be coupled anaerobically to the reduction of nonphysiological dyes, thereby ecluding a monooxygenase type reaction. Both the CO and formate oxidase activities comigrated with small vesicles into the 30% layer of a discontinuous sucrose gradient. No free formate could be demonstrated during the oxidation of CO; in addition, the CO and formate oxidase activities differed in respect to storage and heat stability, substrate saturation values, and sensitivities against inhibitors. The following reaction was therefore proposed for the oxidation of CO:

$$CO + H_2O \longrightarrow CO_2 + 2e^- + 2H^+ \qquad (3)$$

A soluble menadione reductase from *A. eutrophus* has been partially purified.[86] The enzyme is NADH specific, stimulated by FMN (25%), and inhibited by coumarol. The natural quinone from *A. eutrophus* (Q_8) did not serve as an electron acceptor, analogous to the results obtained for the solubilized membrane-bound hydrogenase.[73]

C. Quinones

A. eutrophus contains both ubiquinone Q_8 and vitamin K_2.[83,87] The relative concentrations vary considerably according to culture conditions. Vitamin K_2 is apparently induced under oxygen limitation. Cells grown under excess oxygen contained only trace amounts of menaquinone.[47] Ubiquinone Q_{10} has been recently detected in *P. carboxydovorans*.[88] Evidence for the in vivo function of quinones in the H_2 oxidation was obtained by extraction (or inactivation) and reconstitution experiments with *A. eutrophus*. H_2 oxidase activity and coupled phosphorylationare strongly diminished and the enzyme activity can be restored by addition of Q_8 or vitamin K_2.[82,83]

D. Cytochromes

All hydrogen bacteria tested so far contain membrane-bound cytochromes of the *b*- and *c*-type, together with *o*- and *a*-type oxidases.[82-84,89-91] In some cases, low-temperature spectra revealed the presence of multiple *c*- or *b*-type cytochromes.[89,90] With the exception of the coryneform bacterium 11/x which induces a cytochrome *d* under autotrophic growth conditions,[90] the transfer of cells from heterotrophic to autotrophic growth does not significantly alter the cytochrome pattern. Stationary phase cells of autrophically grown *A. eutrophus* and *Pseudomonas pseudoflava* contain additional cytochrome *d*, probably induced by oxygen limitation.[47,91] Soluble cytochrome *c* is commonly found in high-speed supernatants; in *A. eutrophus*, the location of the soluble cytochrome c_{550} was found to be periplasmic.[47] The soluble cytochrome c_{550} from *A. ruhlandii* was partially purified ($E_o' = 0.08V$ at pH 7.8)[92], and three *c*-type cytochromes were purified from butanol-extracted cells of *A. eutrophus*.[93] The amino acid composition of the acidic cytochrome *c* is not markedly different from other bacterial cytochrome *c*. Unfortunately the cellular location of this cytochrome is not known.

Recently, a hemoglobin-like pigment has been purified and characterized from autotrophically grown *A. eutrophus*.[94] The protein (mol wt 43,000) contains two prosthetic groups, noncovalently bound FAD and protoheme. Reversible oxygen binding was demonstrated with the NADH-reduced form. The protein catalyzes the reduction of dyes, cytochrome *c*, and oxygen by NADH. Its cytoplasmic nature and the low oxygen turnover number in vitro indicate it is not involved in terminal respiration but may act as an oxygen store within the cell.

V. ENERGY CONSERVATION AND ELECTRON FLOW

The flow of electrons from H_2, NADH, and succinate to oxygen has been of constant interest. The questions raised were: how do electrons from hydrogen couple to the electron transport chain, which carriers are involved, and how many coupling sites are present? Unfortunately, the main body of evidence in the cell-free system is derived from experiments with ultrasonically disrupted cells that were not fractionated into membranes and soluble proteins. Since most of the work has been done with *P. saccharophila* and *A. eutrophus*, organisms containing both a soluble and membrane-bound hydrogenase, the oxidation of hydrogen consists of complex reactions. This is further complicated by the presence of soluble cytochrome *c* that can accept electrons either from the soluble hydrogenase[61] or an NADH-cytochrome *c* reductase[86] and by the presence of an NADH-dependent menadione reductase.[86] These reactions sup-

Table 2
P/O RATIOS IN CELL-FREE EXTRACTS
(AUTOTROPHIC GROWTH)

Substrate	Pseudomonas saccharophila	Paracoccus denitrificans	Alcaligenes eutrophus	
H$_2$	0.73	1.57	1.60	0.52
NADH	1.22	0.95	1.80	0.02
Succinate	0.15	1.00	0.91	0.18
Ascorbate	0.00[a]	0.37[a]	0.63	N.D.[b]
Reference	95	98	102	82

[a] Ascorbate plus cytochrome c.
[b] N.D., not determined.

posedly feed electrons into the respiratory chain in vitro, but not necessarily in vivo, and thereby bypass the first segment, the membrane-bound hydrogenase. These considerations also apply to P/O ratios determined in these systems. The problems encountered with cell-free preparations can be overcome by determination of respiratory proton-translocation by whole cells and molar growth yields. These methods have successfully elucidated the degree of energy conservation in two hydrogen bacteria, *A. eutrophus* and *P. denitrificans*.

A. Organisms
1. Pseudomonas saccharophila

Phosphorylation coupled to the aerobic oxidation of various electron donors was studied, using both heterotrophic and autotrophic cells.[84,95,96] Cell-free extracts from succinate-grown cells coupled the aerobic oxidation of NADH, succinate, and ascorbate to ATP generation. The data show that the oxidation was efficiently coupled with ADP phosphorylation, indicating the participation of all three energy-conservation sites in vitro. In the case of autotrophically grown cells, no phosphorylation of ADP coupled to the oxidation of ascorbate could be detected (Table 2). Therefore, Site 3 was supposed to be nonfunctional under autotrophic growth conditions. Recently Burnell et al.[97] presented evidence that the sidedness of membrane vesicles from *P. denitrificans* depended on the culture conditions. Alterations in the orientation of membrane vesicles might well influence the detection of energy coupling at Site 3, especially as electrons for this segment of the respiratory chain are provided for by nonphysiological dyes or mammalian cytochrome *c*. Succinate oxidation, as well as hydrogen oxidation, were only poorly coupled to ATP generation.

Of particular interest is the coupling of hydrogen via a hydrogenase to the respiratory chain. NAD showed no effect on stimulation of H$_2$ oxidation, and the addition of a NADH trapping system did not significantly inhibit oxidation of hydrogen or coupled phosphorylation.[96] These results eliminate NADH as an intermediate in hydrogen oxidation, in agreement with the observation that the membrane-bound hydrogenase from *P. saccharophila* does not reduce NAD.[48] Electrons from hydrogen were thought to bypass the NADH dehydrogenase and to couple with the respiratory chain at the level of cytochrome *b*. Thus, the failure of Site 1 specific inhibitors (rotenone and atebrin) to eliminate phosphorylation linked to hydrogen, but not the NADH associated ATP generation, led to the conclusion that coupling Site 2 is solely involved in H$_2$ oxidation, while NADH oxidation generates ATP at both coupling Sites 1 and 2. Uncoupled electron flow from hydrogen to cytochrome *b*, however, would mean constant energy spoilage, considering the redox potential of the H$_2$/2H$^+$ couple of E'_o = −0.42 V.

2. Paracoccus denitrificans

The oxidation of hydrogen, NADH, succinate, and ascorbate is coupled to the formation of ATP in cell-free extracts from autotrophically grown cells.[98] Phosphorylation was significantly decreased by specific respiratory inhibitors, atebrin for NADH oxidase, antimycin A for the succinate oxidase, and cyanide for ascorbate oxidation. All three coupling sites were considered to be functional. This result was confirmed by the measurement of respiratory-driven proton translocation in whole cells. Assuming an H^+/ATP ratio of 2, the experimentally determined H^+/O ratio of 8 linked to endogenous respiration of unstarved cells was interpreted to indicate the involvement of Sites 1, 2, and 3, plus a functional proton-translocating transhydrogenase which constitutes Site 0.[16,99] A recent study[100] on the magnitudes of the protonmotive force and the phosphorylation potential reaffirms the function of three proton translocating sites; however, deriving a H^+/ATP ratio of 3, in agreement with results based on measurements of H^+/O ratios in intact cells and growth yields.[101]

Whereas the oxidation of NADH in crude cell extracts was markedly inhibited by rotenone (10 μM), hydrogen oxidation was relatively insensitive, although the associated phosphorylation was partially (30%) decreased. In contrast to *P. saccharophila*, the rates of H_2 oxidation and phosphorylation were lowered by the addition of NAD and a NADH trap which led to the postulation of endogenous NAD as an intermediate able to couple to exogenously added NAD.[98] This, however, is not compatible with the inability of the membrane-bound hydrogenase to reduce NAD, and the absence of a soluble NAD-reducing hydrogenase in this organism.[53] A multiple entry of electrons from hydrogen into the electron transport chain at the level of NADH, cytochrome *b*, and *c* was proposed from inhibitor studies.[98] A recent report on hydrogen oxidation, however, provides evidence for an exclusively direct interaction of the membrane-bound hydrogenase with the electron transport chain.[81] Hydrogen-ferricyanide oxidoreductase activity is inhibited by antimycin A, but not by rotenone and potassium cyanide. It is likely that hydrogenase interacts with the respiratory chain between the rotenone and antimycin A-sensitive sites, and reduces ferricyanide by way of cytochrome *c*.[81]

3. Alcaligenes eutrophus

The entry of electrons from hydrogen into the respiratory chain has been a matter of debate. Hydrogen oxidation in cell-free preparations does not involve NADH as an obligate intermediary carrier both in strain H 20[82] and strain H 16,[83] whereas mediation by NAD and NADH dehydrogenase was postulated for the type strain (ATCC 17697).[102] In the latter case, the soluble NAD-reducing hydrogenase present in the crude cell extract might well have channeled electrons via NADH into the respiratory chain.

Inhibitor studies with HQNO and antimycin A showed the H_2 oxidase to be very sensitive to HQNO and insensitive to antimycin A, whereas NADH or succinate oxidation was strongly affected only by antimycin A. Bongers[82] explained this phenomenon by an abbreviated chain for the flow of hydrogen electrons ($H_2 \rightarrow$ cytochrome *b* $\rightarrow O_2$) which contains one coupling site in the HQNO-sensitive first segment. Pfitzner[83] postulated two parallel chains for H_2 and NADH oxidation in the cytochrome *b* cytochrome *c* segment with different inhibitor sensitivities. Similar results were obtained with *P. facilis* and *C. autotrophicum* 14 g.[103] A more-detailed investigation of the respiratory chain has recently been reported for strain H 16.[104] The organism contains two cytochromes 553 (E_0', 150 and 264 mV) and two cytochromes 562, one of high (418 mV) and one of lower (96 mV) midpoint potential. The study of cytochrome reduction and reoxidation using various substrates and cyanide to discriminate between inhibitor-sensitive and inhibitor-insensitive oxidases demonstrates a branched electron transport chain. The high-potential cytochrome 562 appears to be identical with the carbon monoxide-binding cytochrome o and to be the oxidase of a relatively cyanide-insensitive pathway with high affinity to oxygen (K_m, 3 μM). Ascorbate (+ TMPD), in contrast to hydrogen, NADH, and succinate exclusively reduces the cyanide-sensitive cytochrome a (E'_o, 307 mV) branch of the respiratory chain. A tentative scheme for the electron flow in *A. eutrophus* is shown in Figure 1.

Oxidative phosphorylation coupled to the oxidation of various substrates has been studied, using both cell-free preparations (P/O ratios) and whole cells (H⁺/O ratios and growth yields). The oxidation of hydrogen, NADH, succinate, and ascorbate by crude extracts from autotrophically grown cells was found to be coupled to ATP generation (Table 2), indicating the involvement of three functional coupling sites in vitro.[102] A tight coupling between electron transport and energy generation was also indicated by the ability of CCCP to greatly stimulate oxidase ctivities in lysozyme prepared membrane vesicles.[47] These results are confirmed by whole-cell proton translocation data and growth yield measurements. Cells grown heterotrophically or under H_2/CO_2 exhibited H⁺/O ratios (g equiv. H⁺/g·atom 0) approaching 8 for endogenous substrate, commensurate with the involvement of four coupling sites if H⁺/ATP equals 2.[105-107] The presence of a membrane-bound transhydrogenase constituting the proton-translocating segment at Site 0 was indicated by the NAD-specific stimulation of NADPH oxidase activity and by the ATP-dependent reduction of NADP by NADH.[107] Furthermore, the possible existence of four proton-translocating sites was strengthened by data of H⁺/O ratios exhibited by starved cells which were loaded with defined substrates capable of donating electrons to membrane-bound carriers at different levels in the electron transport chain.[108] Assuming that the maintenance energy requirement is independent of the growth rate, determinations of $Y_{O_2}^{Max}$ (grams cells per mole O_2) from energy-limited cultures allow the calculation of N (moles ATP per moles O_2), provided the value for Y_{ATP}^{Max} (grams cells per moles ATP) is known. N was calculated for *A. eutrophus* from experimentally determined $Y_{O_2}^{Max}$ (lactate limitation), assuming a Y_{ATP}^{Max} value of 8.4 and found to equal 5.9 which is consistent with three functional coupling sites in vivo.[108] The involvement of a coupling site in electron transfer between cytochrome b and oxygen was also shown by molar growth yied measurements. The succinate-fumarate molar growth yield difference of 12 g cells per mole substrate was interpreted to reflect the additional energy (1 ATP) obtained from succinate oxidation via cytochrome b.[109]

Less information is available at present about the number of coupling sites involved in the oxidation of hydrogen by whole cells or membrane preparations. Hydrogen electrons for the respiratory chain can theoretically be provided directly (membrane-bound enzyme) or indirectly via NADH (soluble enzyme). The isolation of a mutant strain which lacks the membrane-bound hydrogenase activity demonstrates the existence of the NADH-mediated hydrogen oxidation in vivo.[80] Partial inhibition of hydrogen oxidation by amytal in whole cell suspensions of *A. ruhlandii* points to a participation of a NADH-mediated reaction on the overall H_2 oxidation in this organism.[89]

Proton translocation experiments with autotrophic cells of *A. eutrophus* yielded H⁺/O ratios of 6 to 7 with hydrogen as electron donor, indicating three functional coupling sites in vivo.[110] Theoretical calculations of energy-substrate consumption[111] and of the ATP requirement for synthesis of cell material from CO_2 imply that three functional coupling sites are commensurate with the lower limit of the gas consumption ratios observed experimentally. The lowest O/C value determined for *A. eutrophus* (O/C, 2.5) means that 7.5 mol ATP are available for the assimilation of 1 mol CO_2, provided hydrogen oxidation proceeds with a P/O ratio of 3. Considering a theoretical value

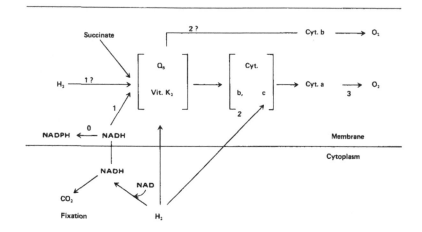

FIGURE 1.   Tentative scheme of electron flow in *Alcaligenes eutrophus* H 16. Postulated coupling sites are indicated by numbers.

of 7.9 mol ATP needed to synthesize cell material from 1 mol CO_2,[112] catabolic and anabolic reactions seem to be tightly coupled under optimal growth conditions. The function of three coupling sites during growth on hydrogen might indicate the membrane-bound hydrogenase quinone segment constitutes a proton-translocating site. Surprisingly, the extensive literature on proton translocation in *P. denitrificans* does not contain data on H^+/O ratios for autotrophic cells loaded with hydrogen. This organism, lacking a soluble NAD-reducing hydrogenase, would be especially suited to test this possibility.

B. Reverse Electron Flow

In chemoautotrophic bacteria, NADH is in especially high demand for the fixation of CO_2. Although the redox potential of hydrogen (E'_0, -0.42 mV) permits an energy-independent reduction of NAD; hydrogen bacteria which possess no soluble NAD-reducing enzyme have to reduce NAD by an ATP-consuming reversal of electron flow. ATP-dependent reduction of NAD in crude extracts of *A. eutrophus* was observed with succinate, ascorbate, or reduced mammalian cytochrome *c* as electron donors.[102] Respiratory inhibitors, uncoupling agents, and oligomycin effectively inhibit this process. The ATP-dependent reduction of NADP by NADH proved the existence of a proton-translocating Site 0.[106]

C. Coupling of H_2 Oxidation to CO_2 Assimilation

Unfavorable growth conditions, e.g., limitation in the energy-consuming reactions lead to "uncoupled" growth in hydrogen bacteria. The energy-providing reaction (H_2 + ½O_2 → H_2O) proceeds uncontrolled by the actual needs for anabolic purposes. Optimal growth, therefore, seems to be controlled by energy supply, e.g., by some step in the energy generating process.[17,35]

A respiratory control mechanism via ADP/ATP was proposed by Bartha[113] to explain the increase in gas consumption by respiring cell suspensions (resting or growing) upon exposure to CO_2 in the H_2/O_2 gas mixture. Withdrawal of CO_2 increases the intracellular ATP concentration by 30%, and the NADH/NAD ratio rises from 0.5 to 1.4,[114] suggesting, too, that the degree of NAD reduction and/or the phosphate potential is responsible for the respiratory control effect.[113] The stimulation of the H_2 oxidase by CO_2, reported for particulate cell-free preparations from *A. eutrophus* H

16 (see Reference 7) could, however, not be verified with lysozyme prepared membrane suspensions from the same strain.[78] This CO_2 effect is not a unique feature of all hydrogen bacteria. An extremely sensitive response to CO_2 is observed with *A. eutrophus* and *A. ruhlandii*,[31,113] both organisms with two hydrogenases. H_2 oxidation by a mutant strain of *A. eutrophus*, defective in the NAD-reducing hydrogenase, was recently reported to be insensitive to CO_2.[80] This mutant strain, in contrast to the wild type, and a number of hydrogen bacteria, containing only the membrane-bound hydrogenase, exhibit exceptionally high rates of H_2 oxidation in the absence of CO_2. Therefore, the lack of the CO_2 enhancement effect in these organisms might be due to a respiratory chain that oxidizes hydrogen already at its maximum capacity. It remains to be shown whether the high H_2 oxidase activity reflects the necessity of reverse electron transport for NAD reduction. Energy is seemingly wasted under uncoupled growth conditions, and the primary biochemical events leading to this situation are not known. There are possible explanations for the observed phenomena during uncontrolled hydrogen oxidation:

1. Coupling sites are still functional, but ATP is either degraded by an ATPase or energy-spilling reactions or used for increased maintenance purposes.[113]
2. Not all three coupling sites are functional. Under low iron supplementation, *A. eutrophus* exhibits $H^+/0$ ratios of 3 to 4, indicating the loss of one site.[110]
3. Electrons flow via an abbreviated chain which is not coupled to oxidative phosphorylation. The observed uncoupled activity may or may not be a function of the growth rate.[35,113]

Recently, the study of synchronous heterotrophic cell cultures of *A. eutrophus* indicated that the extent of respiratory control varies during the cell cycle.[115,116] There are two groups of temporally linked events, those that occur maximally at 0.4 and 0.9 of a cycle (respiratory activity and ATPase activity) and those that occur maximally at 0.1 to 0.15 and 0.55 to 0.6 of a cycle (stimulation of respiration by CCCP, inhibition of ATPase activity by dicyclohexylcarbodiimide, $t_{1/2}$ of proton conductance). These transitions likely reflect changing energy requirements at different stages in the cell cycle. No significant oscillations were detected in the $H^+/0$ ratios exhibited by cells from various stages in the cell cycle.

D. Nitrate Respiration

Growth using nitrate as an electron acceptor has been reported for *P. denitrificans*,[111] *A. eutrophus*,[117] *P. pseudoflava*,[91] and *P. saccharophila*.[84] With hydrogen as electron donor, anaerobic growth of *P. denitrificans* is very poor unless yeast extract is added. Under these mixotrophic conditions, about two thirds of the cell carbon is derived from yeast extract and one third from CO_2.[118] Nitrate respiration with hydrogen was reported to be very modest for *A. eutrophus*[117] and *P. pseudoflava*.[91] Autotrophic cells of *P. facilis* completely reduce nitrate to nitrite at the expense of hydrogen. However, the process is not coupled to the assimilation of CO_2.[30,76] During anaerobic growth on fructose, two growth phases were observed for *A. eutrophus*. During the first phase, nitrate is reduced to nitrite which accumulates in the medium. In the second phase, nitrite is reduced with the formation of gaseous nitrogen which is the sole product of denitrification in this organism.[117] In contrast to *P. denitrificans*,[119] the dissimilatory membrane-bound nitrate reductase in *A. eutrophus* is also present in aerobically grown cells which reduce nitrate to nitrite without adaptation. The enzyme is, however, strongly repressed by ammonium ions. The nitrite reductase system is induced only under anaerobic growth conditions, as is the case with *P. deni-*

trificans.[117,120] Surprisingly, cells grown with fructose and nitrate contain an active membrane-bound H_2 to NO^-_3 oxidoreductase, probably due to residual hydrogenase activity in fructose-grown cells.[117]

Data on the characterization of enzymes involved in nitrate respiration are available only for *P. denitrificans.* The purified nitrate reductase is a molybdenum containing Fe-S protein.[121] The soluble nitrite reductase contains bo th types *c* and *d* hemes and exhibits oxidase activity in addition to its nitrite reductase activity.[122] Evidence derived from spectrophotometry and differential inhibition led to the conclusion that the nitrate reductase interacts with the respiratory chain at the leve of the *b*-type cytochromes without involvement of the *c*-type cytochromes. Membrane vesicles that coupled the anaerobic oxidation of NADH and succinate to ATP generation, however, failed to synthesize ATP when ascorbate/TMPD was the electron donor.[46] Analogous results were obtained with crude extracts of *P. saccharophila* grown heterotrophically as well as autotrophically with nitrate as electron acceptor.[84] Anaerobic heterotrophic growth of *A. eutrophus* resulted in the production of a cytochrome *d* (in addition to cytochrome aa_3 and *o*), with a concomitant loss of proton translocation at Site 3 and the disappearance of membrane-bound transhydrogenase activity when nitrate was used as electron acceptor.[110]

VI. REGULATION

It is likely that the natural habitat of hydrogen bacteria supplies both organic substrates and hydrogen for growth. The capacity to use the two types of substrates concomitantly is of general ecological importance. The advantage of growing on hydrogen and using this electron donor for additional energy generation during mixotrophic growth poses regulatory problems because mixotrophic metabolism involves dissimilation of substrates which do not share common catabolites. The predominantly catabolic function of the tricarboxylic acid cycle in heterotrophic cells is transformed into an exclusively biosynthetic function during chemolithotrophy. Regulatory phenomena as discussed below, e.g., repression of hydrogenase activity by organic compounds and repression of catabolic enzymes by hydrogen, have to be overcome under mixotrophic growth conditions. Very little biochemical data are available concerning these regulatory effects in hydrogen bacteria. The subject has been extensively reviewed.[6-8] As no substantially new results have been reported since then, the literature will be surveyed briefly.

A. Hydrogenase Synthesis

A study of the literature on hydrogenase synthesis leaves the reader with little insight and in a state of confusion. All hydrogen bacteria show a diminished capacity for hydrogen oxidation when cultured on organic substrates. The residual hydrogenase activity varies considerably, depending on the species, the organic carbon source, oxygen concentration, and the growth phase. For a number of hydrogen bacteria, the enzyme(s) seems to be strictly inducible by autotrophic growth conditions.[25-28,54] An inhibitory effect of high oxygen concentration on enzyme synthesis has been observed with both heterotrophic[6] and autotrophic cultures.[25,28,54] Enzyme synthesis can either occur under an H_2/O_2 atmosphere[9,123] or needs CO_2 in addition.[79,113] Organic substrates either stimulate,[28] inhibit,[9,26,54] or prevent[123] enzyme synthesis. Depending on the organic carbon source metabolized, *A. eutrophus* may have residual hydrogenase activities ranging from 0 to 40% of the autotrophic value.[8,79,124] Obviously neither the exact physiological conditions nor the molecular mechanism of hydrogenase synthesis are yet known.

B. Mixotrophy

In the absence of CO_2, hydrogen is able to completely suppress the formation of catabolic enzymes involved in the degradation of various organic compounds.[6,8] Under these conditions without CO_2, the cells stop growing.[28] With the exception of *P. denitrificans* and *A. autotrophicum*, this "hydrogen effect" has been reported for every strain tested until now. The inhibitory effect of hydrogen on the chemoorganic metabolism can be overcome by the addition of CO_2 to the H_2/O_2 gas atmosphere. Growth under these mixotrophic conditions results in higher cell yields than those obtained with either auto- or heterotrophic cultures. The concomitant utilization of hydrogen and an organic substrate has been reported for *A. eutrophus*,[8,80,125] *P. denitrificans*,[118] *P. facilis*,[125] *Mycobacterium*,[9] and a coryneform bacterium 11/x.[27] Unfortunately, studies on growth rates and cell yields under mixotrophic conditions have been rarely supplemented with data on substrate consumption, enzyme activities, and CO_2 fixation rates. Mixotrophically grown cells are reported to contain lowered hydrogenase and ribulose bisphosphate carboxylase activities compared to the value for autotrophic cells. The hydrogenase activity in mixotrophic cells of *Mycobacterium gordonae* can, however, by far exceed the enzyme levels of autotrophic cells, depending on the organic carbon source metabolized. In this organism, hydrogen also facilitates the degradation of normally unfavorable organic substrates.[9]

The extent to which both inorganic and organic substrates serve as energy and/or carbon sources has not been investigated. Experimental evidence for the fixation of CO_2 concomitant with the assimilation of organic compounds into cell material has been so far only shown for *P. denitrificans* growing mixotrophically with yeast extract.[118] It is doubtful whether the sharp regulatory effects described above for batch cultures are applicable to the growth of hydrogen bacteria in nature. A situation in which concentrations of nutrients are low or limiting (e.g., continuous culture) may well lead to a peaceful metabolic interaction of the two different pathways of substrate consumption and assimilation.

VII. GENETIC STUDIES

The study of energy conservation has been greatly facilitated by use of mutant strains.[126] Problems of metabolic regulation in hydrogen bacteria, e.g., hydrogenase synthesis and the hydrogen and CO_2 enhancement effects, can be much more easily approached by application of genetic methodology. Therefore, during the last few years, efforts have concentrated on the establishment of genetic transfer systems in these bacteria.

A number of bacteriophages are now known for hydrogen bacteria.[127,128] Temperate phages were isolated for *P. pseudoflava*,[12] although transduction has not yet been reported.[129] Reh and Schlegel[130] were the first to establish a genetic transfer system for *N. opaca* 1 b. The ability to grow chemoautotrophically can be transferred from this organism to various chemoorganoheterotrophic *Nocardia* strains and to *Corynebacterium hydrocarboclastus*. Genetic information is thought to be transferred by conjugation. Frequent spontaneous loss of autotrophy and the ability to cure strains by mitomycin C, ethidiumbromide, and rifampicin treatment indicated the extrachromosomal location of the autotrophic marker. These genetic results were supplemented by enzymatic studies:[131]

1.   In the donor strain as well as in the recombinants, the synthesis of the three key enzymes for autotrophic growth (soluble NAD-reducing hydrogenase, phosphoribulokinase, and ribulose bisphosphate carboxylase) are under coordinate control.

2.   Enzymes of the donor and recombinants show identical K_m values and their cellular location and gel electrophoretic properties are the same.

These results are in accordance with the conclusions drawn from genetic studies. In *N. opaca*, the genetic information for autotrophic growth is transferred *en bloc* and is localized on a plasmid. All strains tested contain one or more plasmids, regardless of culture condition.[132] Recently, evidence has been presented for the extrachromosomal location of genetic information for the hydrogenase in *P. facilis.*[133]

VIII. CONCLUSION

Chemolithotrophic bacteria are distinguished by their use of unconventional electron donors for their respiratory chains. The advantage of being able to exist in habitats which are low or completely devoid of organic material is paid for by an exceptionally low energy yield from substrate oxidation. The majority of chemoautotrophic microorganisms do not only have to expend extra energy for NAD reduction by reverse electron transport, but also consume more energy in carbon synthesis from CO_2 than heterotrophic bacteria.[105] Compared to the rest of their chemoautotrophic relatives, hydrogen bacteria lead a life of luxury, especially those which are able to reduce NAD directly by hydrogen without involvement of the respiratory chain. They combine simplicity of nutrient requirement with exceptionally high autotrophic growth rates and a tight and effective coupling between energy generation and CO_2 assimilation. Their chemoautotrophic enzyme equipment is the least sophisticated among chemoautotrophic bacteria. The heterotrophic respiratory chain can be transformed into a chemoautotrophic one without impairment of the ATP yield simply by inserting a hydrogenase into the cytoplasmic membrane. Therefore, hydrogen bacteria do not possess exotic respiratory chains, but as is frequently pointed out,[15,16,106] *A. eutrophus* and especially *P. denitrificans* contain pseudomitochondrial electron transport chains characterized by the presence of a membrane-bound transhydrogenase (Site 0) and a high-potential cytochrome *c* (Site 3) essential for the existence of four proton-translocating segments.[106]

As anticipated by their aerobic hydrogen metabolism, hydrogen bacteria possess the most stable and oxygen-insensitive hydrogenases thus far reported.

The ease with which heterotrophy is converted to autotrophy might be an explanation of why the ability to grow on hydrogen and CO_2 is so abundant among aerobic bacteria of widely varying taxonomic position. The concomitant exploitation of two potential energy and carbon sources under mixotrophic conditions, as is likely to occur in nature, provides a selective advantage over strict chemoheterotrophic bacteria. The extrachromosomal location of the genetic information for chemolithoautotrophy thus far reported for two strains together with the failure to isolate strict autotrophic hydrogen bacteria lends itself to speculate that chemolithoautotrophy with hydrogen is just a clever acquisition to the "normal" chemoorganoheterotrophic cell metabolism.

IX. ACKNOWLEDGMENT

I wish to express my thanks to all colleagues who have made their results and manuscripts available to me prior to publication. I am very grateful to Dr. William A. Toscano for reading the manuscript.

REFERENCES

1. **Yamagata, S. and Nakamura, H.**, Uber die Hydrogenase, nebst einer Bemerkung uber den Mechanismus der bakteriellen Knallgasreaktion, *Acta Phytochim.*,10, 297, 1938.
2. **Walker, C. C. and Yates, M. G.**, The hydrogen cycle in nitrogen-fixing *Azotobacter chroococcum*, *Biochimie*, 60, 225, 1978.
3. **Bothe, H., Distler, E., and Eisbrenner, G.**, Hydrogen metabolism in blue-green algae, *Biochimie*, 60, 277, 1978.
4. **Aragno, M., Walther-Mauruschat, A., Mayer, F., and Schlegel, H. G.**, Micromorphology of Gram-negative hydrogen bacteria. I. Cell morphology and flagellation, *Arch. Microbiol.*, 114, 93, 1977.
5. **Walther-Mauruschat, A., Aragno, M., Mayer, F., and Schlegel, H. G.**, Micromorphology of Gram-negative hydrogen bacteria. II. Cell envelope, membranes, and cytoplasmic inclusions, *Arch. Microbiol.*, 114, 101, 1977.
6. **Schlegel, H. G. and Eberhardt, U.**, Regulatory phenomena in the metabolism of knallgas bacteria, in *Advances in Microbial Physiology*, Vol. 7, Rose, A. H. and Tempest, D. W., Eds., Academic Press, London, 1972, 205.
7. **Schlegel, H. G.**, Mechanisms in chemo-autotrophy, in *Marine Ecology*, Vol. 2 (Part 1), Kinne, O., Ed., John Wiley & Sons, New York, 1975, 9.
8. **Rittenberg, S. C.**, The roles of exogenous organic matter in the physiology of chemolithotrophic bacteria, in *Advances in Microbial Physiology*, Vol. 3, Rose, A. H. and Tempest, D. W., Eds., Academic Press, London, 1969, 159.
9. **Park, S. S. and De Cicco, B. T.**, Hydrogenase and ribulose diphosphate carboxylase during autotrophic, heterotrophic, and mixotrophic growth of scotochromogenic mycobacteria, *J. Bacteriol.*, 127, 731, 1976.
10. **Calloway, D. H. and Kumar, A. M.**, Protein quality of the bacterium *Hydrogenomonas eutropha*, *Appl. Microbiol.*, 17, 176, 1969.
11. **Bongers, L.**, Life support systems for space missions, *Dev. Ind. Microbiol.*,5, 139, 1964.
12. **Schlegel, H. G.**, Physiology and biochemistry of knallgas bacteria, in *Advances in Comparative Physiology and Biochemistry*, Vol. 2, Lowenstein, O., Ed., Academic Press, New York, 1966, 185.
13. **Schlegel, H. G.**, The physiology of hydrogen bacteria, *Antonie van Leeuwenhoek J. Microbiol. Serol.*, 42, 181, 1976.
14. **Haddock, B. A. and Jones, C. W.**, Bacterial respiration, *Bacteriol. Rev.*, 41, 47, 1977.
16. **John, P. and Whatley, F. R.**, The bioenergetics of *Paracoccus denitrificans*, *Biochim. Biophys. Acta*, 463, 129, 1977.
17. **Bongers, L.**, Some aspects of continuous cultures of hydrogen bacteria, *Dev. Ind. Microbiol.*, 11, 241, 1970.
18. **Goto, E., Kodama, T., and Minoda, Y.**, Isolation and culture conditions of thermophilic hydrogen bacteria, *Agric. Biol. Chem.*, 41, 685, 1977.
19. **Stouthamer, A. H.**, Energetic aspects of the growth of micro-organisms, in *Microbial Energetics*, Haddock, B. A. and Hamilton, W. A., Eds., Cambridge University Press, London, 1977, 285.
20. **Bartha, R. and Ordal, E. J.**, Nickel-dependent chemolithotrophic growth of two Hydrogenomonas strains, *J. Bacteriol.*,89, 1015, 1965.
21. **Repaske, R. and Repaske, A. C.**, Quantitative requirements for exponential growth of *Alcaligenes eutrophus*, *Appl. Environ. Microbiol.*, 32, 585, 1976.
22. **Ruhland, W.**, Beitrage zur Physiologie der Knallgasbakterien, *Jahrb. Wiss. Bot.*, 63, 321, 1924.
23. **Repaske, R. and Mayer, R.**, Dense autotrophic cultures of *Alcaligenes eutrophus*, *Appl. Environ. Microbiol.*, 32, 592, 1976.
24. **Berndt, H., and Wölfle, D.**, Hydrogenase: its role as electron generating enzyme in the nitrogen fixing hydrogen bacterium *Xanthobacter autotrophicus*, in *Hydrogenases: their catalytic activity, structure and function*, Schlegel, H. G. and Schneider, K., Eds., E. Goltze, Gottingen, Federal Republic of Germany, 1978, 327.
25. **Packer, L. and Vishniac, W.**, Chemosynthetic fixation of carbon dioxide and characteristics of hydrogenase in resting cell suspensions of *Hydrogenomonas ruhlandii* nov. spec., *J. Bacteriol.*, 70, 216, 1955.
26. **Schneider, K., Rudolph, V., and Schlegel, H. G.**, Description and physiological characterization of a coryneform hydrogen bacterium, strain 14 g, *Arch. Mikrobiol.*,93, 179, 1973.
27. **Canevascini, G. and Eberhardt, U.**, Chemolithotrophic growth and regulation of hydrogenase formation in the coryneform hydrogen bacterium strain 11/x, *Arch. Microbiol.*, 103, 289, 1975.
28. **Aragno, M. and Schlegel, H. G.**, Physiological characterization of the hydrogen bacterium *Aquaspirillum autotrophicum*, *Arch. Microbiol.*, 116, 221, 1978.

29. **Repaske, R., Ambrose, C. A., Repaske, A. C., and De Lacy, M. L.**, Bicarbonate requirement for elimination of the lag period of *Hydrogenomonas eutropha*, *J. Bacteriol.*, 107, 712, 1971.
30. **Schatz, A.**, Uptake of carbon dioxide, hydrogen and oxygen by *Hydrogenomonas facilis*, *J. Gen. Microbiol.*, 6, 329, 1952.
31. **Eberhardt, U.**, On chemolithotrophy and hydrogenase of a Gram-positive Knallgas bacterium, *Arch. Mikrobiol.*, 66, 91, 1969.
32. **Kanai, R., Miyashi, S., and Takamiya, A.**, Knall-gas reaction-linked fixation of labelled carbon dioxide in an autotrophic *Streptomyces*, *Nature (London)*, 188, 873, 1960.
33. **Lechtmann, M. D., Goldner, B. H., and Canfield, J. H.**, Studies with aerobic hydrogen oxidizing bacteria for application to a regenerative life support system, *Dev. Ind. Microbiol.*, 5, 299, 1964.
34. **Kodama, T., Igarashi, V., and Minoda, Y.**, Material balance and efficiency of energy conversion for the autotrophic growth of a hydrogen bacterium, *Agric. Biol. Chem.*, 39, 83, 1975.
35. **Bongers, L.**, Energy generation and utilization in hydrogen bacteria, *J. Bacteriol.*, 104, 145, 1970.
36. **Schuster, E. and Schlegel, H. G.**, Chemolithotrophes Wachstum von Hydrogenomonas H16 im Chemostaten mit elektrolytischer Knallgaserzeugung, *Arch. Mikrobiol.*, 58, 380, 1967.
37. **Voitovich, Y.V., Ponomarev, I. P., Artyukh, N. G., Kesler, T. G., and Sid'ko, F. Y.**, The efficiency of free energy utilization by hydrogen-oxidizing bacteria, *Mikrobiologiya*, 42, 14, 1973.
38. **Jüttner, R.**, Autotrophe Massenkultur von *Alcaligenes eutrophus* Stamm H 16 — PHB⁻⁴ im Chemostaten, Ph.D. thesis, University of Göttingen, Germany, 1977.
39. **Andreesen, M.**, Personal communication, 1975.
40. **Namsarev, B. B., Nozhevnikov, A. N., and Zavarzin, G. A.**, Utilization of formic acid by hydrogen bacteria, *Mikrobiologiya*, 40, 772, 1975.
41. **Friedrich, C. G., Bowien, B., and Friedrich, B.**, Formate and oxalate metabolism in *Alcaligenes eutrophus*, *J. Gen. Microbiol.*, 115, 185, 1979.
42. **Zavarzin, G. A. and Nozhevnikova, A. N.**, Aerobic carboxydobacteria, in *Microbial Ecology*, Vol. 3, Kinne, O., Ed., John Wiley & Sons, New York, 1977, 305.
43. **Meyer, O. and Schlegel, H. G.**, Reisolation of the carbon monoxide utilizing hydrogen bacterium *Pseudomonas carboxydovorans*(Kistner) comb. nov., *Arch. Microbiol.*, 118, 35, 1978.
44. **Kalacheva, G. S.**, Phospholipids of hydrogen bacteria, *Upr. Biosint. Vodorodnykh Bakt. Drugihk, Khemoautotrofo*, Tezisy Vses. Soveshch, Fedorova, Ya. V., Kesler, T. G., and Shavkun, I. V., Eds., Akad. Nauk. SSSR, Sib. Otd. Krasnoyarsk, U.S.S.R., 1976, 50.
45. **Kimura, H. and Futai, M.**, Effects of phospholipids on L-lactate dehydrogenase from membranes of *Escherichia coli*: activation and stabilization of the enzyme with phospholipids, *J. Biol. Chem.*, 253, 1095, 1978.
46. **John, P. and Whatley, F. R.**, Oxidative phosphorylation coupled to oxygen uptake and nitrate reduction in *Micrococcus denitrificans*, *Biochim. Biophys. Acta*, 216, 342, 1970.
47. **Probst, I. and Schlegel, H. G.**, Respiratory components and oxidase activities in *Alcaligenes eutrophus*, *Biochim. Biophys. Acta*, 440, 412, 1976.
48. **Bone, D. H.**, Localization of hydrogen evolving enzymes in *Pseudomonas saccharophila*, *Biochem. Biophys. Res. Comm.*, 3, 211, 1960.
49. **Wittenberger, H. and Repaske, R.**, Studies on hydrogen oxidation in cell-free extracts of *Hydrogenomonas eutropha*, *Biochim. Biophys. Acta*, 47, 542, 1961.
50. **Eberhardt, U.**, Uber das Wasserstoff-aktivierende System von *Hydrogenomonas* H 16.I. Verteilung der Aktivitat auf zwei Zellfraktionen, *Arch. Mikrobiol.*, 53, 288, 1966.
51. **Vishniac, W. and Trudinger, P. A.**, Carbon dioxide fixation and substrate oxidation in the chemosynthetic sulfur and hydrogen bacteria, *Bacteriol. Rev.*, 26, 168, 1962.
52. **Atkinson, D. E. and McFadden, B. A.**, The biochemistry of Hydrogenomonas. I. The hydrogenase of *Hydrogenomonas facilis* in cell free preparations, *J. Biol. Chem.*, 210, 885, 1954.
53. **Schneider, K. and Schlegel, H. G.**, Localization and stability of hydrogenases from aerobic hydrogen bacteria, *Arch. Microbiol.*, 112, 229, 1977.
54. **Aggag, M. and Schlegel, H. G.**, Studies on a Gram-positive hydrogen bacterium, *Nocardia opaca* 1 b.I. Description and physiological characterization, *Arch. Mikrobiol.*, 88, 299, 1973.
55. **Aggag, M. and Schlegel, H. G.**, Studies on a Gram-positive hydrogen bacterium *Nocardia opaca* 1b.III. Purification, stability and some properties of the soluble hydrogen dehydrogenase, *Arch. Microbiol.*, 100, 25, 1974.
56. **Bone, D. H., Bernstein, S., and Vishniac, W.**, Purification and some properties of different forms of hydrogen dehydrogenase, *Biochim. Biophys. Acta*, 67, 581, 1963.
57. **Bone, D.**, Inhibitor, isotopic and kinetic studies on hydrogen dehydrogenase, *Biochim. Biophys. Acta*, 67, 589, 1963.
58. **Repaske, R.**, The electron transport system of *Hydrogenomonas eutropha*. I. Diphosphopyridine nucleotide reduction by hydrogen, *J. Biol. Chem.*, 237, 1351, 1962.

59. Repaske, R. and Dans, C. L., A factor for coupling NAD to hydrogenase in *Hydrogenomonas eutropha, Biochem. Biophys. Res. Comm.*, 30, 136, 1968.
60. Pfitzner, J., Linke, H. A. B., and Schlegel, H. G., Eigenschaften der NAD-spezifischen Hydrogenase aus *Hydrogenomonas* H 16, *Arch. Microbiol.*, 71, 67, 1970.
61. Schneider, K. and Schlegel, H. G., Purification and properties of soluble hydrogenase of *Alcaligenes eutrophus* H 16, *Biochim. Biophys. Acta*, 452, 66, 1976.
62. Schneider, K. and Cammack, R., Soluble hydrogenase from *Alcaligenes eutrophus*, an iron-sulfur flavoprotein, in *Hydrogenases: Their Catalytic Activity, Structure and Function*, Schlegel, H. G. and Schneider, K., Eds., E. Goltze, Gottingen, Federal Republic of Germany, 1978, 221.
63. Schneider, K. and Schlegel, H. G. Identification and quantitative determination of the flavin component of hydrogenase from *Alcaligenes eutrophus, Biochem. Biophys. Res. Commun.*, 84, 564, 1978.
64. Schneider, K., Cammack, R., Schlegel, H. G., and Hall, D. O., The iron-sulfur centers of soluble hydrogenase from *Alcaligenes eutrophus, Biochim. Biophys. Acta*, 578, 445, 1979.
65. Gruzinskii, I. V., Gogotov, I. N., Bechina, E. M., and Semenov, Ya. V., Hydrogenase activity of hydrogen-oxidizing bacteria *Alcaligenes eutrophus, Mikrobiologiya*, 46, 625, 1977.
66. Pichukova, E. E., Varfolomeer, S. D., and Berezin, I. V., Oxygen as a stabilizer of hydrogenase of hydrogen bacterium *Alcaligenes eutrophus* Z-1, *Dokl. Acad. Nauk SSSR*, 236, 1253, 1977.
67. Schneider, K. and Schlegel, H. G., The NAD-reducing soluble hydrogenase from *Alcaligenes eutrophus*, in *Abstr. Int. Symp. Microbial Growth on C₁ Compounds*, Pushkino, SSSR, 1977, 98.
68. Rao, K. K. and Hall, D. O., Hydrogen production from isolated chloroplasts, in *Photosynthesis in Relation to Model Systems*, Barber, J., Ed., Elsevier, Amsterdam, 1979, 300.
69. Packer, L. and Vishniac, W., The specificity of a diphosphopyridine nucleotide-linked hydrogenase, *Biochim. Biophys. Acta*, 17, 153, 1955.
70. Aggag, M., unpublished data, 1974.
71. Wilson, J. E., Ambiquitous enzymes: variation in intracellular distribution as a regulatory mechanism, *Trends Biochem. Sci.*, 3, 124, 1978.
72. Schink, B. and Schlegel, H. G., Hydrogen metabolism in aerobic hydrogen-oxidizing bacteria, *Biochimie*, 60, 297, 1978.
73. Schink, B. and Schlegel, H. G., The membrane-bound hydrogenase of *Alcaligenes eutrophus*. I. Solubilization, purification, and biochemical properties, *Biochim. Biophys. Acta*, 567, 315, 1979.
74. Eminova, E. E. and Romanova, A. K., Hydrogenase activity of the thermophilic hydrogen-oxidizing bacterium *Pseudomonas thermophila, Mikrobiologiya*, 46, 505, 1977.
75. Schink, B., unpublished data, 1978.
76. Schatz, A. and Bovell, C., Jr., Growth and hydrogenase activity of a new bacterium, *Hydrogenomonas facilis, J. Bacteriol.*, 63, 87, 1952.
77. Atkinson, D. E., The biochemistry of *Hydrogenomonas*. IV. The inhibition of hydrogenase by oxygen, *J. Biol. Chem.*, 218, 557, 1956.
78. Probst, I., unpublished data, 1977.
79. Eberhardt, U., Uber das Wasserstoff-aktivierende System von *Hydrogenomonas* H 16. II. Abnahme der Aktivitat bei heterotrophem Wachstum, *Arch. Microbiol.*, 54, 115, 1966.
80. Schink, B. and Schlegel, H. G., Mutants of *Alcaligenes eutrophus* defective in autotrophic metabolism, *Arch. Microbiol.*, 117, 123, 1978.
81. Sim, E. and Vignais, P. M., Hydrogenase activity in *Paracoccus denitrificans*. Partial purification and interaction with the electron transport chain, *Biochimie*, 60, 307, 1978.
82. Bongers, L., Phosphorylation in hydrogen bacteria, *J. Bacteriol.*, 93, 1615, 1967.
83. Pfitzner, J., Uber das Elektronentransport system bei *Hydrogenomonas eutropha* Stamm H 16, *Zentralbl. Bakteriol. Parasitenk. Infektionskr. Hyg., Abt. 1, Orig. Reihe A*, 200, 396, 1972.
84. Ishaque, H., Donawa, A., and Aleem, M. I. H., Energy coupling mechanisms under aerobic and anaerobic conditions in autotrophically grown *Pseudomonas saccharophila, Arch. Biochem. Biophys.*, 159, 570, 1973.
85. Meyer, O. and Schlegel, H. G., Oxidation of carbon monoxide cell extracts of *Pseudomonas carboxydovorans, J. Bacteriol.*, 137, 811, 1979
86. Repaske, R. and Lizotte, C. L., The electron transport system of *Hydrogenomonas eutropha*. II. Reduced nicotinamide adenine dinucleotide-menadione reductase, *J. Biol. Chem.*, 240, 4777, 1965.
87. Lester, R. L. and Crane, F. L., The natural occurrence of coenzyme Q and related compounds, *J. Biol. Chem.*, 234, 2169, 1959.
88. Meyer, O., personal communication, 1978.
89. Packer, L., Respiratory carriers involved in the oxidation of hydrogen and lactate in a facultative autotroph, *Arch. Biochem. Biophys.*, 78, 54, 1958.
90. Bernard, U., Probst, I., and Schlegel, H. G., The cytochromes of some hydrogen bacteria, *Arch. Mikrobiol.*, 95, 29, 1974.

91. Auling, G., Reh, M., Lee, C. M., and Schlegel, H. G., *Pseudomonas pseudoflava*, a new species of hydrogen-oxydizing bacteria: its differentiation from *Pseudomonas flava* and other yellow-pigmented, Gram-negative, hydrogen-oxidizing species, *Int. J. Syst. Bacteriol.*, 28, 82, 1978.

92. Bone, D. H., Cytochromes of *Pseudomonas saccharophila*, *Nature (London)*, 197, 517, 1963.

93. Fang, F. S. and Burris, R. H., Cytochrome *c* in *Hydrogenomonas eutropha*, *J. Bacteriol.*, 96, 298, 1968.

94. Probst, I., Wolf, G., and Schlegel, H. G., An oxygen-binding flavohemoprotein from *Alcaligenes eutrophus*, *Biochim. Biophys. Acta*, 576, 471, 1978.

95. Ishaque, M., Donawa, A., and Aleem, M. I. H., Oxidative phosphorylation in *Pseudomonas saccharophila* under autotrophic and heterotrophic growth conditions, *Biochem. Biophys. Res. Comm.*, 44, 245, 1971.

96. Donawa, A., Ishaque, M., and Aleem, M. I. H., Energy conversion in autotrophically grown *Pseudomonas saccharophila,* *Eur. J. Biochem.,* 21, 293, 1971.

97. Burnell, J. N., John, P., and Whatley, F. R., The reversibility of active sulphate transport in membrane vesicles of *Paracoccus denitrificans*, *Biochem. J.*, 150, 527, 1975.

98. Knobloch, K. Ishaque, M., and Aleem, M. I. H., Oxidative phosphorylation in autotrophically grown *Micrococcus denitrificans*, *Arch. Mikrobiol.*, 76, 114, 1971.

99. Scholes, P. B. and Mitchell, P., Respiration-driven proton translocation in *Micrococcus denitrificans*, *J. Bioenerg.*, 1, 309, 1970.

100. Kell, D. B., John, P., and Ferguson, S. J., The protonmotive force in phosphorylating membrane vesicles from *Paracoccus denitrificans*, *Biochem. J.*, 174, 257, 1978.

101. Meijer, E. M., van Versefeld, H. W., van der Beek, E. G., and Stouthamer, A. H., Energy conservation during aerobic growth in *Paracoccus denitrificans*, *Arch. Microbiol.*, 112, 25, 1977.

102. Ishaque, M. and Aleem, M. I. H., Energy coupling in *Hydrogenomonas eutropha*, *Biochim. Biophys. Acta*, 223, 388, 1970.

103. Bernard, U. and Schlegel, H. G., NADH- and H$_2$-oxidation in hydrogen bacteria studied by respiratory chain inhibitors, *Arch. Mikrobiol.*, 95, 39, 1974.

104. Beatrice, M. C. and Chappell, J. B., The respiratory chain of *Hydrogenomonas* H 16, *Biochem. J.*, 178, 15, 1979.

105. Beatrice, M. C. and Chappell, J. B., Respiration-driven proton translocation in *Hydrogenomonas eutropha* H 16, *Biochem. Soc. Trans.*, 2, 151, 1974.

106. Drozd, J. W. and Jones, C. W., Oxidative phosphorylation in *Hydrogenomonas eutropha* H 16 grown with and without iron, *Biochem. Soc. Trans.*, 2, 529, 1974.

107. Jones, C. W., Brice, J. M., Downs, A. J., and Drozd, J. W., Bacterial respiration-linked proton translocation and its relationship to respiratory chain composition, *Eur. J. Biochem.*, 52, 265, 1975.

108. Jones, C. W., Aerobic respiratory systems in bacteria, in *Microbial Energetics*, Haddock, B. A., and Hamilton, W. A., Eds., Cambridge University Press, London, 1977, 23.

109. Bongers, L., Yields of *Hydrogenomonas eutropha* from growth on succinate and fumarate, *J. Bacteriol.*, 102, 598, 1970.

110. Drozd, J. W., Growth energetics in *Alcaligenes eutrophus* H 16, *Proc. Soc. Gen. Microbiol.*, 4, 72, 1977.

111. Fedorova, T. A., Calculation of energy-substrate consumption by hydrogen bacteria, *Mikrobiologiya*, 42, 160, 1973.

112. Hempfling, W. P. and Vishniac, W., Yield coefficients of *Thiobacillus neopolitanus* in continuous culture, *J. Bacteriol.*, 93, 874, 1967.

113. Bartha, R., Physiologische Untersuchungen uber den chemolithotrophen Stoffwechsel neu isolierter Hydrogenomonas-Stamme, *Arch. Mikrobiol.*, 41, 313, 1962.

114. Ahrens, J. and Schlegel, H. G., Zur Regulation der NAD-abhangigen Hydrogenase-Aktivitat, *Arch. Mikrobiol.*, 55, 257, 1966.

115. Edwards, C. and Jones, C. W., Respiratory properties of synchronous cultures of *Alcaligenes eutrophus* H 16 prepared by a continuous-flow size selection method, *J. Gen. Microbiol.*, 99, 383, 1977.

116. Edwards, C., Spode, J. A., and Jones, C. W., The properties of adenosine triphosphatase from exponential and synchronous cultures of *Alcaligenes eutrophus* H 16, *Biochem. J.*, 172, 253, 1978.

117. Pfitzner, J. and Schlegel, H. G., Denitrifikation bei *Hydrogenomonas eutropha* Stamm H 16, *Arch. Mikrobiol.*, 90, 199, 1973.

118. Banerjee, A. K. and Schlegel, H. G., Zur Rolle des Hefeextraktes wahrend des chemolithotrophen Wachstums von *Micrococcus denitrificans*, *Arch. Mikrobiol.*, 53, 132, 1966.

119. Pichinoty, F., L'inhibition par l'oxygene de la denitrification bacterienne, *Ann. Inst. Pasteur Paris*, 109, 248, 1965.

120. Lam, Y. and Nicholas, D. J. D., Aerobic and anaerobic respiration in *Micrococcus denitrificans*, *Biochim. Biophys. Acta*, 172, 450, 1969.

121. Forget, P. and Der Vartanian, D. V., The bacterial nitrate reductases: EPR studies on nitrate reductase A from *Micrococcus denitrificans*, *Biochim. Biophys. Acta*, 256, 600, 1972.

122. Newton, N. A., The two-haem nitrite reductase of *Micrococcus denitrificans*, *Biochim. Biophys. Acta*, 185, 316, 1969.

123. Lindsay, E. M. and Syrett, P. G., The induced synthesis of hydrogenase by *Hydrogenomonas facilis*, *J. Gen. Microbiol.*, 19, 223, 1958.

124. Stukus, P. E. and De Cicco, B. T., Autotrophic and heterotrophic metabolism of *Hydrogenomonas*: regulation of autotrophic growth by organic substrates, *J. Bacteriol.*, 101, 339, 1970.

125. De Cicco, B. T. and Stukus, P. E., Autotrophic and heterotrophic metabolism of *Hydrogenomonas*. I. Growth yields and patterns under dual substrate conditions, *J. Bacteriol.*, 95, 1469, 1968.

126. Cox, G. B. and Gibson, F., Studies on electron transport and energy-linked reactions using mutants of *Escherichia coli*, *Biochim. Biophys. Acta*, 346, 1, 1974.

127. Pootjes, C. F., Isolation of a bacteriophage for *Hydrogenomonas facilis*, *J. Bacteriol.*, 87, 1259, 1964.

128. Rautenstein, Ya. L. Khavina, E. S., Solov'eva, N. Ya., Moskalenko, L. N., and Filatova, A. D., Bacteriophages of hydrogen bacteria and lysogeny in these cultures, *Mikrobiologiya*, 42, 167, 1973.

129. Auling, G., Mayer, F., and Schlegel, H. G., Isolation and partial characterization of normal and defective bacteriophages of Gram-negative hydrogen bacteria, *Arch. Microbiol.*, 115, 237, 1977.

130. Reh, M. and Schlegel, H. G., Chemolithoautotrophie als eine übertragb are, autonome Eigenschaft von *Nocardia opaca* 1b, Nachr. Akad. Wiss. Gottingen. II. Math.-Phys. Kl. 12, 207, 1975.

131. Ecker, C. and Reh, M., unpublished data, 1978.

132. Sartorius, G. and Reh, M., personal communication, 1978.

133. Pootjes, C. F., Evidence for plasmid coding of the ability to utilize hydrogen gas by *Pseudomonas facilis*, *Biochem. Biophys. Res. Comm.*, 76, 1002, 1977.

Chapter 9

RESPIRATORY ELECTRON FLOW IN FACULTATIVE PHOTOSYNTHETIC BACTERIA

D. Zannoni and A. Baccarini-Melandri

TABLE OF CONTENTS

I. INTRODUCTION

In general there are two different ways to deal with a review of a subject that is still a great matter of debate: either to present a mere description of the body of experimental work available in the literature or to focus attention on only a few examples and organize a logical framework within which the most current ideas about this topic can be applied, thereby enabling new suggestions to be made. The second is certainly the more difficult and risky approach, but in our opinion, it also seems to be the most interesting way to describe a process, like respiration in facultative phototrophs, that varies considerably in detail in different organisms and which is also subjected to more environmental parameters than in other bacteria. In order to illustrate why we feel the second approach to be more appropriate and stimulating, a brief outline of the main physiological characteristics of facultative photosynthetic bacteria (Rhodospirillaceae) is necessary.

In the bacterial world, the greatest capacity for adaptation to a range of nutritional conditions is shown by purple photosynthetic bacteria and hydrogen bacteria. Among the former, the nonsulfur purple bacteria (Rhodospirillaceae) are of particular interest because of their remarkable metabolic versatility. In fact, in addition to the phototrophic and photoheterotrophic conditions of growth in anaerobiosis on inorganic (hydrogen) or organic (succinate, malate, etc.) sources of reducing power, many other possibilities of adaptation to different growth conditions are available to this family of prokaryotes: growth as ordinary chemoheterotrophic bacteria in the dark using either oxygen or nitrate[1] as electron acceptors and growth in aerobiosis in darkness on hydrogen and carbon dioxide[2] and in a fermentative manner on a sugar as the sole carbon and energy source.[3-6] This extreme metabolic flexibility is shown in Table 1.

We would also like to point out the great difference between the photosynthetic process in phototrophic bacteria and blue-green algae (Cyanobacteria), which represents the second group of prokaryotic organisms with a photosynthetic apparatus. The latter group, with a few exceptions,[7] performs an oxygenic photosynthesis like that of green plants, whereas phototrophic bacteria carry out a completely anoxygenic process. In Cyanobacteria therefore photosynthetic reducing equivalents are derived from water and are not dependent on the availability of reduced metabolites: this situation is the opposite to that occurring in Rhodospirillaceae, in which reducing equivalents are derived from organic or inorganic exogenous sources and in which a strict interaction between photosynthetic and respiratory reactions must exist. From this feature and because photosynthetic green and sulfur purple bacteria, such as *Chlorobium* and *Chromatium*, do not respire, it follows that the Rhodospirillaceae are the only prokaryotes in which light-dependent and respiratory electron flow and the interaction between these two processes can be investigated comparatively in the same organism under different and extreme conditions of growth.

The most striking morphological change detectable in photosynthetically grown cells in comparison with cells growing aerobically in the dark is the organization of the intracytoplasmic membranes.[8,9] The photosynthetic apparatus (light-harvesting pigments, reaction center bacterio-chlorophyll, and electron carriers) is predominantly localized on intracytoplasmic membranes, known as chromatophores (literally from the Greek, bearing color), and it is well documented in the literature that the oxygen partial pressure is the main factor in regulation of pigment formation.[10-12] A decrease in oxygen partial pressure below a threshold level which seems to be different for different species,[12,13] not only induces the formation of the photosynthetic apparatus, but also stimulates the process of membrane formation, with an enlargement of the intracytoplasmic membrane.[14] An increase of oxygen tension inhibits both formation of photosynthetic apparatus and membrane formation.[15,16]

Table 1
THE ABILITY OF METABOLIC ADAPTATION OF RHODOSPIRILLACEAE

| | Conditions of growth | | | | |
Mode of growth	Carbon source	Source of reducing power	Electron acceptors	Light	Dark
Photoautotrophic	CO_2	$H_2, H_2S, S_2O_3^{--}$	—	+	—
Photoheterotrophic	Organic compounds	Organic compounds	—	+	—
Chemoheterotrophic	Organic compounds	Organic compounds	O_2, NO_3	—	+
Chemoautotrophic	CO_2	H_2	O_2	—	+
Fermentative	Organic compounds	Organic compounds	None, DMSO, TMAO	—	+

Because of the strict and mutual interrelation between the photosynthetic and respiratory apparatuses, one cannot describe the aerobic respiratory system of Rhodospirillaceae and disregard the enormous and very useful amount of data available in the literature on the photosynthetic electron transport system of the same organisms.

On the basis of these considerations, we have selected two organisms, *Rhodopseudomonas capsulata* and *Rhodopseudomonas sphaeroides*, which have been studied in great detail in both their respiratory and photosynthetic functions; this choice does not exclude pertinent references to other species of facultative photosynthetic bacteria. Excellent reviews dealing with many of the topics to be considered in this survey of the facultative photosynthetic bacteria have appeared in the past and recent years,[17-26] but are totally or predominantly focussed on the photosynthetic process operating in these organisms. Since one goal of this review is to emphasize the recent and controversial aspects related to the physiology and the bioenergetics of the redox chain in nonsulfur purple bacteria, we consider the examination of these reviews necessary for a detailed coverage of this topic.

II. RESPIRATORY ELECTRON TRANSPORT CHAIN

Studies on the respiratory function of Rhodospirillaceae and also on its relation to the photosynthetic function have been performed in whole cells, in EDTA-lysozyme-treated cells (spheroplasts), and in membrane fragments obtained either by mechanical breaking of intact cells or by osmotic lysis of spheroplasts. After disruption of the bacterial cells by mechanical treatments (French Pressure Cell, sonication, etc.), inside-out membrane vesicles containing the electron transport carriers associated with the respiratory and photosynthetic electron flow, with an opposite polarity with respect to whole cells,[27-29] can be isolated by differential centrifugation. Moreover, in the recent past, different procedures which give vesicles conserving the same polarity as the whole cell (right-side-out vesicles) have been described.[30,31] The characterization of the properties of these particles[31] in comparison with the chromatophore vesicles (inside-out vesicles) has shown an asymmetrical arrangement of the electon transport system within the membrane.

A feature which emerges from different research approaches and which appears to be common to many genera of the Rhodospirillaceae family is the presence of a branched respiratory electron flow. Branching of respiratory systems, a widespread phenomenon in bacteria,[32-43] can occur at different points in the redox chain, as for example at the level of primary dehydrogenases, in the ubiquinone cytochrome *b* region, or at the level of the terminal oxidases. Although clear demonstrations of branched respiratory chains in nonsulfur bacteria have been reported in detail for only a few genera,[44-46] the presence of different oxidases under different conditions of growth (e.g., *R. sphaeroides*)[47,48] could indicate a possible operativity of alternative pathways of electron transport to oxygen in many other species.

As mentioned above, detailed studies of the nature of the redox carriers and of the organization of the respiratory chain of facultative photosynthetic bacteria started recently and were greatly facilitated by the powerful tool offered for these studies by respiratory deficient mutants, especially those of *R. capsulata* isolated by Marrs and Gest.[49] In order to help the reader throughout this review, we think it appropriate at this point to describe the main characteristics of the respiratin-deficient mutants of *R. capsulata* mentioned in the following paragraphs (Table 2).

A. Respiratory Oxidases

The terminal oxidase systems of aerobically grown Rhodospiillaceae seem to be generally associated with *b*-type cytochromes. Only in *R. sphaeroides* does aerobic growth lead to the development of a cytochrome *c* oxidase, with the same spectroscopic properties as the cytochrome *aa₃* complex of mitochondria.[50] However in membranes from cells growing in the dark in the presence of a low oxygen concentration, a *b*-type oxidase appears to be present in *R. sphaeroides,*[47,48] according to the general statement that *b*-type cytochromes seem to represent the oxidases characteristic of this group of microorganisms.[44,46,51,52]

The chemical definition of cytochrome *b* usually emphasizes its nonreactivity with carbon monoxide or cyanide *in situ,* when bound to the membrane in a functional state.[53] Many of the oxidases detected in facultative photosynthetic bacteria are able to react with carbon monoxide, and furthermore the respiratory electron transport system is sensitive to cyanide. Therefore *b*-type oxidases largely differ from the "classical" *b*-type cytochrome of mitochondrial systems in their autooxidizability and their property of combining with CO.

In *R. capsulata,* the relationship between respiratory activities and specific electron transport pathways has been well characterized using aerobic mutants. In fact the suggestion by Baccarini-Melandri et al.,[45] based upon classical biochemical analysis, that a branched respiratory system operates in aerobic membrane fragments of wild-type *R. capsulata,* has been confirmed independently by the evidence of the operativity in vivo of complementary portions of this branched chain in two mutant strains, M6 and M7, which are impaired in different steps of the respiratory apparatus.[49] These strains, identified by means of the Nadi reaction, are both spontaneous revertants of mutant M5, and are unable to grow aerobically in darkness.[49] The mutant strain, M7, lacks cytochrome *c* oxidase activity, and redox titrations of *b*-type components indicate that it also lacks a high-potential cytochrome, characterized by a midpoint potential at pH 7.0 ($E_{m,7}$) of 410 mV.[51] This latter finding led to the obvious conclusion that cytochrome b_{410} must necessarily be associated with cytochrome *c* oxidase activity in *R. capsulata* membranes.[51] However, the possibility that the absence of cytochrome b_{410} is only accidental cannot be completely excluded in consequence of the fact that the mutagen nitrosoguanidine was used to produce strain M5. As reported by Guerola et al.,[54] it is very likely that more than one mutation is involved in a mutant selected after nitrosoguanidine treatment. Subsequent genetic analysis suppors the idea that one mutation causes the loss of both cytochrome b_{410} and cytochrome *c* oxidase.[125] The discovery of a gene transfer process occurring in *R. capsulata* provides a useful tool for chromosomal mapping.[55-57] The "gene transfer agent" (GTA) system has been used to perform crosses involving M5 mutant (parent strain) and one unlinked marker, e.g., the *rif₁₀* mutation which causes rifampin resistance. The results of recombination tests, analyzed on the basis of classical genetics of merozygotic crosses, seem to be in agreement with the transfer frequency expected for a single mutation in both M6 and M7 strains.[125]

The feature that characterizes the alternative pathway of oxygen consumption of *R.*

Table 2
RESPIRATORY MUTANTS OF *RHODOPSEUDOMONAS CAPSULATA*

Strain designation	Aerobic dark growth		Photosynthetic growth		Lesion
	Malate	Succinate	Malate	Succinate	
M1	−	+	+	+	NADH-deH
M2	−	−	+	−	NADH- and succinate-deH
M3	+	−	+	−	Succinate-deH
M4	+	+	+	+	Lacks cytochrome b_{410}
M5	−	−	+	+	Lacks cytochrome b_{410} and cytochrome b_{260}
M6	+	+	+	+	Lacks cytochrome b_{260}
M7	+	+	+	+	Lacks cytochrome b_{410}
MT113	+	nd	−	−	Lacks cytochrome c_{340}
Y11	+	nd	−	−	Ubiquinone cytochrome *b-c* region

capsulata is its sensitivity to only high concentrations of KCN, as well as carbon monoxide. This respiratory pathway has been tentatively associated with the activity of a *b*-type cytochrome with an $E_{m,7} = 270$ mV.[52] Redox titrations performed on membrane fragments of mutant M7 under an atmosphere of nitrogen and CO have shown a shift of about 100 mV in the mid-point potential of cytochrome b_{270}, suggesting its involvement in the alternative CO-sensitive pathway.[52]

The definitive proof of a correlation between cytochrome b_{270} and the alternative respiratory pathway has been obtained using membranes from mutant M6 endowed with only cytochrome *c* oxidase activity. In this mutant, cytochrome b_{270} is still spectrally present, but because of some kind of lesion not yet characterized, does not bind carbon monoxide and does not react with oxygen.[52] The recent finding that light-induced oxygen consumption normally carried out by membranes of *R. capsulata* is not present in photosynthetic or semiaerobic membranes of mutant M6 has led to the conclusion that the b_{270}-containing branch is also responsible for the light-driven oxygen reduction.[58]

It is not only *R. capsulata* that contains *b*-type components characterized by a high mid-point potential. In a bacteriochlorophyll mutant of *R. sphaeroides* grown aerobically in the dark, two *b*-type cytochromes can be detected with $E_{m,7}$ values of 390 and 255 mV.[59] Their similarity with the two high-potential *b* cytochromes of *R. capsulata* is obvious. Unfortunately, no specific function has been associated with these components, and therefore their function remains obscure. However, the observation that membranes from aerated cells of *R. sphaeroides* GA can perform oxygen consumption driven by light[126] suggests as a working hypothesis the possible implication of cytochrome b_{255}, as well as cytochrome b_{270} of *R. capsulata*, in this process.

The cytochrome *c* oxidase activity in aerobic membranes of *R. sphaeroides* is normally carried out by the cytochrome aa_3 complex which is a well known oxidative system in plant and animal mitochondria. Redox titrations performed at 607 nm (band of absorption of *a*-type cytochromes) showed that two components with midpoint potentials at pH 7.0 of 375 and 200 mV can be identified, and they contribute with a ratio close to unity in forming this spectral band.[50] Moreover, the necessity for copper for synthesis and operativity of the oxidase has been clearly established.

B. *b*- and *c*-Type Cytochromes in the Electron Transport Chain
In addition to the high mid-point potential *b*- and aa_3-type components associated with oxidase activities, many other electron transport carriers have been identified in membranes from nonsulfur purple bacteria. Cytochromes of the *b*-type can usually be

identified by *in situ* redox titrations because of their extremely hydrophobic nature and also because of the difficulty in clearly distinguishing them on the basis of spectral characteristics. In contrast, *c*-type cytochromes, mostly c_2, loosely attached to the cytoplasmic membrane by electrostatic interaction, have been studied extensively *in situ*, as well as in solution.[60-63] However, it is well documented that in solution and the bound form of cytochrome c_2 has markedly different characteristics;[62-64] therefore the redox analysis *in situ* is certainly closer to the in vivo properties of this electron carrier.

There are many similarities among nonsulfur purple bacteria regarding the values of mid-point potentials determined between -100 and $+200$ mV in titrations of *b*-type cytochromes. Three components with $E_{m,7}$ values of 185,44 and -104 mV were reported in aerobic membranes of *R. sphaeroides* by Connelly et al.[64] in addition to the two high mid-point-potential *b*'s identified in aerobically grown cells of an albino mutant (strain V-2) of *R. sphaeroides* (see the section entitled Respiratory Oxidases). Very similar components in the low medium-potential range have been described in both aerobically ($E_{m,7} = 148,56$, and -32) and photosynthetically ($E_{m,7} = 60$ and -25 mV) grown cells of *R. capsulata*.[65,66] However, of these three *b*-type cytochromes, it is only the cytochrome b_{80} of *R. sphaeroides* and b_{60} of *R. capsulata* which are observed to be kinetically competent for photosynthetic electron transport in single-turnover fash experiments.[67,68] Both b_{50} and b_{60} cytochromes are observed to be reduced in the millisecond range in electron transfer reactions which are thought to be part of the light-induced cyclic electron flow. (The actual details of these reactions are described more extensively in the section entitled Ubiquinone Cytochrome *b-c* Oxidoreductase System). The pK of this *b*-type cytochrome in the reduced form, studied in dark titrations and in titrations of the flash-induced absorbance change, is 7.4.[69]

In conditions of steady-state NADH respiration performed by membranes of *R. capsulata*, the inhibition of the cytochrome *c* oxidase activity by 5.10^{-5} *M* cyanide causes a very small increase of the level of reduction of cytochrome *b*.[65] This observation led Zannoni et al.[65] to suggest the possibility that cytochrome b_{60} is reoxidized through both cytochrome *c* and cytochrome b_{260} containing pathways in aerobic vesicles of *R. capsulata*.

The functional role of a cytochrome *b* having a mid-point potential at pH 7.0 around 150 mV has been previously associated with the reduction of flash-activated ferricytochrome c_2.[67] However, its standard redox titration indicates an *n*-value of unity, and therefore this mathematical resolution dissociates cytochrome b_{150} with the kinetically identified electron donor to cytochrome c_2 (see the section entitled Ubiquinone Cytochrome *b-c* Oxidoreductase System) which titrates as an $n = 2$ component.[70] In addition, the possibility that cytochrome b_{150} does not represent a functional redox carrier in vivo has been raised by Dutton and Prince,[70] since its content in chromatophores is extremely variable, from undetectable to 35% of the total *b*-type cytochrome complement. This last observation can however be partially clarified. Previous studies on aerobic membranes of *R. capsulata* have shown that cytochrome b_{150} shifts its mid-point potential toward higher value (180 mV) in redox titrations performed under an atmosphere of nitrogen/carbon monoxide.[52] In addition the periplasmic localization of cytochrome cc', a typical CO-binding pigment of nonsulfur purple bacteria,[60] (see also below) exhibiting a broad absorbing band in the 552 to 560-nm range and therefore interfering with cytochrome *b* absorption spectrum, has also been clearly established in aerobic cells of *R. capsulata*.[52] These two observations seem to suggest (1) the nature of cytochrome b_{150} as a CO-binding pigment and (2) give a reasonable explanation for its variable concentration in membrane vesicles. In fact in line with the assumption that cytochrome b_{150} and cytochrome cc' coincide, a marked decrease in the amount for both cytochrome cc' and cytochrome b_{150} has been observed during the formation of inside-out membrane vesicles by means of French pressure treatment of spheroplasts.[52]

In support of this view, redox titration analysis in aerobic membranes of a nonphotosynthetic mutant (MT113) deficient in cytochrome of c type have shown that the almost total lack of CO-binding pigment is associated with a parallel drastic decrease of cytochrome b_{150}. However, the respiratory electron flow in MT113 does not present any deficiency which would be directly related to the lack of cytochrome b_{150}, thus its physiological role still remains obscure. In this connection, of the three cytochromes species resolved potentiometrically at 560 nm minus 540 nm in membranes from *Rhodospirillum rubrum* ($E_{m,7}$ = 168, −5, and −103 mV), the component with a midpoint potential of −5 mV (n = 1) and alpha-max 562 nm, has been proposed to be "bound cytochrome cc' by Jackson and Dutton.[67] This suggestion does not appear to be correct because of the following considerations: the mid-point potential of cytochrome cc' in the solution is about 0 mV, and it seems to be extremely unlikely that its $E_{m,7}$ in the associated state remains the same. In fact, it is quite clear that the mid-point potential of electron transport carriers can be considerably affected by the lipid environment of the membrane.[71-73] Moreover, the similarities between the values of the relative contributions to the over-all measured absorbance change of b-type components in *R. capsulata*, *R. sphaeroides*, and *R. rubrum* are so pronounced that on account of the recent observation of the MT113 mutant mentioned above ($E_{m,7}$ of the CO-binding pigment, cyt cc' *in situ*, suggested to be + 150 mV), the most reasonable analogy seems to compare cytochrome b_{150} of *R. capsulata* with cytochrome b of *R. sphaeroides* and *R. rubrum*, exhibiting similar mid-point potentials. Thus, it follows that the best candidate to be "bound cytochrome cc'" in membranes from *R. rubrum* is probably cytochrome b_{168}.

Regarding c-type cytochromes, their presence in nonsulfur purple bacteria seems to be mainly related to only one component, cytochrome c_2, which defines a characteristic class of electron transport proteins.

Cytochrome c_2 can be distinguished from other c-type electron carriers, i.e., mitochondrial cytochrome c and bacterial cytochrome c_{551}, by amino acid sequence and by its redox potential[63,74] The *R. sphaeroides* c_2 contained in membrane vesicles has a redox potential of 290 mV at pH 7.0, while *R. capsulata* c_2 shows a mid-point potential close to 340 mV.[65-67] These values, determined by means of *in situ* redox titrations, are generally very different from the $E_{m,7}$ calculated for the cytochrome in solution as well as their pH dependence. Bound cytochromes c_2 of *R. sphaeroides* and *R. rubrum* are nearly pH independent between 5 and 11, whereas the redox potential of the pure cytochrome shows a pronounced pH dependence.[63] Thus, it seems evident that the interaction between membrane and cytochrome modifies the properties of cytochrome c_2.[63]

Rapid light-induced oxidation of cytochrome c_2 has been observed in several of Rhodospirillaceae, e.g., *R. rubrum*, *R. capsulata*, *R. sphaeroides*, but only in aerobic membranes from *R. capsulata* has its role as physiological donor to the cytochrome c oxidase system been unequivocally defined by an immunological approach.[75] In addition the absence of cytochrome c_2 in membranes from the photosynthetic mutant *R. capsulata* strain MT113 has been shown to be associated with the lack of ascorbate-DCPIP or DAD oxidase activity.[127] This observation seems to give conclusive support to the role of cytochrome c_2 in both aerobic and light-induced electron flow of *R. capsulata*.

Another c-type cytochrome, only detectable spectrophotometrically by means of redox titrations, has been found in aerobic membranes from *R. capsulata* and *R. sphaeroides*. Its mid-point potential is very close to 100 mV and represents less than 20%

of the total c-type cytochrome complement.[64,65] Recently a light particulate fraction of membranes from *R. sphaeroides* isolated by differential centrifugation and enriched in cytochrome c-type content has been described by Barrett et al.[76] In these vesicles, cytochrome c_2 and c_{100} are present in a ratio of approximately 2:3. However, no specific role has been assigned to cytochrome c_{100}, and its function remains unknown. Nevertheless, an interesting connection between c_{100} and b_{150} arises from the redox titration analysis of MT113 mutant. As observed for cyt b_{150} also cyt c_{100} react with CO.[126] In addition, as described above, the amount of b_{150} in MT113 is about 80% less than in wild-type membranes, but surprisingly c_{100} also seems to be repressed. The possible explanation of these parallel decreases is to suggest the presence of a single redox component absorbing broadly in the 552- to 560-nm region. Indeed the cytochrome cc' exhibits a broad band at 550 to 563 nm in a reduced state as reported by Bartsch.[77] Its spectrum resembles that of myoglobin, and under acid denaturing conditions, it reverts to a typical hemochromogen spectrum of a c-type heme. Because of these spectral characteristics, the possibility of detecting this cytochrome in both titrations of b- and c-type carriers performed at 551 and 561 nm is evident and could also explain the observed difference in the mathematical resolution of the $E_{m,7}$ in redox titrations performed at 551 to 540 or 561 to 570 nm.

C. Ubiquinone Cytochrome b-c Oxidoreductase System

The quinone complement of photosynthetically and aerobically grown cells of Rhodospirillaceae is in large excess over the other electron transport components. Ubiquinone-10 (UQ-10) alone or in combination with rhodoquinone is the most representative quinone species in these bacteria.[78]

The participation of this carrier in the light-dependent electron flow of nonsulfur purple bacteria has been firmly established by different experimental approaches.[79-83] The change from phototrophic to heterotrophic conditions of growth does not alter the nature or the type of quinone species present; however, its total amount is largely influenced by the parameters of growth. Although studies on the role of UQ-10 in respiration have been performed in some detail only in *Rhodopseudomonas palustris*[84] and *R. capsulata*,[65] there are no reasons to question its functional role in respiratory electron transport of other species of Rhodospirillaceae.

The location and the mechanism of the reactions in which quinones are involved, however, is still a debated matter in bioenergetics.[85]

It is quite evident at this point that several components present in membranes from aerobically grown cells show similarities, if not identity, in their chemical, thermodynamic, and immunological properties with components involved in light-dependent electron flow. This is particularly true for the ubiquinone cytochrome b- cytochrome c oxidoreductase electron transport pathway of membrane fragments from *R. capsulata* and *R. sphaeroides*.

However, although photosynthetic and aerobic apparatuses seem to have many similarities in common, the enormous quantity of data concerning the sequence of redox carriers involved which has been accumulated in recent years is mainly related to the light-induced electron flow. In fact, cyclic electron flow can be driven by a single photochemical event, and this affords certain experimental advantages. This process involves the production of one reducing equivalent delivered to ubiquinone and at the same time provides an oxidizing equivalent on cytochrome c_2. Both these reactions are effectively complete within a millisecond. The electron on the ubiquinone then enters the cytochrome b-c oxidoreductase system, and this in turn reduces the ferricytochrome c_2. Even if the presence of an "alternative" pathway of oxygen consumption in *R. capsulata* appears to complicate the sequence of reactions involved in the ubiquinone cytochrome *b-c* region, there is no objective reason to believe that the ubiquinone cytochrome *b-c* electron transport pathway in aerobic and photosynthetic membranes is different.

Therefore, much of the data available on photosynthetic membranes can be reasonably transferred to aerobic membranes, considering that the redox components involved are probably the same. The light-induced redox changes of cytochrome b_{50} have been studied in *R. sphaeroides* GA by Petty and Dutton.[69] These authors examined the relationship between the pH dependence of the mid-point potential of this *b*-type component, which has a pK at 7.4, and the proton release with or without antimycin A present, as a function of pH and have proposed that cytochrome b_{50} can be considered a proton carrier. In the absence of antimycin A, one proton seems to be released into the inner phase during the electron flow from cytochrome b_{50} to cytochrome c_2.

Analysis of the re-reduction of light-generated ferricytochrome c_2 of *R. capsulata* and *R. sphaeroides* seems to suggest the presence of a component-mediating electron flow between cytochrome b_{50} and cytochrome c_2.[68,70] This component, named Z in the absence of any clearly assignable optical spectrum, has an oxidation-reduction midpoint potential of $+155$ mV at pH 7.0 and *n* value of 2. From pH 5 to pH 11, the mid-point varies by -60 mV/pH.[70] Since Z titrates as a two-electron species, this indicates that Z could exist under Z/ZH• and ZH•/ZH₂ forms.[86,70] Recent data has shown that the kinetics of ferricytochrome c_2 reduction ollows that of a second-order process, where the rate of reaction varies with the concentration of both ferricytochrome c_2 and ZH₂.[87]

Analysis of the redox changes of cytochromes *b* and *c* under continuous illumination, performed in ubiquinone-extracted and reconstituted membranes from *R. capsulata* Ala pho⁺, led Baccarini-Melandri and Melandri[83] to suggest a role for ubiquinone on the oxidizing site of cytochrome *b*, in addition toits function on the reducing side of the same cytochrome. This observation has also been confirmed in single-turnover flash experiments with the same experimental approach.[88] Detailed analysis by Takamya et al.[89] of the half time of re-reduction of cytochrome c_2 induced by a single-turnover flash, indicate that of the total quinone pool present in *R. sphaeroides* (25 to 30 ubiquinone per reaction center), only one or two molecules of quinone per reaction center are essential for rapid cytochrome c_2 re-reduction. Ubiquinone appears at present to be the most likely candidate as the immediate electron donor to ferricytochrome *c* ; one or two molecules of ubiquinone therefore play the role of Z in the ubiquinone cytochrome *b*-cytochrome *c* electron transport pathway.[90] The redox mechanism involving Z still remains largely unknown, and indeed none of the current schemes satisfactorily explain all the kinetic data available.

Nevertheless, the recent finding of an antimycin, oxidant-induced reduction of cytochrome b_{50}, a well-known phenomenon in animal[91] and plant mitochondria[92] as well as in *Paracoccus denitrificans*,[93] has patially clarified the possible redox reaction of Z[94] in chromatophores from *R. sphaeroides* R26 (blue-green mutant). In bacterial membranes, this phenomenon occurs only if Z is reduced before he oxide pulse. The simplest explanation of this effect is that oxidation of cytochrome c_2 by ferricianide is followed by its re-reduction, generating ZH• from ZH₂. The ZH• form generated is unstable and then reduces cytochrome b_{50}. However, it has been shown that in the presence of antimycin the kinetics of cytochrome *b* reoxidation is inconsistent with a rapid re-reduction of the photooxidized cytochrome c_2. These opposite observations have been recently explained by examining the effect of antimycin following subsaturating flashes in order to reduce only partially via UQH• the cytochrome b_{50} and to maintain its availability for reaction with Z•.[90] In effect, if cytochrome b_{50} is reduced before light activation, there is no rapid re-reduction of photoxidized c_2 at any flash intensity because no more cytochrome b can be reduced by $ZH^\bullet$, but with Z reduced and b oxidized, some significant reduction, antimycin insensitive, after a subsaturating pulse takes place. In the presence of antimycin, it clearly appears that cytochrome b_{50} can be reduced by both $UQH^\bullet$ and $ZH^\bullet$, but whether this reaction also occurs in the uninhibited state remains to be established.

Studies on a recently isolated nonphotosynthetic mutant of *R. capsulata* (strain Y11) that has a block in electron transport between the ubiquinone b region and cytochrome c_2 should give information that may help to resolve the puzzle of the ubiquinone cytochrome b-c_2 oxidoreductase system. *R. capsulata* Y11 is able to grow aerobically in accord with the presence of a branched respiratory system, but neither NADH nor succinate can drive the reduction of exogenous cytochrome c.[95] Experimental evidence indicates that in *R. capsulata* R126 (green derivative of Y11), no b-type component appears to be reduced following a single saturating flash of light, in contrast to the wild-type strain.[25] Moreover, even if these chromatophores are poised at Eh = 100 mV (c_2 and Z reduced, b_{50} should be oxidized) no re-reduction of cytochrome c_2 is observed.

III. ENERGY CONSERVATION IN RESPIRATORY ELECTRON TRANSPORT

ATP synthesis associated with NADH and succinate oxidation in membrane fragments from heterotrophically grown cells of Rhodospirillaceae has been known for a long time.[96,97]

Although, especially in recent years, a rather large amount of information has been accumulated on the nature of the redox carriers involved in respiration of Rhodospirillaceae, very little is known about the efficiency of this process. The main difficulties arise from the fact that the classical methods of studying the efficiency of energy conservation, such as P/O ratios and ADP/O ratios, cannot be easily applied with membrane vesicles from facultatives phototrophs. In fact P/O ratios are invariably low, as in many other bacterial systems, possibly due to the contamination of open membrane fragments and right-side-out vesicles which could possibly exhibit an uncoupled electron flow.

Unfortunately, in intact cells, direct measurements of ATP synthesis have proven to be difficult because of the impermeability of the plasma membrane to ATP and ADP and the turnover time for ATP. Therefore exogenous ADP cannot be phosphorylated due to the topology of the ATP synthetase complex which is located on the cytoplasmic side.[29] This enzyme complex which presents many chemical and physical characteristics similar to those of analogous energy transducing systems also appears to be the coupling enzyme involved in photosynthetic ATP synthesis (for a review on coupling factors of Rhodospirillaceae, see Melandri and Baccarini-Melandri.)[98] Insight into the efficiency of the respiratory chain of facultative photosynthetic bacteria may possibly be obtained by in vivo measurements of the true molar growth yields with respect to molecular oxygen (Y_{o2}) or dissimilated carbon source $Y_{substrate}^{max}$) and hence the efficiency of aerobic energy conservation.[99] The use of concentrations of KCN inhibiting selectively at the level of the two main oxidases and of utilizing the two mutants strains of *R. capsulata*, M6 and M7, could probably give specific information on the energy conservation efficiency of the different pathways.

Because of these technical difficulties and in the absence of direct determination of P/O ratios in intact cells, three main approaches have been used to localize the energy coupling sites of respiration in membrane fragments of heterotrophically grown facultative photosynthetic bacterial cells: (1) measurement of phosphate esterification; (2) distribution of permeant anions; (3) quenching of amine fluorescence.

Each method has its merits and its shortcomings. Moreover, not all three approaches have been applied to any one of the redox chains described in this review.

A. P/O Ratios and Membrane Energization

ATP formation driven by NADH and succinate oxidation has been unequivocally demonstrated in membranes from aerobically grown cells of *R. capsulata*. Baccarini-Melandri et al.[45] reported P/O ratios of 0.45 and 0.15 for NADH and succinate respiration, respectively, in membrane fragments from *R. capsulata* harvested in the stationary phase of growth. Very similar values have been shown by Saunders and Jones[100] in phototrophically grown *Rhodopseudomonas viridis*. Measurements of phosphate esterification performed in membranes from *R. capsulata* in the presence of respiratory inhibitors and artificial electron donors and acceptors suggest that there is a branched redox chain, the two pathways of oxygen reduction having different efficiency of phosphorylation.[45]

These P/2e⁻ ratio measurements, correlated with the observation of membrane energization by means of fluorescent amines, have indicated a first energy-conserving step between NADH dehydrogenase and ubiquinone and a second energy-conserving step in the ubiquinone cytochrome *b-c* region.

The same scheme of redox chain associated with respiratory energy conservation has been described by Isacv et al.[101] in *R. rubrum*. In this case, the method applied was based on the use of an artificial phospholipid membrane as a selctive electrode for penetrating anions.

In both *R. capsulata* and *R. rubrum* (Figure 1), the energy state generated by NADH oxidation appears to be only slightly relaxed by antimycin, in agreement with the presence of a branched respiratory electron transport chain in these two organisms. In addition, the quenching of atabrine induced by NADH oxidation with UQ1 as an artificial electron acceptor in the presence of KCN and antimycin demonstrates unequivocally the existence of the first energy conserving site for NADH oxidation in *R. capsulata*. Moreover, a second coupling site between succinate dehydrogenase and cytochrome *c* has been suggested by the quenching of atabrine fluorescence, depending on reduction of oxidized-diaminodurene ($E_o' = 245$ mV) in aerobic membranes of *R. capsulata*.[45]

A controversial point is related to the presence of a third coupling site between cytochrome c_2 and oxygen. This partial reaction is generally carried out using ascorbate and DAD or TMPD as artifical electron donors. Although the mechanism of these two artificial components as electron mediators for cytochrome *c* oxidase activity seems to be the same, their reaction relative to the energy conservation of this partial respiratory activity is completely different. In fact during oxidation, TMPD becomes a stable radical losing one electron, while DAD loses two electrons and two protons. It follows that the reoxidation of reduced-DAD is always associated with proton translocation, supporting the concept of artificial loops as proposed by Trebst and Hauska[102] and based on the chemiosmotic hypothesis. Indeed experiments with membrane chromatophores of *R. capsulata* and other bacterial systems demonstrate that TMPD oxidation is not coupled, but oxidation of DAD is coupled.[103] However, the accumulative uptake of nutrients driven by oxidation of reduced TMPD by vesicles of many other bacterial systems has been shown.[104,105] In accord with Hauska et al.,[103] the most reasonable explanation of these discrepancies may be the presence of different polarities in different bacterial preparations. In the former case, ATP formation would reflect the presence of right-side-out vesicles, while the latter case would resemble in-

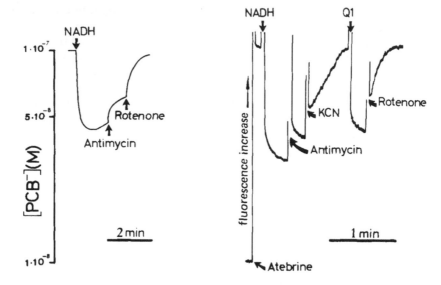

FIGURE 1..   Membrane energization dependent on NADH respiration. (A) Uptake of phenyl diacarbaundecaborate anion (PCB⁻) induced by adition of NADH (1 mM) to an aerobic suspension of membranes from *Rhodospirillum rubrum.* (Adapted from Isaev, P. U., Liberman, E. A., Samuilov, V. D., Skulachev, P. V., and Isofina, L. M., *Biochim. Biophys. Acta,* 216, 22, 1970.) (B) Quenching of atebrine fluorescence induced by addition of NADH (0.2 mM) to an aerobic suspension of membranes from *R. capsulata.* (Adapted from Baccarini-Melandri, A., Zannoni, D., and Melandri, B. A., *Biochim. Biophys. Acta,* 314, 298, 1973. In both experiments antimycin and rotenone were present at concentrations fully inhibiting electron flow; in (B) KCN was added at a concentration of 4 mM, and ubiquinone-1 (Q1) was added at 40 μM.

side-out vesicles. In fact as shown by Hellingwerf et al.,[106] membrane vesicles from the facultative photosynthetic bacterium *R. sphaeroides,* with the same orientation as the cytoplasmic membrane in whole cells (right-side out), actively transport amino acids under aerobic dark respiration in the presence of ascorbate-TMPD.

The association of proton translocating activity through cytochrome oxidase aa_3 in mammalian mitochondria originally reported by Wikström et al.[107] is at present a debated and controversial topic.[108,109] Due to the lack of direct measurements in facultative photosynthetic bacteria, it seems to us wise at present to eave open the problem of proton translocation and associated energy conservation between cytochrome c_2 and oxygen. It is interesting that a direct correlation has been found by Jones et al.[110] between the presence of a membrane-bound cytochrome of *c* type and the presence of three proton translocating segments in various species of bacteria, independent of the kinds of terminal oxidases present.

Although more experimental work appears to be necessary on energy conservation in respiratory electron transport and on the exact topology of the redox components involved, we have formulated one of the possible minimal scheme of respiration utilizing some observations reported above (see the section entitled Ubiquinone Cytochrome b-c Oxidoreductase System) on the light-driven electron transport. This attempt, illustrated in Figure 2A, refers for the sake of simplicity to the mutant strain of *R. capsulata* lacking the alternative oxidase (M6 strain). In this case, no branching operates, and cytochrome c_2 is reoxidized through a *b*-type oxidase (cytochrome b_{410}).

However, the presence of a branch at the ubiquinone cytochrome *b* level in many members of Rhodospirillaceae seems to complicate this simple scheme of the energy conservation process linked to succinate oxidation. It is not yet well established that succinate oxidation through the alternative pathway of oxygen consumption may generate energy associated with ATP synthesis. Low KCN concentration (5.10^{-5} M) has been shown to inhibit succinate oxidase activity by at least 50% in aerobically grown cells of *R. capsulata*, reducing the rate of oxygen consumption to extremely low values. No phosphorylation activity has been observed in these experimental conditions, suggesting the idea of a nonphosphorylating step through the succinate-alternative oxidase pathway. Recently, experiments in our laboratory on membranes from M7 mutant cells harvested in the exponential phase of growth and exhibiting a relatively high succinate oxidase activity indicated the presence of an energy conserving step in the ubiquinone cytochrome b region. This observation can be reconciled with previously reported data on wild-type membranes treated with low KCN, in the light of the marked different rates of succinate oxidase activities observed in the wild-type strain in comparison with the M7 strain.[45] This result suggests an interaction of ZH_2, either with c_2, as in Figure 2A or with the alternative pathway of oxygen consumption as in Figure 2B. If the oxidation of ZH_2 through cytochrome c_2 is the necessary first reaction involving the electron flow between cytochrome b and cytochrme c, no energy conservation with succinate as electron donor should be present in M7 membrane fractions because there is no oxidized partner for ZH_2. The alternative pathway in aerobic membranes could provide, in the absence of cytochrome c oxidase activity, a second "oxidizing equivalent" for ZH_2.

In order to account for a protonmotive Q cycle, the alternative oxidase (cytochrome b_{260}) is presented in the scheme (Figure 2B) as localized on the phase of the membrane opposite to the other oxidase, i.e., on the phase of the membran facing the periplasmic space in whole cells. This arrangement, previously suggested by Garland et al.[111] (see also Jones)[112] takes into account the difference in the efficiency of energy conservation associated with the two terminal pathways.

IV. BRANCHED CHAINS, PHYSIOLOGICAL ASPECTS AND SOME EVOLUTIONARY SPECULATION

Do the branched chains of facultative photosynthetic bacteria represent an important step in the development of more accurate and efficient control of the energy conservation associated with the respiratory electron flow? The data concerning the sequence of the electron transport components unequivocally demonstrate the widespread presence of branched chains in bacteria.[32-43] In spite of these observations, the branched nature of the respiratory system has rarely been interpreted in terms of its physiological significance.[33] The experimental growth conditions are often so different from the natural environment in which that particular bacterium is found that the physiological interpretation of data involves extensive extrapolation. A possible significant reason for the presence of alternate oxidases could be shown by the observation that in many bacteria their synthesis is observed as a response to a lowering of the oxygen concentration. The most important and in our knowledge the best documented example of induction of alternate oxidases in response to physiological growth conditions has been found in the aerobic bacterium *Azotobacter vinelandii*, in which the oxygen level must be kept to critical values in order to maintain the nitrogen fixation process.[113]

An accurate analysis of the respiratory activities and associated phosphorylation in membranes from *R. capsulata* cells grown with a low oxygen tension in the light in order to simulate the natural environment of this bacterium has been performed by Zannoni et al.[114] A comparison between dark-aerobic grown and light-semiaerobic-

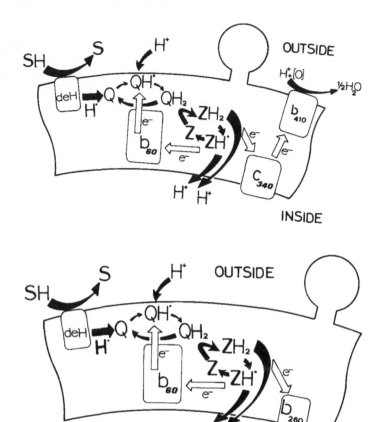

FIGURE 2.   Possible mechanisms of respiratory electron transport in membranes chromatophores from *Rhodopseudomonas capsulata*: (A) strain M6
and (B) strain M7. Inside is inside chromatophores or outside plasma membrane. Outside is outside chromatophores or inside plasma membrane. Further explanation is in the text.

grown cells led to the conclusion that under the latter conditions of growth, the electron flow between NADH-dehydrogenase and UQ-10 (operatively defined as the first
site of phosphorylation) is completely repressed. In other words in semiaerobic conditions in the light, conditions probably close to those present in nature, where the energy
charge of the cell is kept rather high by light-dependent ATP synthesis, part of the
respiratory chain, i.e., the NADH-dehydrogenase-UQ-10-alternative pathway, may assume the function of dissipating reducing power.

   This dissipative process does not seem to be the only physiological explanation for
the presence of a branched chain in facultative photosynthetic bacteria. The isolation
of a mutant of *R. rubrum* (F11) deficient in photooxidase activity and unable to grow
anaerobically in the light[115,116] (although normal rates of endogenous photophosphorylation have been shown in membrane chromatophores) led Gimenez et al.[117] to suggest
a possible redox-regulatory mechanism associated with the alternative pathway of oxygen consumption.

   This hypothesis, previously suggested by Nishimura and Chance for *R. rubrum*[118]
cells, has recently been confirmed[117] by showing that a low oxygen level stimulated the

initiation of photosynthetic growth in wild-type cells, whereas high aeration was necessary in F11 mutant strain. In the latter case, the mutant appears to replace the normal light-induced oxygen consumption pathway, with a new oxidase sensitive to low cyanide and with low oxygen affinity. In our opinion, another rationale for the necessity of an alternative oxidase activity might be the need for oxidizing carriers on the low-potential side of the cyclic electron transport pathway, i.e., Ubiquinone cytochrome *b* region. The ability for phototrophic growth of the aerobic mutant *R. capsulata* M6 endowed only with cytochrome *c* oxidase activity and lacking the pathway associated with the light induced oxygen uptake seems to blur the suggestion of an analogous function for the branched apparatus of *R. capsulata*. On the other hand, the observation that the aerobic mutant M5 blocked in both oxidases needs a minimal medium of growth supplemented with sodium ascorbate shows that some redox-reaction not yet explained may cause the necessity for a branched chain, even for the light-induced electron flow of *R. capsulata*.

It may be appropriate to consider the hypothesis that alternate oxidases, normally cyanide resistant, are primarily a defense mechanism against naturally occurring respiratory inhibitors and therefore play an important selective advantage. The ability to produce antibiotics and toxins is widespread among fungi, bacteria, and higher plants, and soil samples commonly contain chemicals which inhibit at least moderately the growth of bacteria and fungi.[119] Moreover, many well-known inhibitors of the cytochrome system (e.g., antimycin and rotenone) are natural products, possibly reflecting the evolutionary conservatism of the cytochrome system, which makes it an attractive target for nonspecific widespread growth retardants.

The high degree of genetic conservatism among procarotic organisms is widely documented by the close structural and sequential similarities of *c*-type cytochromes from purple nonsulfur bacteria, *P. denitrificans*, and mitochondria.[120-122] These findings have encouraged speculation about the bacterial energy metabolism. The possibility that plasmids conferring antibiotic resistance can be transferred between distantly related bacteria and because these plasmids can be integrated into the bacterial chromosome seems to exclude any suggestion about traditional phylogenetic trees for bacteria.[123] Conclusions made by analysis of single molecular components appear to be an unreliable guide and an oversimplification of the evolutionary process.

Nevertheless, it should be pointed out that studies on the aerobic respiratory systems of some Rhodospirillaceae species show that there are amazing similarities with another bacterial system, *P. denitrificans*, which is one of the few bacterial organisms with a respiratory chain very similar to mitochondria. The similarities between some Rhodospirillaceae and *P. denitrificans* are not only related to the nature of the redox carriers involved, but also to the sensitivity of the respiratory system to electron transport inhibitors, like rotenone, antimycin A, and cyanide and to the energy-transducing inhibitors, like oligomycin and aurovertin. John and Whatley[124] have previously suggested that mitochondria may have evolved from an endosymbiosis of *P. denitrificans* with a fermentative anaerobic host cell. In view of the similarities reported above, it might be speculated in turn that the same philogenetic implications could be extended to some of Rhodospirillaceae.

V. ACKNOWLEDGMENTS

We are deeply indebted to B. L. Marrs, St. Louis University and B. A. Melandri, University of Bologna for the stimulating discussion and for critically reading the manuscript.

This work was partially supported by grant no. 77.01412.04 from Consiglio Nazionale delle Ricerche Italy and by a long-term North Atlantic Treaty Organization (N.A.T.O.) fellowship to D. Zannoni.

REFERENCES

1. **Katoh, T.**, Nitrate reductase in the photosynthetic bacterium *Rhodospirillum rubrum, Plant Cell Physiol.*, 4, 199, 1963.
2. **Madigan, M. and Gest, H.**, Chemoautotrophyc growth of *Rhodopseudomonas capsulata* in darkness on $H_2:CO_2:O_2$, *J. Bacteriol.*, in press.
3. **Uffen, R. L. and Wolfe, R. S.**, Anaerobic growth of purple non-sulfur bacteria under dark conditions, *J. Bacteriol.*, 104, 462, 1970.
4. **Schön, G. and Biedermann, M.**, Bildung fluchtiger Saurenbei der Vergarung von Pyruvat und Fructose in anaerober DunkelKultur von *Rhodospirillum rubrum, Arch. Microbiol.*, 85, 77, 1972.
5. **Jungermann, K. and Schön, G.**, Pyruvate formate lyase in *Rhodospirillum rubrum Ha,* adapted to anaerobic dark conditions, *Arch. Microbiol.*, 99, 109, 1974.
6. **Yen, H. C. and Marrs, B. L.**, Growth of *Rhodopseudomonas capsulata* under anaerobic dark conditions with dimethyl sulfoxide, *Arch. Biochem. Biophys.*, 181, 411, 1977.
7. **Stanier, R. Y. and Cohen-Bazire, G.**, Phototrophic prokaryotes: the Cyanobacteria, *Ann. Rev. Microbiol.*, 31, 225, 1977.
8. **Oelze, J., Biedermann, M., Freund-Molbert, E., and Drews, G.**, The bacteriochlorophyll content and protein composition of chromatophores of *Rhodospirillum rubrum* during morphogenesis of the photosynthetic apparatus, *Arch. Mikrobiol.*, 66, 154, 1969.
9. **Oelze, J., Biedermann, M., and Drews, G.**, Die morphogenese der photosynthese apparates von *Rhodospirillum rubrum, Biochim. Biophys. Acta*, 173, 436, 1969.
10. **Cohen-Bazire, G., Sistrom, W. R., and Stanier, R. Y.**, Kinetics studies of pigment synthesis by nonsulfur purple bacteria, *J. Cell. Comp. Physiol.*, 49, 25, 1957.
11. **Lascelles, J.**, The synthesis of enzymes concerned in bacteriochlorophyll formation in growing cultures of *Rhodopseudomonas sphaeroides, J. Gen. Microbiol.*, 28, 599, 1962.
12. **Biedermann, M., Drews, G., Marx, R., and Schroder, J.**, Der Einfluss des Sauerstoffpartialdruckees und der antibiotica actinomycin und puromycin auf das wachstum die synthese von bacteriochlorophyll und die thylakoidmorphogenese in dunkelkulturen von *Rhodospirillum rubrum, Arch. Mikrobiol.*, 56, 133, 1967.
13. **Dierstien, R. and Drews, G.**, Nitrogen-limited continuous culture of *Rhodopseudomonas capsulata,* growing photosynthetically or heterotrophically under low oxygen tension, *Arch. Microbiol.*, 99, 117, 1974.
14. **Tauschel, H. D. and Drews, G.**, Thylakoidmorphogenese bei *Rhodopseudomonas palustris, Arch. Mikrobiol.*, 59, 381, 1967.
15. **Lampe, H. H. and Drews, G.**, Die Differenzierung des Membransystems von *Rhodopseudomonas capsulata* hinsichtlich seiner photosynthetischen und respiratorischen Functionen, *Arch. Mikrobiol.*, 84, 1, 1972.
16. **Dierstein, R. and Drews, G.**, Control of composition and activity of the photosynthetic apparatus of *Rhodopseudomonas capsulata* grown in ammonium-limited continuous culture, *Arch. Microbiol.*, 106, 227, 1975.
17. **Lascelles, J.**, The bacterial photosynthetic apparatus, *Adv. Microb. Physiol.*, 2, 1, 1968.
18. **Baltscheffsky, H., Baltscheffsky, M., and Thore, H.**, Energy conversion reactions in bacterial photosynthesis, *Curr. Top. Bioenerg.*, 4, 273, 1971.
19. **Oelze, J. and Drews, G.**, Membranes of photosynthetic bacteria, *Biochim. Biophys. Acta*, 265, 209, 1972.
20. **Parson, W. W.**, Bacterial photosynthesis, *Ann. Rev. Microbiol.*, 28, 41, 1974.
21. **Parson, W. W. and Cogdell, R. S.**, The primary photochemical reaction of bacterial photosynthesis, *Biochim. Biophys. Acta*, 416, 105, 1975.

22. **Gromet Elhanan, Z.,** Electron transport and photophosphorylation in photosynthetic bacteria, in *Encyclopedia of Plant Physiology,* Vol. 5, Trebst, A. and Avron, M., Eds., Springer-Verlag, Berlin, 1977, 637.

23. **Jones, O. T. G.,** Electron transport and ATP synthesis in the photosynthetic bacteria, in *Microbial Energetics,* Haddock, B. A. and Hamilton, W. A., Eds., Cambridge University Press, London, 1977, 151.

24. **Drews, G.,** Structure and development of the membrane system of photosynthetic bacteria, *Curr. Top. Bioenerg.,* 8, 161, 1978.

25. **Marrs, B. L.,** Mutations and genetic manipulations as probes of bacterial photosynthesis, *Curr. Top. Bioenerg.,* 8, 261, 1978.

26. **Baccarini-Melandri, A. and Zannoni, D.,** Photosynthetic and respiratory electron flow in the dual functional membrane of facultative photosynthetic bacteria, *J. Biomemb. Bioenerg.,* in press.

27. **Von Stedingk, L. V. and Baltscheffesky, H.,** The light induced reversible pH change in chromatophores from *Rhodospirillum rubrum, Arch. Biochem. Biophys.,* 117, 400, 1966.

28. **Scholes, P., Mitchell, P., and Moyle, J.,** The polarity of proton translocation in some photosynthetic microorganisms, *Eur. J. Biochem.,* 8, 450, 1969.

29. **Prince, R. C., Baccarini-Melandri, A., Hauska, G. A., Melandri, B. A., and Crofts, A. R.,** Asymmetry of an energy transducing membrane. The location of cytochrome c_2 in *Rhodopseudomonas sphaeroides* and *Rhodopseudomonas capsulata, Biochim. Biophys. Acta,* 387, 212, 1975.

30. **Hochman, A., Fridberg, I., and Carmeli, C.,** The location and function of cytochrome c_2 in *Rhodopseudomonas capsulata* membranes, *Eur. J. Biochem.,* 58, 65, 1975.

31. **Michels, P. A. M. and Konings, W. N.,** Structural and functional properties of chromatophores and membranes vesicles from *Rhodopseudomonas sphaeroides, Biochim. Biophys. Acta,* 507, 353, 1978.

32. **Jurtshuk, P., Mueller, T. J., and Accord, N. C.,** Bacterial terminal oxidases, *CRC Crit. Rev. Microbiol.,* 3, 399, 1975.

33. **White, D. C. and Sinclair, P. R.,** Branched electron transport systems in bacteria, *Adv. Microb. Physiol.,* 5, 173, 1971.

34. **Hill, G. C. and Cross, G. A. M.,** Cyanide-resistant respiration and branched cytochrome system in *kinetoplastidae, Biochim. Biophys. Acta,* 305, 590, 1973.

35. **Cheah, K. S.,** Properties of electron transport particles from *Halobacterium cutirubrum, Biochim. Biophys. Acta,* 180, 320, 1969.

36. **Jones, C. W. and Redfearn, E. R.,** Electron transport in *Azotobacter vinelandii, Biochim. Biophys. Acta,* 113, 467, 1966.

37. **Jones, C. W. and Redfearn, E. R.,** The cytochrome system of *Azotobacter vinelandii, Biochim. Biophys. Acta,* 143, 340, 1967.

38. **Appleby, C. A.,** Electron transport system of *Rhizobium japonicum.* II. *Rhizobium* haemoglobin, cytochromes and oxidases in free-living (cultured) cells. *Biochim. Biophys. Acta,* 172, 88, 1969.

39. **Pudek, M. R. and Bragg, P. D.,** Inhibition of the respiratory chain oxidases of *Escherichia coli, Arch. Biochem. Biophys.,* 164, 682, 1974.

40. **Weston, J. A. and Knowles, C. J.,** The respiratory system of the marine bacterium *Beneckea natrigiens.* I. Cytochrome composition, *Biochim. Biophys. Acta,* 333, 228, 1974.

41. **Jones, M. V. and Hughes, D. E.,** The oxidation of nicotinic acid by *Pseudomonas ovalis.* Chester. The terminal oxidase, *Biochem. J.,* 129, 755, 1972.

42. **Oka, T. and Arima, T.,** Cyanide resistance in *Achromobacter.* I. Induced formation of cytochrome a_2 and its role in cyanide-resistant respiration, *J. Bacteriol.,* 90, 744, 1965.

43. **Sinclair, P. R. and White, D. C.,** Effect of nitrate, fumarate and oxygen on the formation of the membrane-bound electron transport system of *Haemophilus influenzae, J. Bacteriol.,* 101, 365, 1970.

44. **Taniguchi, S. and Kamen, M. D.,** The oxidase system of heterotrophycally-grown *Rhodospirillum rubrum, Biochim. Biophys. Acta,* 96, 395, 1965.

45. **Baccarini-Melandri, A., Zannoni, D., and Melandri, B. A.,** Energy trasduction in photosynthetic bacteria. VI. Respiratory sites and energy conservation in membranes from dark-grown cells of *Rhodopseudomonas capsulata, Biochim. Biophys. Acta,* 314, 298, 1973.

46. **King, M. T. and Drews, G.,** The respiratory electron transport system of heterotrophycally-grown *Rhodopseudomonas palustris, Arch. Microbiol.,* 102, 219, 1975.

47. **Sasaki, T., Mokotawa, Y., and Kikuchi, G.,** Occurrence of both a-type and o-type cytochromes as the functional terminal oxidases in *Rhodopseudomonas sphaeroides, Biochim. Biophys. Acta,* 197, 284, 1970.

48. **Whale, F. R. and Jones, O. T. G.,** The cytochrome system of heterotrophycally grown *Rhodopseudomonas sphaeroides, Biochim. Biophys. Acta,* 223, 146, 1970.

49. **Marrs, B. L. and Gest, H.,** Genetic mutations affecting the respiratory electron transport system of the photosynthetic bacterium *Rhodopseudomonas capsulata, J. Bacteriol.,* 114, 1052, 1973.

50. Saunders, V. A. and Jones, O. T. G., Properties of the cytochrome a-like material developed in the photosynthetic bacterium *Rhodopseudomonas sphaeroides, Biochim. Biophys. Acta*, 333, 439, 1974.

51. Zannoni, D., Baccarini-Melandri, A., Melandri, B. A., Evans, E. H., Prince, R. C., and Crofts, A. R., Energy transduction in photosynthetic bacteria. IX. The nature of cytochrome c oxidase in the respiratory chain of *Rhodopseudomonas capsulata, FEBS Lett.*, 48, 152, 1974.

52. Zannoni, D., Melandri, B. A., and Baccarini-Melandri, A., Energy transduction in photosynthetic bacteria. XI. Further resolution of cytochromes of b type and CO-sensitive oxidase present in the respiratory chain of *Rhodopseudomonas capsulata, Biochim. Biophys. Acta*, 449, 386, 1976.

53. Lemberg, R. and Barret, J., *The Cytochromes*, Academic Press, New York, 1973, 8.

54. Guerola, N., Ingraham, J. L., and Cerda-Olmedo, E., Induction of closely linked mutations by nitrosoguanidine, *Nature (London) New Biol.*, 230, 122, 1971.

55. Marrs, B. L., Genetic recombination in *Rhodopseudomonas capsulata, Proc. Natl. Acad. Sci. U.S.A.*, 71, 971, 1974.

56. Yen, H. C. and Marrs, B. L., A map of genes for carotenoid and bacteriochlorophyll biosynthesis in *Rhodopseudomonas capsulata, J. Bacteriol.*, 126, 619, 1976.

57. Marrs, B. L., Genetic and bacteriophage, in *The Photosynthetic Bacteria*, Clayton, R. K. and Sistrom, W. R., Eds., Plenum Press, New York, 1978.

58. Zannoni, D., Jasper, P., and Marrs, B. L., Light-induced oxygen uptake as a probe of electron transport between respiratory and photosynthetic components in membranes of *Rhodopseudomonas capsulata, Arch. Biochim. Biophys.*, 191, 625, 1978.

59. Saunders, V. A. and Jones, O. T. G., Detection of two further b-type cytochromes in *Rhodopseudomonas sphaeroides, Biochim. Biophys. Acta*, 396, 439, 1975.

60. Horio, T. and Kamen, M. D., Bacterial cytochromes. II. Functional aspects, *Ann. Rev. Microbiol.*, 24, 399, 1970.

61. Kakuno, T., Hosoi, K., Higuti, T., and Horio, T., Electron and proton transport in *Rhodospirillum rubrum* chromatophores, *J. Biochem. (Tokyo)*, 74, 1193, 1973.

62. Prince, R. C. and Dutton, P. L., The pH dependence of the oxidation-reduction midpoint potential of cytochrome c_2 in vivo, *Biochim. Biophys. Acta*, 459, 573, 1977.

63. Pettigrew, G. W., Bartsch, R. G., Meyer, T. E., and Kamen, M. D., Redox potential of the photosynthetic bacterial cytochromes c_2 and the structural bases for variability, *Biochim. Biophys. Acta*, 503, 509, 1978.

64. Connelly, J. L., Jones, O. T. G., Saunders, V. A., and Yates, D. W., Kinetic and thermodynamic properties of membrane-bound cytochromes of aerobically and photosynthetically grown *Rhodopseudomonas sphaeroides, Biochim. Biophys. Acta*, 292, 644, 1973.

65. Zannoni, D., Melandri, B. A., and Baccarini-Melandri, A., Energy transduction in photosynthetic bacteria. X. Composition and function of the branched oxidase system in wild type and respiration deficient mutants of *Rhodopseudomonas capsulata, Biochim. Biophys. Acta*, 423, 413, 1976.

66. Evans, E. H. and Crofts, A. R., *In situ* characterisation of photosynthetic electron transport in *Rhodopseudomonas capsulata, Biochim. Biophys. Acta*, 357, 89, 1974.

67. Dutton, P. L. and Jackson, J. B., Thermodynamic and kinetic characterisation of electron transfer components *in situ* in *Rhodopseudomonas sphaeroides* and *Rhodospirillum rubrum, Eur. J. Biochem.*, 30, 495, 1972.

68. Evans, E. H. and Crofts, A. R., A thermodynamic characterization of the cytochromes of chromatophores from *Rhodopseudomonas capsulata, Biochim. Biophys. Acta*, 357, 78, 1974.

69. Petty, K. M. and Dutton, P. L., Ubiquinone-cytochrome b electron and proton transfer: a functional pK on cytochrome b_{50} in *Rhodopseudomonas sphaeroides* membranes, *Arch. Biochem. Biophys.*, 172, 346, 1976.

70. Prince, R. C. and Dutton, P. L., Single and multiple reactions in the ubiquinone-cytochrome b-c_2 oxidoreductase of *Rhodopseudomonas sphaeroides*: the physical-chemistry of the major electron donor to cytochrome c_2 and its coupled reaction, *Biochim. Biophys. Acta*, 462, 731, 1977.

71. Ernster, L., Lee, I. Y., Norling, B., and Person, B., Studies with Ubiquinone-depleted submitochondrial particles. Essentially of ubiquinone for the interaction of succinate dehydrogenase, NADH dehydrogenase and cytochrome b, *Eur. J. Biochem.*, 9, 299, 1969.

72. Ericinska, M., Oshino, N., and Chance, B., The b cytochromes in succinate-cytochrome c reductase from pigeon breast mitochondria, *Arch. Biochem. Biophys.*, 157, 431, 1973.

73. Goldberger, R., Pumphrey, A., and Smith, A., Studies on the electron-transport system. XLVI. On the modification of the properties of cytochrome b, *Biochim. Biophys. Acta*, 58, 307, 1962.

74. King, M. T. and Drews, G., The function and localization of ubiquinone in the NADH- succinate-oxidase system of *Rhodopseudomonas palustris, Biochim. Biophys. Acta*, 305, 230, 1973.

75. Baccarini-Melandri, A., Jones, O. T. G., and Hauska, G., Cytochrome c_2 — An electron carrier shared by respiratory and photosynthetic electron transport chain of *Rps. capsulata, FEBS Lett.*, 86, 151, 1978.

201

76. **Barrett, J., Hunter, C. N., and Jones, O. T. G.**, Properties of a cytochrome c-enriched light parti-culate fraction isolated from the photosynthetic bacterium *Rps. sphaeroides, Biochem. J.*, 174, 267, 1978.

77. **Bartsch, R. G.**, Spectroscopic properties of purified cytochromes of photosynthetic bacteria, in *Bac-terial Photosynthesis*, Gest, H., San Pietro, A., and Vernon, P. L., Eds., Antioch Press, Yellow Springs, Ohio, 1963, 475.

78. **Maroc, J., Klerk, H. D., and Kamen, M. D.**, Quinones of *Athiorodaceae, Biochim. Biophys. Acta*, 162, 621, 1968.

79. **Clayton, R. K.**, Evidence for the photochemical reduction of coenzyme Q in chromatophores of photosynthetic bacteria, *Biochem. Biophys. Res. Commun.*, 9, 49, 1962.

80. **Takamya, K. and Takamya, A.**, Light-induced reactions of ubiquinone in photosynthetic bacterium *Chromatium D, Plant Cell Physiol.*, 8, 719, 1967.

81. **Bales, H. and Vernon, P.**, Effect of reduced 2,6-Dichlorophenol indophenol upon the light-induced absorbancy changes in *Rhodospirillum rubrum* chromatophores, in *Bacterial Photosynthesis*, Gest, H., San Pietro, A., and Vernon, L. P., Eds., Antioch Press, Yellow Springs, Ohio, 1963, 269.

82. **Horio, T., Horiuti, Y., Yamamoto, N., and Nishikawa, K.**, Light-influenced ATP-ase activity: bac-terial in *Methods in Enzymology*, San Pietro, A., Ed., Academic Press, New York, 1972, p. 24, p. 96.

83. **Baccarini-Melandri, A. and Melandri, B. A.**, A role for ubiquinone-10 in the b-c₂ segment of the photosynthetic bacterial electron transport chain, *FEBS Lett.*, 80, 459, 1977.

84. **King, M. T. and Drews, G.**, The function and localization of ubiquinone in the NADH and succinate oxidase systems of *Rhodopseudomonas palustris, Biochim. Biophys. Acta*, 305, 230, 1973.

85. **Mitchell, P.**, Possible molecular mechanism of the protonmotive function of cytochrome systems, *J. Theor. Biol.*, 62, 327, 1976.

86. **Crofts, A. R., Crowther, D., and Tierney, G. V.**, Electrogenic electron transport in photosynthetic bacteria, in *Electron Transport and Oxidative Phosphorylation*, Quagliarello, E., Papa, S., Slater, E. C., Palmieri, F., and Siliprandi, N., Eds., North-Holland, Amsterdam, 1975, 233.

87. **Prince, R. C., Bashford, C. L., Takamiya, K., van den Berg, W. H., and Dutton, P. L.**, Second order kinetics of the reduction of cytochrome c₂ by the Ubiquinone cytochrome b-c₂ oxidoreductase of *Rhodopseudomonas sphaeroides, J. Biol. Chem.*, 253, 4137, 1978.

88. **Bowyer, J. B., Baccarini-Melandri, A., Melandri, B. A., and Crofts, A. R.**, The role of ubiquinone-10 in cyclic electron transport in *Rhodopseudomonas capsulata* Ala pho⁺: effects of lyophilization and extraction, *Z. Naturforsch.*, in press.

89. **Takamiya, K., Prince, R. C., and Dutton, P. L.**, Thermodynamic and functional heterogeneity among the ubiquinones of *Rps. sphaeroides*, in *Frontiers of Biological Energetics*, Dutton, P. L., Leigh, J. S., and Scarpa, A., Eds., Academic Press, in press.

90. **Prince, R. C., van den Berg, W. H., Takamiya, K., Bashford, L. C., and Dutton, P. L.**, A single ubiquinone plays a central role in electron flow through the ubiquinone-cytochrome b-c₂ oxidored-uctase, in *Frontiers of Biological Energetics*, Dutton, P. L., Leigh, J. S., and Scarpa, A., Eds., Academic Press, in press.

91. **Wikstrom, M. K. F.**, The different cytochrome b components in the respiratory chain of animal mitochondria and their role in electron transport and energy conservation, *Biochim. Biophys. Acta*, 301, 155, 1973.

92. **Storey, B. T.**, The respiratory chain of plant mitochondria. XIII. Redox state changes of cytochrome b₅₆₂ in mung bean seedling mitochondria treated with Antimycin A, *Biochim. Biophys. Acta*, 267, 48, 1972.

93. **John, P. and Papa, S.**, Rapid oxygen-induced reduction of b-type cytochromes in *Paracoccus deni-trificans, FEBS Lett.*, 85, 179, 1978.

94. **Dutton, P. L. and Prince, R. C.**, Equilibrium and disequilibrium in the ubiquinone-cytochrome b-c₂ oxidoreductase of *Rps. sphaeroides, FEBS Lett.*, 91, 15, 1978.

95. **La Monica, R. F. and Marrs, B. L.**, Studies on a mutant of *Rps. capsulata* which is deficient in electron transport between cytochrome b and c, in *Abstr. Int. Conf. Primary Electron Transport and Energy Transduction in Photosynthetic Bacteria*, Brussels, ThB3, 1976.

96. **Geller, D. M.**, Oxidative phosphorylation in extracts of *Rhodospirillum rubrum, J. Biol. Chem.*, 247, 2945, 1962.

97. **Klemme, J. M. and Schlegel, H. G.**, Untersuchungen zum cytochrom-oxidase-system aus anaerob in licht und aerob im dunkeln gewachsenen zellen von *Rhodopseudomonas capsulata, Arch. Mikrobiol.*, 68, 326, 1969.

98. **Melandri, B. A., and Baccarini-Melandri, A.**, Coupling factors ATPases from photosynthetic bac-teria, *J. Bioenerg.*, 8, 109, 1976.

99. **Strouthamer, A. H.**, Determination and significance of molar growth yields, in Methods in Micro-biology, Vol. 1, Narrio, J. R. and Ribbons, D. W., Eds., Academic Press, London, 1969, 629.

100. **Saunders, V. A. and Jones, O. T. G.**, Oxidative phosphorylation and effects of aerobic conditions in *Rps. viridis*, *Biochim. Biophys. Acta*, 305, 581, 1973.

101. **Isaev, P. U., Liberman, E. A., Samvilov, V. D., Skulachev, P. V., and Tsofina, L. M.**, Conversion of biomembrane-produced energy into electric form. III. Chromatophores of *Rhodospirillum rubrum*, *Biochim. Biophys. Acta*, 216, 22, 1970.

102. **Trebst, A. and Hauska, G.**, Energie Konservierung in der photosynthetischen membrane der Chloroplasten, *Naturwissenschaften*, 61, 308, 1974.

103. **Hauska, G., Trebst, A., and Melandri, B. A.**, Artificial conservation in the respiratory chain. No native coupling site between cytochrome *c* and oxygen, *FEBS Lett.*, 73, 257, 1977.

104. **Burnell, J. N., John, P., and Mhatley, F. R.**, The reversibility of active sulphate transport in membrane vesicles of *Paracoccus denitrificaus*, *Biochem. J.*, 150, 527, 1975.

105. **Nichols, W. W. and Hamilton, W. A.**, The transport of D-lactate by membrane vesicles from *Paracoccus denitrificaus*, *FEBS Lett.*, 65, 107, 1976.

106. **Hellingwerf, K. J., Michels, P. A. M., Dorpema, J. W., and Konings, W. N.**, Transport of amino acids in membrane vesicles of *Rhodopseudomonas sphaeroides* energized by respiratory and cyclic electron flow, *Eur. J. Biochim.*, 55, 397, 1975.

107. **Wikström, M. K. F.**, Proton pump coupled to cytochrome *c* oxidase in mitochondria, *Nature (London)*, 266, 271, 1977.

108. **Moyle, J. and Mitchell, P.**, Cytochrome *c* oxidase is not a proton pump, *FEBS Lett.*, 88, 268, 1978.

109. **Wikström, M. and Krab, K.**, Cytochrome *c* oxidase is a proton pump, *FEBS Lett.*, 91, 8, 1978.

110. **Jones, C. W., Brice, J. M., Downs, A. J., and Drozd, J. W.**, Bacterial respiration-linked proton translocation and its relationship to respiratory-chain composition, *Eur. J. Biochem.*, 52, 265, 1975.

111. **Garland, P. B., Clegg, P. A., Boxer, D., Downie, J. A., and Haddock, B. A.**, Proton translocating nitrate reductase of *Escherichia coli*, in *Electron Transfer Chains of Oxidative Phosphorylation*, E. Quagliariello, Papa, S., Palmieri, F., Slater, E. C., and Siliprandi, Eds., North-Holland Amsterdam, 1975, 351.

112. **Jones, C. W.**, Aerobic respiratory system in bacteria, in *Microbial Energetics*, Haddock, B. A. and Hamilton, W. A., Eds., Cambridge University Press, London, 1977, 23.

113. **Jones, C. W., Brin, J. M., Wright, V., and Acknell, B. A. C.**, Respiratory protection of nitrogenase in *Arotobacter vinelandii*, *FEBS Lett.*, 29, 77, 1973.

114. **Zannoni, D., Melandri, B. A., and Baccarini-Melandri, A.**, The branched respiratory system of the facultative photosynthetic bacterium *Rps. capsulata*, in *Functions of Alternative Terminal Oxidases*, Degn, H. Lloyd, D., and Hill, G., Eds., Pergamon Press, Oxford, 1978, p. 49, p. 169.

115. **Del Valle-Tascon, D., Gimenez, G., and Ramirez, J. M.**, Photooxidative system of *Rhodospirillum rubrum*. I. Photooxidations catalyzed by chromatophores isolated from a mutant deficient in photooxidative activity, *Biochim. Biophys. Acta*, 459, 76, 1977.

116. **Del Valle-Tascon, D. Gimenez-Gallego, G., and Ramirez, J. M.**, Light-dependent ATP formation in a non-phototrophyc mutant of *Rhodospirillum rubrum* deficient in oxygen photoreduction, *Biochim. Biophys. Res. Commun.*, 66, 514, 1975.

117. **Gimenez-Gallego, G., Del Valle-Tescon, D., and Ramirez, J. M.**, A possible physiological function of the oxygen-photoreducing system of *Rhodospirillum rubrum*, *Arch. Microbiol.*, 109, 119, 1976.

118. **Nishimura, M. and Chance, B.**, Studies on the electron transfer system in photosynthetic bacteria. III. Spectroscopic studies on cytochrome system, in Studies on Microalgae and Photosynthetic Bacteria, Japanese Society of Plant Phys., Eds., The University of Tokyo, Japan, 1963, 239.

119. **Whittaker, R. H. and Feeny, P. P.**, Allelochemis: chemical interaction between species, *Science*, 171, 757, 1971.

120. **Dus, K., Sletten, K., and Kamen, D.**, Cytochrome c_2 of *Rhodospirillum rubrum* II. Complete amino acid sequence and phylogenetic relationship, *J. Biol. Chem.*, 243, 5507, 1968.

121. **Timkovich, R. and Dickerson, R. E.**, The structure of *Paracoccus denitrificans* cytochrome *c* 550, *J. Biol. Chem.*, 251, 4003, 1976.

122. **Dickerson, R. E., Takano, T., Einsemberg, D., Samson, L., Cooper, A., and Margoliash, E.**, Ferrycytochrome *c*. I. General features of the horse and bonito proteins at 2.8 A resolution, *J. Biol. Chem.*, 246, 1511, 1971.

123. **Ambler, R. P.**, Bacterial cytochrome *c* and molecular evolution, *Syst. Zool.*, 22, 554, 1973.

124. **John, P. and Whatley, F. R.**, *Paracoccus denitrificans* and the evolutionary origin of the mitochondrion, *Nature (London)*, 254, 495, 1975.

125. **Marrs, B.**, personal communication.

126. **Zannoni, D.**, unpublished observations.

127. **Zannoni, D. and Marrs, B.**, in preparation.

INDEX

A

a, cytochrome, see Cytochrome a

Absorbance changes, see Absorption spectra

Absorption spectra
bacteriorhodopsin, I: 65
carotenoids, I: 61
coenzyme F_{420}, I: 164—165, 168, 170
cyclic electron transport chain, I: 61—62
cytochrome a_1, I: 156
cytochrome b_1, I: 144, 147
cytochrome o, I: 139—140, 146—156
cytochromes, aerobic respiratory chain, I: 117,
120—123
factor F_{342}, I: 182—183
factor F_{430}, I: 181—182
glycine reductase selenoprotein, II: 55—56
membrane-bound components, I: 40
proline reductase, II: 61

Acceptors
electron, ammonia oxidation and, see also
specific electron acceptors by name, II:
94—95
hydrogen, see Hydrogen, acceptors

Acetate
formation of, I: 6
fumarate and, II: 3, 7
glycine and, II: 51—52, 56
growth yields and, I: 6, 9, 13, 17
methane, reduction to, I: 177—180
nitrate and, II: 34, 37
oxidation, II: 66, 72
pyruvate and, II: 7
uncoupler of oxidative phosphorylation, I: 13

^{14}C-Acetate, production from ^{14}C-glycine, assay,
glycine reductase, II: 54

Acetobacter
pasteurianum, cytochrome a_1, I: 121
peroxidans, cytochromes, I: 138
species, cytochromes, I: 138, 159
suboxydans
cytochrome o, I: 139, 141, 148, 155
strain ATCC 621, cytochrome o, I: 155
strain IAM 1828, cytochrome o, I: 155

Acetobacterium woodii, acetate, I: 178, 180

Acetophilic methanogens, methane production
from acetate by, I: 178

Acetyl-CoA, pyruvate and, II: 72

Acetylene, inhibition of ammonia and methane
oxidation by, II: 108

Achromobacter sp., cytochromes, I: 156

Acidification process, II: 89—90

Acid-induced ATP synthesis, II: 117

Acid-labile sulfide
membranes of Escherichia coli, I: 120
nitrate reductase and, II: 24

Acid-labile sulfur, sulfur-oxidizing bacteria
studies, see also Labile sulfur, II: 124, 126

Acidophilic bacteria, electrochemical proton
gradient, I: 45

Acids, weak, see Weak acids

ACMA, solute translocation studies, I: 65—66

Acrylate Co A-esters, hydrogen acceptors, II: 3

Action spectra
cytochrome a_1, I: 141—142
cytochrome d, I: 141—142
cytochrome o, I: 141—142, 147—149, 152

Activation, sulfate, II: 67

Active center, nitrate reductase, II: 23

Active transport systems
nitrate respiration, II: 38
solute translocation studies, I: 34, 36

Adenine nucleotide pools, cell cycle studies, I: 97

Adenosine diphosphate, see ADP

Adenosine phosphosulfate, see APS

Adenosine triphosphatase, see ATPase

Adenosine triphosphate, see ATP

Adenylate kinase, sulfur oxidation and, II: 124

Adenylate pools, cell cycle studies, I: 94, 100
energy charge, I: 100

ADP
ATP and, I: 3—4, 6, 9, 12, 97; II: 66, 172
cell cycle studies, I: 97, 100
fumarate reduction and, II: 10, 13
glycine reductase and, II: 54, 62
phosphorylation, II: 169, 192

ADP/O ratios, facultative photosynthetic
bacteria studies, II: 192

ADP sulfurylase, sulfur oxidation and, II: 124

Aerobacter aerogenes, cytochrome o, I: 139, 141

Aerobic electron transport systems, I: 43—50
electrochemical proton gradient, I: 45—46, 48
quinones and, I: 48—50
solute transport coupled to, I: 46—48

Aerobic growth, Staphylococcus sp., II: 148

Aerobic metabolism, hydrogen and, II: 2

Aerobic respiratory chain, composition of and
activity of, see Respiratory chain, aerobic

Aerobic steady-state levels, cytochrome
reduction, I: 127

Age, cell, see Cell, age

δ-ALA, see δ-Aminolevulinic acid

Alanine
carriers, solute translocation studies, I: 43, 73
uptake, I: 64

Alcaligenes autotrophicum
hydrogenases, II: 164
hydrogen effect, II: 175

Alcaligenes eutrophus
ATP, II: 171—172
carbon dioxide, II: 172—173
carbon monoxide, II: 162
cell breaking, II: 163
cell cycle studies, I: 92, 95, 98, 104—105
cytochromes, II: 168, 170—171
dehydrogenases, II: 167—168
energy conservation and electron flow, II:
168—174
coupling of H_2 oxidation to CO_2
assimilation, II: 172—173

B

synthesis of membrane components, I: 102

PM6
    classification, I: 189
    trimethylamine oxidation, I: 194
species
    cytochrome *o*, I: 152—153
    hydrogen, II: 160
stearothermoplilus, nitrate reductase, II: 20
subtilis
    aro D, menaquinone and, I: 49—50
    cytochromes, I: 138, 153
    respiratory chain, I: 48—49; II: 114
    strain JB69, cytochrome *o*, I: 153
    strain PC1219, cytochrome *o*, I: 153
    strain W23, cytochrome c_{553}, I: 44
    synchronous cultures, cell cycle studies, I: 93,
        101—104, 107
    synthesis of membrane components, I: 102
    transport systems, I: 43—44, 48—50, 71—73
Bacteria, see specific bacteria by class, genus, or
        type
Bacterial cell, see Cell
Bacterial membrane, see Membrane
Bacteriochlorophyll
    cyclic electron transport and, I: 59, 62
    facultative photosynthetic bacteria studies, II:
        184, 187
    synthesis, I: 106
Bacteriophaeophytin, cyclic electron transport
        and, I: 59, 62
Bacteriophages, hydrogen bacteria, II: 175
Bacteriorhodopsin, I: 43, 64—66
    absorption spectrum, I: 65
    molecular weight, I: 64—65
    purple and red membranes, separation of, I: 64
Bacterium
    4B6
        classification, I: 189
        formaldehyde oxidation, I: 192
        methylamine oxidation, I: 196
        trimethylamine oxidation, I: 195
    5B1, classification, I: 189
    C2A1
        classification, I: 189
        methylamine oxidation, I: 196
        trimethylamine oxidation, I: 195
    growth inhibitors, II: 197
    5H2
        classification, I: 189
        formaldehyde oxidation, I: 192
        tetramethylammonium oxidation, I: 194
    species 1, formate oxidation, I: 193
    Type-M, trimethylamine oxidation, I: 195
Bacterium coli, see also *Escherichia coli*, II: 50
Bacteroides
    fragilis
        cytochromes, II: 145
        fumarate, II: 9, 145
        glucose, II: 4, 6, 8—9
        growth yields, II: 146
        heme, II: 144

hemin, II: 9
    protoheme, II: 144—146
    subspecies *distanosis*, protoheme, II: 144
    subspecies *fragilis*, protoheme, II: 144
    subspecies *thetaiotomicron*, protoheme, II:
        144
    sugar catabolism, II: 145
hypermagas, protoheme, II: 144
megaterium, transport systems, I: 72
melaninogenicus, protoheme, II: 143, 145
oralis, protoheme, II: 144
ruminicola
    B₁4 strain, heme, II: 146
    GA20 strain, protoheme, II: 144, 146
    glucose, II: 146
    heme, II: 146
    protoheme, II: 143—146
    strain 23, protoheme, II: 144, 146
    species, protoheme, II: 142—147, 153
Balanced growth, cell cycle studies, I: 90—91,
        96—97
Bases, weak, see Weak bases
Batch cultures
    ammonia-oxidizing bacteria studies, II: 89—90,
        92—93, 95
    growth yield studies, I: 2
    hydrogen bacteria studies, II: 175
Beggiatoaceae, cytochrome *o* studies, I: 148
Benzyl viologen
    heme-requiring bacteria studies, II: 145
    hydrogen bacteria studies, II: 167
Bifidobacterium longum, cytochromes, II: 151
Bioenergetics, I: 88
Biogenesis, membrane, I: 88, 91, 101—102, 104
Biological nitrogen cycle, ammonia-oxidizing
        bacteria studies, II: 88—89, 106
Bisulfite
    reduction of, II: 69—71
    reductase, see Bisulfite reductase
    thiosulfate reduced to, II: 71
Bisulfite reductase, sulfate reduction and, II:
        69—71, 77, 79
    hydrogenase and, II: 79
    localization, II: 77
Bisulfite reductase II, sulfate reduction and, II:
        70—71, 77
    localization, II: 77
Black Sea isolate JR1, coenzyme F_{420}, I: 167
Blue-green algae, see also Cyanobacteria
    intracytoplasmic membranes in, I: 213
    photosynthetic process, II: 184
Blue-green mutant, *Rhodopseudomonas
        sphaeroides* (R26) antimycin, II: 191—192
Bordetella pertussis, hematin, II: 152
Bovine rhodanese, sulfur-oxidizing bacteria
        studies, II: 130—131
Branched respiratory chains, see Respiratory
        chains, branched
Braun's lipoprotein, function of, I: 35
Breaking, cell, II: 163

C

209

aerobic respiratory chain studies, I: 124
ammonia-oxidizing bacteria studies, II: 102,
104—105
facultative photosynthetic bacteria studies, II:
193
growth yield studies, I: 7
solute translocation studies, I: 34—36, 38—40,
68—69, 73
sulfur-oxidizing bacteria studies, II: 117, 131
Chemiostatic theory and models
formate-fumarate electron transport, II: 14—15
sulfate reduction-phosphorylation coupling, II:
79—80
Chemoautolithotrophic organisms, see also
Hydrogen bacteria, II: 160, 176
Chemoautotrophs
ammonia-oxidizing, see Ammonia-oxidizing
chemoautotrophic bacteria; specific
bacteria by name
facultative photosynthetic bacteria studies, II:
185
hydrogen bacteria studies, II: 172, 175—176
methanogenic, see Methanogenic bacteria
reverse electron flow studies, II: 105
sulfur-oxidizing, see Sulfur-oxidizing
chemoautotrophic bacteria
Chemoheterotrophic bacteria
facultative photosynthetic bacteria studies, II:
184—185
hydrogen bacteria studies, II: 176
Chemolithotrophs
growth yield studies, I: 2, 25—27
hydrogen bacteria studies, II: 174, 176
sulfur oxidation by, II: 114, 125—126
Chemoorganotrophic growth, hydrogen bacteria,
II: 160, 175—176
Chemostat cultures
ammonia-oxidizing bacteria studies, II: 93, 95,
106
cell cycle studies, I: 98—99
growth yield studies, I: 4—6, 8—9, 11—13,
15—16, 18, 21—25, 27—28
hydrogen bacteria studies, II: 162
Chlorate, reduction to chlorite, I: 107
Chlorate-resistant mutants, molybdenum cofactor
and, II: 28, 38—39
Chlorella fusca, growth yield studies, I: 27
Chlorin, studies of, II: 140, 153
Chlorobiaceae, cyclic electron transfer systems, I:
59
Chlorobium
limicola, APS reductase, II: 124
species, respiration in, II: 184
4-Chloro-7-nitrobenzofurazan, see Nbf-Cl
2-Chloro-6-(trichloromethyl)pyridine, ammonia
oxidation and, II: 88
Chromatiaceae, cyclic electron transfer systems,
I: 59
Chromatium
species
respiration, II: 184
rhodanese, II: 103

lyase reaction, fumarate and, II: 4
Citrobacter sp., nitrate, II: 36
Clocks, cell cycle studies, I: 98
vinosum, transport systems, I: 62
Chromatophores
ammonia-oxidizing bacteria studies, II: 89
definition, II: 184
facultative photosynthetic bacteria studies, II:
184—185, 193, 196
solute translocation studies, I: 42, 60—63
pH gradients generated in, I: 62
Chromosome replication, membrane synthesis
and, I: 101
CH₃-S-CoM methylreductase, methonogenic
bacteria studies, I: 171—176, 181
ATP requirements, I: 171—172
coenzyme M analogues, I: 173—175
components, I: 171, 181
inhibitors, I: 172—174
RPG effect, I: 174—176
Citrate
fermentation, II: 3—4, 6, 8—9
Clostridial ferredoxin, heme-requiring bacteria
studies, II: 145
Clostridium
caproicum, Stickland reaction, II: 50—51
formicoaceticum, fumarate, II: 3, 5, 8—9, 15
HF, see Clostridium, sticklandii
lentoputrescens, Stickland reaction, II: 51—52
pasteurianium, transport systems, I: 66
perfringens, nitrate, II: 36
species
nitrate, II: 37
Stickland reaction, II: 50—62
sporogenes
glycine reductase, II: 53, 58
Stickland reaction, II: 50—51, 53, 58
sticklandii
glycine reductase, II: 51—57, 62
proline reductase, II: 57—62
Stickland reaction, II: 51—62
tertium, nitrate, II: 36
thermoaceticum
formate dehydrogenase, II: 52
Stickland reaction, II: 52
valerianicum, Stickland reaction, II: 50—51
p-CMS, fumarate reduction and, II: 11
Coenzyme F₄₂₀, methogenic bacteria studies, I:
163—170
absorption spectra, I: 164—165, 168, 170
fragments, I: 165, 169
reduction, stoichiometry, I: 165—166
side chains, I: 165—166
structure of, I: 165, 168
Coenzyme M, methanogenic bacteria studies, I:
170—177, 181
acetate reduction by, I: 178—179
analogues, I: 173—175
C₁ and C₂ derivatives, I: 176—177
CDR factor, I: 177, 181
CH₃-S-CoM methylreductase, see CH₃-S-CoM
methylreductase

Fumarate-glycerol oxidoreductase pathway,
    anaerobic growth, I: 116
Fumarate reductase
    electron transport chain, II: 10—12
    heme-requiring bacteria studies, II: 145—146
    impermeable dyes and, II: 12—13, 15
    localization, I: 56—57; II: 12—15
    membrane bound, II: 3
    methanogenic bacteria studies, I: 167, 181
    sidedness in *Vibrio succinogenes*, II: 12—15
    solute translocation studies, I: 55—57, 59
    synthesis, I: 55
Fungus
    growth inhibitors, II: 197
    heterotrophic, ammonia oxidation studies, II:
        88
Futile cycles, ATP turnover, I: 12—13, 27

G

β-Galactoside carrier, solute translocation studies,
    I: 72—73
Gas uptake, hydrogen bacteria studies, II:
    161—162
Genetic studies, hydrogen bacteria, II: 175—176
Gene transfer agent system, facultative
    photosynthetic bacteria studies, II: 186
Gluconate
    formation of, I: 6
    growth yield studies, I: 6, 16—21
Glucose
    fermentation, II: 2, 4, 6—9, 28, 146
    growth yield studies, I: 4—6, 8—13, 15—26
    oxidation, II: 148
    uptake mechanism, I: 25, 28
    utilization rates, I: 6
Glutamate, carbon and energy source, II: 37
Glutamate synthase, glutamine synthetase and, I:
    18
Glutaminase, ATP and, I: 12
Glutamine synthetase
    ATP and, I: 12, 18
    glutamate synthase and, I: 18
Glutamine uptake, ATP and, I: 56
Glutathione, reduced, sulfur-oxidizing bacteria
    studies, II: 119—121, 129—130
Glutathione reductase, sulfur-oxidizing bacteria
    studies, II: 119
Glycerol
    fermentation of, II: 4, 6
    fumarate reduction and, II: 9
    growth yield studies, I: 4—5, 7—9, 17—19,
        21—22
Glycerol-fumarate oxidoreductase pathway,
    anaerobic growth, I: 116
L-Glycerol-3-phosphate dehydrogenase
    aerobic respiratory chain studies, I: 116—119,
        130—131
    localization, I: 130—131
α-Glycerophosphate, cell cycle studies, I: 100

L-α-Glycerophosphate
    cell cycle studies, I: 104—105
    solute translocation studies, I: 54—56, 66
α-Glycerophosphate dehydrogenase, cell cycle
    studies, I: 105
L-α-Glycerophosphate dehydrogenase, solute
    translocation studies, I: 43, 55—58
α-Glycerophosphate oxidase, aerobic respiratory
    chain studies, I: 119—120
Glycine
    acetate and, II: 51—52, 56
    ammonia and, II: 52
    deamination, II: 52—53, 56, 62
    hydrogen acceptor, II: 50
    reduction, II: 51—57
        chemical reactions, II: 51—52
        metal requirements, II: 52—53
        reductase, see Glycine reductase
Glycine reductase
    assays of, II: 53—54
    fraction C, II: 57, 62
        molecular weight, II: 57
    protein A, see Glycine reductase, selenoprotein
    protein B, II: 54, 57, 62
        molecular weight, II: 57
        tritium exchange assay, II: 54
    purification and properties of, II: 54—57
    reaction, stereochemistry, II: 51—53
    selenoprotein, II: 52—57, 62
        absorption spectrum, II: 55—56
        molecular weight, II: 55
        physical and chemical properties, table, II:
            55
    Stickland reaction, II: 51—57, 62
Glycogen
    growth yield studies, I: 5, 9—10
    synthesis, I: 9—10
Gramicidin, phosphorylation and, II: 10, 79
Green bacteria, photosynthesis, II: 184
Green derivative, *Rhodopseudomonas capsulata*
    (R126), cytochrome, II: 192
Group translocation systems, classification, I: 37
Growing zones, cell cycle studies, I: 107—108
    equatorial, I: 108
Growth, I: 2—13
    ATP and, see also ATP, synthesis, growth yield
        studies, I: 7—13, 26—27
    balanced, see Balanced growth
    cell volume, linear, during cell cycle, I: 100
    measurable parameters, I: 2—7
    membrane, see Membrane, growth
    rate
        ammonia and, II: 90—93, 106
        growth yields and, I: 3—4, 13—17, 22—25,
            29; II: 77—79
        oxygen consumption rate and, I: 23
        specific, II: 90, 160
        thermophilic hydrogen bacterium, highest
            specific, II: 160
    requirements, hydrogen bacteria, II: 160—162
    uncoupled, hydrogen bacteria, II: 162,
        172—173

220   *Diversity of Bacterial Respiratory Systems*

K

K, vitamin, see Vitamin K
K_{12}, *Escherichia coli*, see *Escherichia coli*, K_{12}
    strain
2-Ketoglutarate, formation of, I: 6
Klebsiella
    aerogenes
        ATP/O ratio, I: 20
        cell cycle studies, I: 98—99
        growth yield studies, I: 4—10, 12—13,
            15—22, 24—26, 28
        nitrate, II: 20—30, 32, 36—37
        nitrate reductase, II: 20—30
        P/O ratio, I: 20
        respiratory chain, I: 19—20; II: 28—30
        synchronous cultures, cell cycle studies, I:
            98—99
    pneumoniae,
        fumarate, II: 10
        nitrate, II: 28, 32
        nitrate reductase, II: 28
    species, polysaccharide synthesis, I: 10
Knall gas reaction, hydrogen oxidation, II: 160

L

Labile sulfur, sulfur-oxidizing bacteria studies, II:
    120
    acid-labile, II: 124, 126
Lactate
    fermentation, II: 4, 6, 8—9
    glucose fermentation and, II: 2
    growth yield studies, I: 17, 19, 21—22
    nitrate and, II: 35—37
    oxidation, II: 72, 150
D-Lactate
    facilitated secondary transport system, I: 44
    formation of, I: 6
    growth yield studies, I: 6, 9, 13
    oxidation of, I: 44, 47, 50
    solute translocation studies, I: 44—47, 58—59
Lactate dehydrogenase, sulfate reduction and, II:
    72—73
D-Lactate dehydrogenase
    aerobic respiratory chain studies, I: 116—119
    cytochrome *o* studies, I: 140
    localization, I: 44, 130—131
    pyruvate reductase and, I: 13
    solute translocation studies, I: 43, 58
    synthesis of, I: 104
L-Lactate dehydrogenase
    aerobic respiratory chain studies, I: 116—119
    localization, I: 130—131
    solute translocation studies, I: 43
D-Lactate oxidase
    aerobic respiratory chain studies, I: 119—120
    cytochrome *o* studies, I: 143
Lactic acid bacteria, heme-requiring, studies of,
    II: 149—153

Lactobacillus
    brevis, cytochromes, II: 151
    L-lactate dehydrogenase, I: 130—131
    menaquinone-6, II: 77
    NADH dehydrogenase, I: 44, 130
    nitrate reductase, I: 51—53; II: 22—24, 27,
        29, 33, 35, 41—42
    nitrite reductase, II: 29, 33, 35
    oxygen reduction, I: 130
    respiratory chain, I: 43, 116
    plantarum
        cytochromes, II: 151
        pseudocatalase, II: 150
    species
        catalase, II: 150
        protoheme, II: 150—151
    strain 30a, pyruvate, II: 61
Lactose
    carrier, solute translocation studies, I: 72—73
    permeases, I: 104
    transport of, I: 49, 52—54, 57—59
    uptake of, I: 68
Leerlauf oxidation, hydrogen, II: 161
Leucine, translocation of, I: 66
Leuconostoc
    mesenteroides
        catalase, II: 150
        cytochromes, II: 151
        hematin, II: 151
        pseudocatalase, II: 150
    species, pseudocatalase, II: 150
Light-dependent electrochemical proton gradient
        generation, I: 65—66
Light-induced electron transport, I: 59, 61—64;
        II: 184, 187, 194—197
    ATP synthesis coupled to, I: 59
Linear growth, volume, during cell cycle, I: 100
Lineola longa, synchronous cultures, cell cycle
        studies, I: 93
Lipids
    diffusion of, cell cycle studies, I: 107
    synthesis of, I: 101—103
    vesicles, primary transport systems, I: 42—43
Localization, membrane
    autotrophic marker, hydrogen bacteria studies,
        II: 175—176
    carriers, electron and hydrogen, I: 44; II:
        75—77, 102—103
    cytochrome synthesis, I: 109
    cytoplasmic membrane, I: 35
        APS reductase, II: 77
        ATPase, I: 35, 116
        ATP sulfurylase, II: 77
        bisulfite reductase, II: 77
        cytochrome *b*, II: 77
        cytochrome cc_3, II: 77
        electron carriers, I: 44
        ferredoxin, II: 77
        flavodoxin, II: 77
        formate dehydrogenase, I: 52; II: 12—15
        fumarate reductase, I: 56—57; II: 12—15
        L-glycerol-3-phosphate dehydrogenase, I:

Membrane-bound transhydrogenase, hydrogen
  bacteria studies, II: 174, 176
Membrane-permeable ions, distribution of, I:
  39—40
Membrane-permeable weak acids or bases, uptake
  and distribution of, I: 38, 40—41, 45,
  63—64
Membrane vesicles (cytoplasmic)
  components, I: 41
  facultative photosynthetic bacteria studies, II:
    185, 193—194
  functions, I: 42
  inside-out-oriented, isolation of, I: 61
  isolation, I: 41, 61
  localization at, see Localization, membrane
    vesicle
  model system for transport, I: 42
  oxidative phosphorylation, I: 42
  protonmotive force studies, I: 125—126
  right-side- out-oriented, isolation of, I: 61
Menadione
  heme-requiring bacteria studies, II: 143
  hydrogen bacteria studies, II: 166
Menadione reductase, hydrogen bacteria studies,
  II: 168
Menaquinone
  aerobic respiratory chain studies, I: 117,
    119—120
  APS reductase and, II: 68
  heme-requiring bacteria studies, II: 146—147,
    149
  nitrate reductase and, II: 39
  fumarate reductase and, I: 56
  respiratory chain, role in, I: 48—49
  solute translocation studies, I: 48—49, 56
Menaquinone-6, sulfate reduction and, II: 75, 77
  localization, II: 77
Menaquinone-7, solute translocation studies, I: 49
Menaquinone-8, aerobic respiratory chain studies,
  I: 117, 119—120
Mercaptan, coenzyme M analogue, I: 174—175
Mesoheme, heme-requiring bacteria studies, II:
  147
Mesoporphyrin, heme-requiring bacteria studies,
  II: 146—147
Mesosomal-like membranes, hydrogen bacteria,
  II: 163
Messenger RNA, outer membrane proteins and,
  I: 101
Metabolism
  adaptivity of, facultative photosynthetic
    bacteria, II: 184—185
  aerobic, hydrogen and, II: 2
  anaerobic, see Anaerobic metabolism
  energy, cell cycle studies, I: 100
  oscillations, synchronous cultures, I: 99
  overflow, I: 5—7
  sulfur, II: 114
Metal-binding agents, ammonia oxidation and,
  II: 94—95
Metals, see also specific metals by name
  glycine reductase requirements for, II: 52—53
  hydrogen bacteria studies, II: 161, 166

nitrate reductase activity and, II: 28—31
Methane
  ammonia oxidation inhibited by, II: 96
  ammonia-oxidizing bacteria studies, II: 96,
    100—101
  growth yields, I: 213—214
  hydroxylation of, II: 96
  methanogenic bacteria studies, see
    Methanogenic bacteria; specific bacteria by
    name
  methanotrophic bacteria studies, see
    Methanotrophic bacteria, specific bacteria
    by name
  methylotrophic bacteria studies, see
    Methylotrophic bacteria; specific bacteria
    by name
  oxidation, see also Methane-oxidizing bacteria,
    I: 203—206, 209, 213; II: 94—95, 97, 108
    inhibition of, II: 108
    reactions involved in, table, I: 205—206
  P/O ratio, I: 209
Methane monooxygenase
  ammonia-oxidizing bacteria studies, II: 95—97,
    100—101, 106
  energy conservation, II: 100
  inhibitors of, II: 100—101
  methylotrophic bacteria studies, I: 203—206,
    211—212, 214
Methane-oxidizing bacteria, see also specific
    bacteria by name
  ammonia oxidation studies, II: 88, 95—97, 100,
    106, 108
  hydroxylamine oxidized by, II: 100—101
Methanobacterium
  AZ, coenzyme F_{420}, I: 167
  formicicum
    coenzyme F_{420}, I: 170
    coenzyme M, I: 172
  G, coenzyme F_{420}, I: 167
  mobile, growth factor, I: 182
  M. o. H.
    ATP synthesis, I: 180—181
    coenzyme F_{420}, I: 163—167
    coenzyme M, I: 172
    hydrogenase, I: 163
  omelianskii, ATP synthesis, I: 180—181
  species
    coenzyme F_{420}, I: 170
    hydrogen, II: 2
  thermoautotrophicum
    ATP synthesis, I: 181
    coenzyme F_{420}, I: 163—164, 167
    coenzyme M, I: 171—173, 176—177
    factor F_{430}, I: 182
    factor F_{432}, I: 183
    hydrogenase, I: 163
Methanobrevibacter ruminantium
  coenzyme F_{420}, I: 164—167
  coenzyme M, I: 172
  hydrogenase, I: 163, 165
  strain M-1, coenzyme F_{420}, I: 167
  strain PS
    coenzyme F_{420}, I: 164

227

Methylotrophic bacteria, respiration in, I:
188—214; II: 94, 96—97, 107
ammonia oxidation studies, II: 94, 96—97, 107
classification, I: 188—190
cytochromes, I: 196—204, 208—209, 211, 213;
II: 96—97
general discussion, I: 188
growth yields, I: 200, 212—214
methanotrophs, I: 188, 203—212
electron transport systems and energy
transduction, I: 208—212
primary oxidation pathways, I: 203—208
nonmethane-utilizing, I: 188—203
electron transport systems and energy
transduction, I: 197—203
primary oxidation pathways, I: 188—197
physiological aspects, I: 212—214
P/O ratios, I: 200—202, 213—214
types, I: 188—190
Methylreductase, CH$_3$-S-CoM, see CH$_3$-S-CoM
methylreductase
Mg^{2+}-Ca;2$^+$-activated membrane-bound ATPase,
I: 57, 97, 126
Michaelis constant
ammonia oxidation, II: 88—89, 100—101, 106
cytochrome 562, II: 171
hydrogen, II: 166, 176
NADH dehydrogenase, II: 107
nitrite oxidation, II: 104
thiosulfate, II: 128
Microbial growth yield values, see Growth yields,
microbial
Micrococcus
denitrificans
cytochromes, I: 155; II: 77
nitrate reductase, II: 24—26
nitrite reductase, II: 123
halodenitrificans, nitrate reductase, II: 20
lysodeiktus, transport systems, I: 72
pyogenes var. albus, cytochomre o in, I: 139
species, nitrate reductase, II: 21—22
Midpoint oxidation-reduction potential
cytochrome a$_1$, I: 121
cytochrome aa$_3$, II: 187
cytochrome b, II: 186—188, 191
cytochrome b$_1$, I: 121
cytochrome b$_{50}$, II: 191
cytochrome b$_{270}$, II: 187
cytochrome b$_{556}$, I: 121, 127—128
cytochrome b$_{558}$, I: 121
cytochrome b$_{562}$, I: 127
cytochrome c$_2$, I: 60; II: 189
cytochrome cc$'$, II: 189—190
cytochrome d, I: 122
cytochrome o, I: 122
P$_{870}$, electron donor, I: 59—60
Minicell-forming mutant, Escherichia coli,
cytochrome segregation studies, I: 109
Mitochondria
cytochrome c localized in, II: 77
cytochrome oxidases, I: 145
oxidative phosphorylation, see Oxidative

phosphorylation, mitochondrial
respiration in, I: 97, 100
state 3, I: 97, 100
state 4, I: 97, 100
Mixed-function oxidases, cytochrome o, I: 138,
148
Mixobacter strain AL1, synchronous cultures, cell
cycle studies, I: 96, 103
Mixotrophic growth, hydrogen bacteria studies,
II: 160, 174—176
definition, II: 160
Mixotrophic sulfur-oxidizing chemoautotrophs,
classification, II: 114
MK, fumarate and, II: 10—11, 15
Molar growth yields, see Growth yields
Molecular weight
APS reductase, II: 68
bacteriorhodopsin, I: 64—65
copper-containing protein, I: 204
cytochrome b$_1$, I: 121
cytochrome c, autooxidizable, I: 198
cytochrome c$_1$, II: 72
cytochrome c$_{551}$, II: 74
cytochrome cc$_3$, II: 74
cytochrome o, I: 149
desulforedoxin, II: 75
ferredoxin, II: 75
flavodoxin, II: 75
formate, I: 193
glycine reductase
fraction C, II: 57
protein B, II: 57
selenoprotein, II: 55
hemoglobin-like pigment, hydrogen bacteria,
II: 168
iron flavoprotein, I: 204
membrane-bound hydrogenase, II: 166
methanol, I: 192
methylamine, I: 196
NADH dehydrogenase, I: 118
nitrate reductase, II: 21—22
proline reductase, II: 60
rubredoxin, II: 75
Molybdate
glycine reductase and, II: 52
nitrate reductase and, II: 39
Molybdenum
cofactor, nitrate reductase and, II: 27—28,
39—41
formate and, II: 10—11
nitrate reductase and, I: 51; II: 24, 26—29,
39—41
Monocarboxylic acids, transport of, I: 71
Morphology, membrane, hydrogen bacteria, II:
163
MPS, solute translocation studies, I: 44
Mycobacterium
gordonae, hydrogen, II: 175
phlei
cytochrome o, I: 153—154
respiratory chain, II: 114
species

O

nitrite, Michaelis constant, II: 104
nitroxyl, II: 99—101, 104
polythionate, II: 114
primary, pathways, methyltrophic bacteria, I: 188—197, 203—208
 graphic representation, I: 191
pyruvate, II: 72
redox indicator dyes, II: 53
reduction and, pyrophosphatase, regulation, II: 67
substrates, see Substrate, oxidation
succinate, see Succinate, oxidation
sulfide, see Sulfide, oxidation
sulfite, see Sulfite, oxidation
sulfur, see Sulfur, oxidation; Sulfur-oxidizing chemoautotrophic bacteria
tetramethylammonium, I: 194
tetrathionate, II: 128
thiosulfate, see Thiosulfate, oxidation
TMPD, II: 193
trimethylamine, I: 194—195
trimethylsulfonium chloride, I: 193—194
Oxidative phosphorylation
ATPase-deficient mutant, *Escherichia coli*, I: 97
ATP synthesis and, I: 123, 126; II: 114, 117
growth yield studies, I: 4, 9, 11, 13, 19
heme-requiring bacteria studies, II: 146—147, 151—153
hydrogen bacteria studies, II: 162, 171, 173
hydroxylamine oxidation coupled to, II: 102
membrane vesicles, I: 42
mitochondrial, fumarate, II: 10
nitrate reduction and, II: 32—35, 37
site of, I: 101
sulfur-oxidizing bacteria studies, II: 114, 122, 124—125, 127, 131
uncouplers of, I: 13; II: 94—95, 104, 117—118
Oxido-reduction loops, see Transmembrane oxidation-reduction loops
α-Oxoglutarate, heme-requiring bacteria studies, II: 144
2-Oxoglutarate, formation of, I: 6
2-Oxoglutarate dehydrogenase, NADH and, I: 4
Oxygen
 consumption, rate of, see Consumption rates, oxygen
 cytochrome c_2 coupled to, II: 193—194
 growth and growth yield studies, I: 2—15, 17—21, 23—25, 29
 growth yields, see Growth yields, oxygen
 hemoproteins and, see Hemoproteins, oxygen reactive
 hydrogen acceptor, II: 37
 nitrate respiration and, II: 38
 reduction, localization of, I: 130
 respiratory chain, II: 29—31
 sensitivity to, hydrogen bacteria, II: 161, 166—168, 174, 176
 uptake, I: 3—4, 20—22, 92, 94, 97—98, 104; II: 89, 93, 95, 100—101, 106

ammonia-dependent, II: 100—101
inhibition of, II: 93
rate of, II: 95, 106
Oxygenase
 electrons transported to, II: 104
 heme-requiring bacteria studies, II: 140

P

P582, see Pigment P582
P_{870}, electron donor molecule, cyclic electron transport systems, I: 59—61
 midpoint oxidation-reduction potential, I: 59—60
Palmitic acid, labeling of phospholipids with, I: 107
Paracoccus
 denitrificans
  antimycin, II: 191
  cell breaking, II: 163
  coproporphyrinogenase, II: 142
  cytochromes, I: 155; II: 197
  energy conservation and electron flow, II: 169—170
  growth yield studies, I: 7, 17, 19—20
  hydrogen, II: 160, 163—164, 167, 169—170, 172—176
  hydrogenases, II: 164, 167, 170
  mixotrophy, II: 175
  NADH, II: 170
  nitrate, II: 20, 23—24, 29—33, 35, 173—174
  nitrate reductase, II: 20, 23—24, 29—31
  nitrite, II: 93
  proton translocation, I: 124; II: 170, 172
  respiratory chain, I: 20, II: 29—31, 114, 176, 197
  transport systems, I: 67, 72
 species, hydrogen, II: 160
α-Parinaric acid, mobility, I: 101
Passive diffusion systems, solute translocation, I: 36
Passive secondary transport systems, see Secondary transport systems, passive
Pasteurella pestis, hematin, II: 152
PCB⁻, solute translocation studies, I: 40
Pediococcus
 cerevisiae, pseudocatalase, II: 150
 species
  catalase, II: 150
  pseudocatalase, II: 150
Penicillin, nitrate reductase and, I: 107
Pentachlorophenol, sulfate reduction-phosphorylation uncoupled by, II: 79
PEP, heme-requiring bacteria studies, II: 145
PEP carboxykinase, heme-requiring bacteria studies, II: 145
PEP-glucose phosphotransferase system, glucose uptake and, I: 25, 28
Peptidoglycan, cell wall and, I: 35
Pericytoplasmic membrane, methylotrophic

Polyvinylpyrrolidone, synchronous culture
    preparation, I: 93
Porphins, heme-requiring bacteria studies, II:
    140, 147
Porphobilinogen, heme-requiring bacteria studies,
    II: 143—145, 147
Porphyrin, heme-requiring bacteria studies, II:
    140, 147
Porphyrinogen, heme-requiring bacteria studies,
    II: 148
Potassium
    growth yield studies, I: 6, 11, 21
    solute translocation studies, I: 38—39, 66—67,
        72
Primary oxidation pathways, methyltrophic
    bacteria, see also oxidation, I: 188—197,
        203—208
    graphic representation, I: 191
Primary transport systems, I: 37, 42—66
    aerobic, see Aerobic electron transport systems
    anaerobic, see Anaerobic electron transport
        systems
    bacteriorhodopsin, I: 43, 64—66
    classification, I: 37
    cyclic, see Cyclic electron transport systems
    lipid vesicles, I: 42—43
Product formation, rates of, I: 6
Proline
    δ-aminovalerate and, II: 50, 57—59
    hydrogen acceptor, II: 50
    reduction of, II: 57—62
        chemical reaction, II: 57—58
        reductase, see Proline reductase
    uptake, ATP and, I: 56
Proline reductase
    absorption spectrum, II: 61
    assays of, II: 58—60
    molecular weight, II: 60
    NADH and, II: 58, 60—62
    purification and properties of, II: 60—62
    pyruvate and, II: 58, 60—61
    reaction, stoichiometry, II: 58—59
    selenium and, II: 61
    Stickland reaction, II: 57—62
Propionate, fumarate reduction and, II: 2—4
Propionibacterium
    acidi-propionici, nitrate, II: 35—37
    freuden reicheii
        fumarate, II: 8—9
        glucose, II: 8—9
        glycerol, II: 9
        lactate, II: 8—9
    pentosaceum, nitrate, II: 35—36
Propionic acid bacteria, fermentation by, II: 4,
    6—9
Propylene oxide, production from propylene, II:
    96, 98
Protaminobacter ruber
    formaldehyde oxidation, I: 192
    methanol oxidation, I: 192
Proteins
    carrier, see Carrier proteins
    copper-containing, see Copper, protein

containing
    cytoplasmic, soluble, II: 77
    flavo-, see Flavoprotein
    formation of, I: 6
    hydrogen bacteria studies, II: 160, 163—168
    iron- sulfur, see Iron-sulfur proteins
    protoheme-containing, II: 144
    seleno-, see C-lycine reductase, selenoprotein
    siroheme, II: 69—71
    synthesis of, I: 101—103
Proteus
    mirabilis
        chlorate-resistant mutant, II: 39
        hem mutants, II: 40
        nitrate, II: 20—22, 33, 36, 39—41
        nitrate reductase, II: 20—22, 39—41
    rettgeri
        citrate, II: 3—4, 6, 8
        fumarate, I: 56; II: 3—5, 7—9
        glucose, II: 4, 7—9
        pyruvate, II: 9
    vulgaris, cytochromes, I: 139, 156
Protoheme
    bacteria requiring, II: 142—152
    definition, II: 140
    formation and structure, II: 142—143
    heme-requiring bacteria studies, II: 140—153
    hydrogen bacteria studies, II: 168
Protoheme-containing proteins, see also specific
    proteins by name, II: 144
Proton
    conductance, whole cells, oscillations in, I: 104
    conductors, solute translocation studies, I:
        45—46
    extrusion
        aerobic respiratory chain studies, I: 124
        heme-requiring bacteria studies, II: 146—148
        mechanism, solute translocation studies, I:
            60—61
        methylotrophic bacteria studies, I: 200—202,
            209
        nitrate respiration, II: 32—33, 42
    gradient, electrochemical, see Electrochemical
        proton gradient
    motive force, see also Electrochemical proton
        gradient
        aerobic respiratory chain studies, I: 124—
            126, 130
        ammonia-oxidizing bacteria studies, II:
            104—106
        ATP synthesis coupled to, I: 125—126
        heme-requiring bacteria studies, II: 153
        magnitude of, I: 124—126; II: 170
        nitrate reduction, II: 32—33
        Q cycle, I: 128—129; II: 195
        solute translocation studies, I: 36
    potential, formate to fumarate electron
        transport, II: 14—15
    pump function, bacteriorhodopsin, I: 65
    re-entry into cell, chemiosmotic theory, I: 7
    translocation
        ammonia-oxidizing bacteria studies, II:
            102—105, 107

S

U

Milton Keynes UK
Ingram Content Group UK Ltd.
UKHW051950071024
449327UK00026B/2255